Polymeric Materials in Energy Conversion and Storage

Polymeric Materials in Energy Conversion and Storage

Guest Editors
Md Najib Alam
Vineet Kumar

Basel • Beijing • Wuhan • Barcelona • Belgrade • Novi Sad • Cluj • Manchester

Guest Editors

Md Najib Alam
School of Mechanical Engineering
Yeungnam University
Gyeongsan
Korea, South

Vineet Kumar
School of Mechanical Engineering
Yeungnam University
Gyeongsan
Korea, South

Editorial Office
MDPI AG
Grosspeteranlage 5
4052 Basel, Switzerland

This is a reprint of the Special Issue, published open access by the journal *Polymers* (ISSN 2073-4360), freely accessible at: www.mdpi.com/journal/polymers/special_issues/ZI602P23H7.

For citation purposes, cite each article independently as indicated on the article page online and using the guide below:

Lastname, A.A.; Lastname, B.B. Article Title. *Journal Name* **Year**, *Volume Number*, Page Range.

ISBN 978-3-7258-3158-6 (Hbk)
ISBN 978-3-7258-3157-9 (PDF)
https://doi.org/10.3390/books978-3-7258-3157-9

© 2025 by the authors. Articles in this book are Open Access and distributed under the Creative Commons Attribution (CC BY) license. The book as a whole is distributed by MDPI under the terms and conditions of the Creative Commons Attribution-NonCommercial-NoDerivs (CC BY-NC-ND) license (https://creativecommons.org/licenses/by-nc-nd/4.0/).

Contents

About the Editors vii

Preface ix

Vineet Kumar and Md Najib Alam
Polymeric Materials in Energy Conversion and Storage
Reprinted from: *Polymers* 2024, 16, 3132, https://doi.org/10.3390/polym16223132 1

Lu Wei, Junjie Hu, Jiale Wang, Haiyang Wu and Kai Li
Theoretical Analysis of Light-Actuated Self-Sliding Mass on a Circular Track Facilitated by a
Liquid Crystal Elastomer Fiber
Reprinted from: *Polymers* 2024, 16, 1696, https://doi.org/10.3390/polym16121696 5

Laura Hrostea, Anda Oajdea and Liviu Leontie
Impact of PCBM as a Third Component on Optical and Electrical Properties in Ternary Organic
Blends
Reprinted from: *Polymers* 2024, 16, 1324, https://doi.org/10.3390/polym16101324 23

Jungsang Cho, Damon E. Turney, Gautam Ganapati Yadav, Michael Nyce, Bryan R. Wygant and Timothy N. Lambert et al.
Use of Hydrogel Electrolyte in $Zn-MnO_2$ Rechargeable Batteries: Characterization of Safety, Performance, and Cu^{2+} Ion Diffusion
Reprinted from: *Polymers* 2024, 16, 658, https://doi.org/10.3390/polym16050658 35

Alexey R. Tameev, Alexey E. Aleksandrov, Ildar R. Sayarov, Sergey I. Pozin, Dmitry A. Lypenko and Artem V. Dmitriev et al.
Charge Carrier Mobility in Poly(N,N′-bis-4-butylphenyl-N,N′-bisphenyl)benzidine Composites with Electron Acceptor Molecules
Reprinted from: *Polymers* 2024, 16, 570, https://doi.org/10.3390/polym16050570 47

Fuwei Liu, Luyao Gao, Jiajia Duan, Fuqun Li, Jingxian Li and Hongbing Ge et al.
A Novel and Green Method for Preparing Highly Conductive PEDOT:PSS Films for Thermoelectric Energy Harvesting
Reprinted from: *Polymers* 2024, 16, 266, https://doi.org/10.3390/polym16020266 60

Zongsong Yuan, Junxiu Liu, Guqian Qian, Yuntong Dai and Kai Li
Self-Rotation of Electrothermally Responsive Liquid Crystal Elastomer-Based Turntable in Steady-State Circuits
Reprinted from: *Polymers* 2023, 15, 4598, https://doi.org/10.3390/polym15234598 71

Jiawei Xu, Hongwei Hu, Shengtao Zhang, Guanggui Cheng and Jianning Ding
Flexible Actuators Based on Conductive Polymer Ionogels and Their Electromechanical Modeling
Reprinted from: *Polymers* 2023, 15, 4482, https://doi.org/10.3390/polym15234482 90

Jyothsna Surisetty, Mohammadhossein Sharifian, Thomas Lucyshyn and Clemens Holzer
Investigating the Aging Behavior of High-Density Polyethylene and Polyketone in a Liquid Organic Hydrogen Carrier
Reprinted from: *Polymers* 2023, 15, 4410, https://doi.org/10.3390/polym15224410 104

Khulaif Alshammari, Thamer Alashgai, Alhulw H. Alshammari, Mostufa M. Abdelhamied, Satam Alotibi and Ali Atta
Effects of Nd_2O_3 Nanoparticles on the Structural Characteristics and Dielectric Properties of PVA Polymeric Films
Reprinted from: *Polymers* 2023, 15, 4084, https://doi.org/10.3390/polym15204084 118

Bianca K. Muñoz, Andrés González-Banciella, Daniel Ureña, María Sánchez and Alejandro Ureña
Electrochemical Comparison of 2D-Flexible Solid-State Supercapacitors Based on a Matrix of PVA/H_3PO_4
Reprinted from: *Polymers* 2023, 15, 4036, https://doi.org/10.3390/polym15204036 134

Stefania Zappia, Marina Alloisio, Julio Cesar Valdivia, Eduardo Arias, Ivana Moggio and Guido Scavia et al.
Silver Nanoparticle–PEDOT:PSS Composites as Water-Processable Anodes: Correlation between the Synthetic Parameters and the Optical/Morphological Properties
Reprinted from: *Polymers* 2023, 15, 3675, https://doi.org/10.3390/polym15183675 148

Md Najib Alam, Vineet Kumar, Han-Saem Jung and Sang-Shin Park
Fabrication of High-Performance Natural Rubber Composites with Enhanced Filler–Rubber Interactions by Stearic Acid-Modified Diatomaceous Earth and Carbon Nanotubes for Mechanical and Energy Harvesting Applications
Reprinted from: *Polymers* 2023, 15, 3612, https://doi.org/10.3390/polym15173612 163

Paweł Jeżowski and Przemysław Łukasz Kowalczewski
Isinglass as an Alternative Biopolymer Membrane for Green Electrochemical Devices: Initial Studies of Application in Electric Double-Layer Capacitors and Future Perspectives
Reprinted from: *Polymers* 2023, 15, 3557, https://doi.org/10.3390/polym15173557 182

Hao-Xuan Guo, Yuriko Takemura, Daisuke Tange, Junichi Kurata and Hiroyuki Aota
Redox-Active Ferrocene Polymer for Electrode-Active Materials: Step-by-Step Synthesis on Gold Electrode Using Automatic Sequential Polymerization Equipment
Reprinted from: *Polymers* 2023, 15, 3517, https://doi.org/10.3390/polym15173517 196

Md Najib Alam, Vineet Kumar, Taemin Jeong and Sang-Shin Park
Nanocarbon Black and Molybdenum Disulfide Hybrid Filler System for the Enhancement of Fracture Toughness and Electromechanical Sensing Properties in the Silicone Rubber-Based Energy Harvester
Reprinted from: *Polymers* 2023, 15, 2189, https://doi.org/10.3390/polym15092189 204

About the Editors

Md Najib Alam

Dr. Md Najib Alam obtained his B.Sc. and M.Sc. degrees from the University of Kalyani, India, in 2006 and 2008, respectively. He completed his Ph.D. at the same university in 2014, where he addressed numerous challenges related to rubber vulcanization chemistry. Following his Ph.D., Dr. Alam pursued postdoctoral research at Chulalongkorn University, Thailand, from 2015 to 2019. From 2019 to 2022, he served as a research fellow at the School of Mechanical Engineering, Yeungnam University, Republic of Korea. In 2022, he transitioned to the role of assistant professor at the same institution. Throughout his research career, Dr. Alam has published over 50 original research articles in internationally recognized peer-reviewed journals, including the *Journal of Industrial and Engineering Chemistry*, *Composites Science and Technology*, and *Composites Part B: Engineering*. Currently, he is serving as a guest editor of various Special Issues of the MDPI journal *Polymers*. His research focuses on rubber vulcanization, novel rubber composites, stretchable sensors, and smart materials.

Vineet Kumar

Vineet Kumar has been an Assistant Professor at the Department of Mechanical Engineering, Yeungnam University, Gyeongsan-si, Republic of Korea, since 2016. He completed his Ph.D. in the Department of Material Science at the University of Milan-Bicocca, Milan, Italy, in 2014. He finished his Master's of Science in Environment Management from the Forest Research Institute, Dehradun, India, in 2008. He has co-authored more than 75 peer-reviewed articles in reputable SCI journals. He has also served as a Topic Editor for *Polymers* (MDPI) since 2021. Moreover, he has served as a Guest Editor of more than six Special Issues in *Polymers* and *Frontiers in Materials*.

Preface

Recently, global warming and its effect on the environment have been critical issues and the subject of significant discussion in society. To reduce global warming, unnecessary production of heat should be avoided. Heat is a kind of energy that cannot be fully converted to other forms of energy, whereas electrical energy almost always can. Hence, day-by-day, the demand for electrical energy is continuously increasing. However, to meet this demand, energy storage or production devices should be highly efficient.

With continuous advancements in the science and technology of energy devices, polymers have become integral parts of these devices due to the stability of their mechanical and electrical properties. However, specific polymers have specific roles in these systems. Therefore, research on functional polymers or polymer composite materials should be focused on specific applications.

This reprint provides readers with insights into various techniques that can be used to fabricate energy-harvesting and energy storage devices with functional polymers. It offers valuable investigations of the mechanisms of energy conversion and the critical factors for higher efficiency. Additionally, the book discusses alternative polymers that could be beneficial for environmental remediation or have economic advantages. Each chapter thoroughly covers the background of the innovations, presents results and discussions, and concludes with important findings. We are confident that readers will be introduced to several innovative ideas that will be beneficial for their research or academic pursuits.

Md Najib Alam and Vineet Kumar
Guest Editors

Editorial

Polymeric Materials in Energy Conversion and Storage

Vineet Kumar and Md Najib Alam *

School of Mechanical Engineering, Yeungnam University, 280, Daehak-ro, Gyeongsan 38541, Republic of Korea; vineetfri@gmail.com
* Correspondence: mdnajib.alam3@gmail.com

1. Introduction

Energy conversion and storage devices based on polymeric materials are emerging as a promising avenue for renewable power sources. These features are attributed to their versatility, tunable properties, and ease of processing for polymer-based energy materials [1]. Due to their versatile nature, these polymeric materials are currently used in a wide range of applications, such as batteries, fuel cells, solar cells, nanogenerators, and supercapacitors [2]. Moreover, the ability to engineer these materials at the molecular level provides significant advantages in optimizing the performance, efficiency, and longevity of energy systems. Therefore, these materials emerge as promising renewable power sources with high efficiency and durability [3]. There are various advantages to using polymer materials in energy harvesting applications. For example, their unique properties—such as flexibility, lightweight nature, chemical stability, and ease of processing—make them attractive for energy applications. Additionally, these polymers can be chemically modified to enhance specific functionalities like conductivity, ion transport, or mechanical strength, depending on the need [4].

There are various energy conversion applications for polymer-based energy generation systems. Some important applications include solar cells, fuel cells, and energy generators [5,6]. Solar cells utilize polymer-based organic photovoltaic cells that harvest light energy and convert it into electrical energy. For example, organic semiconductors made from conjugated polymers can absorb sunlight and generate electrical energy. These polymer-based harvesting materials are promising due to their lightweight nature, flexibility, low cost, and ease of production via 3D and 4D printing [7]. In addition, fuel cells use polymer-based Nafion membranes (also known as proton exchange membranes) to conduct protons from the anode to the cathode. These membranes are crucial because they separate the fuel and oxidant while allowing ion transport, which is essential for the electrochemical reactions that generate electricity [8]. Finally, energy generators based on piezoelectricity or triboelectricity are frequently used as energy nanogenerators. Composites made from elastomers, piezoelectric materials, or electrically conductive materials are commonly employed in energy harvesting applications [9].

In addition to energy conversion applications, polymeric materials also play a dominant role in energy storage devices. Frequently used materials include those found in batteries and supercapacitors. In lithium-ion batteries and other types of rechargeable batteries, polymers are used as electrolytes, binders, and separators [10]. For example, solid polymer electrolytes (SPEs) are of great interest because they can improve battery safety by replacing liquid electrolytes, which are often flammable. Moreover, polymer-based electrodes offer flexibility and lightweight properties, making battery systems suitable for portable and wearable electronics [11]. For supercapacitors, energy is stored through the accumulation of electrical charges at the interface between an electrode and an electrolyte. Some conductive polymers, such as polyaniline and polypyrrole, are used in supercapacitor electrodes to increase the device's charge storage capacity [12]. These materials exhibit high surface area, high conductivity, and robust electrochemical activity. Keeping these aspects

Citation: Kumar, V.; Alam, M.N. Polymeric Materials in Energy Conversion and Storage. *Polymers* 2024, 16, 3132. https://doi.org/10.3390/polym16223132

Received: 29 October 2024
Accepted: 6 November 2024
Published: 10 November 2024

Copyright: © 2024 by the authors. Licensee MDPI, Basel, Switzerland. This article is an open access article distributed under the terms and conditions of the Creative Commons Attribution (CC BY) license (https:// creativecommons.org/licenses/by/ 4.0/).

in mind, the present editorial summarizes the latest novel articles in the field of polymeric materials for energy conversion and storage. The next section is dedicated to an overview of the latest articles, summarizing the prospects of different polymeric materials for energy applications.

2. Overview of Published Articles

Alam et al. [13] presented an energy harvesting system using rubber composites based on silicone rubber, nano-carbon black (NCB), and molybdenum disulfide (MoS_2). The results reported that at a 17:3 NCB to MoS_2 ratio, the fracture toughness improved by 184% and elongation at break increased by 93% compared to a sample with 20 phr of NCB as the only filler. The results further indicated that the hybrid filler (17:3 NCB to MoS_2 ratio) exhibited an over 100% higher output voltage, reaching up to 3.2 mV compared to NCB as the only filler at 20 phr. Jeżowski et al. [14] developed a robust biopolymer membrane for green electrochemical devices. The results showed that a capacitance of 30 F/g, a resistance of 3 Ω, and a voltage of approximately 1.6 V were achieved. These results support the use of this biopolymer membrane as a sustainable alternative for high-performance storage devices. In another study, Guo et al. [15] reported an interesting concept involving redox-active polymers for capacitors as energy storage devices. Automatic sequential polymerization equipment was used to synthesize redox-active polymers on gold electrodes. The total charge generated varied from 2.17 to 7.14×10^{-4} C. Alam et al. [16] fabricated high-performance composites based on natural rubber, diatomaceous earth, and carbon nanotubes. The results demonstrated robust performance after adding these reinforcing fillers. For example, the fracture toughness was 9.74 MJ/m^3 for the unfilled sample and increased by 484% to 56.86 MJ/m^3 at the 40 phr hybrid sample. The electrical conductivity was around 1.75×10^{-6} S/m, and the output voltage was approximately 25 mV for the composite containing 3 phr CNT.

Zappia et al. [17] fabricated composites based on silver nanoparticles for possible water-processable anodes. The results showed that the viscosity of the composites changed from 14.1 to 0.2 Pa·s. The composites also exhibited remarkable photovoltaic properties in organic cells. Alshammari et al. [18] presented a study on the dielectric properties of PVA-based polymeric films. The results indicated that the electrical conductivity increased from 0.82×10^{-9} S/cm (unfilled) to 6.82×10^{-9} S/cm for the composite sample. Moreover, the relaxation time decreased from 14.2×10^{-5} s (unfilled) to 6.35×10^{-5} s for the composite sample. Another study by Muñoz et al. [19] investigated the electro-mechanical performance of supercapacitors based on PVA and H_3PO_4. The results demonstrated that the composites exhibited high electro-mechanical performance, such as 1.4 F/g and 0.961 W/kg. Additionally, robust performance was reported, with the fabricated device able to recover 96.12% of its original capacitance once the external strain was removed. Surisetty et al. [20] demonstrated the aging behavior of high-density polyethylene (HDPE) and polyketone in liquid organic hydrogen carriers. The aging tests showed that the crystallinity of HDPE at 25 °C was 48.6% after 500 h of aging; however, with an increase in temperature to 60 °C for 500 h, the crystallinity decreased to 46.8%. Yuan et al. [21] presented theoretical studies on the electrothermal response of liquid crystal elastomers for high-performance applications, focusing on various aspects of elastomer-based tunable circuits. Another interesting study by Xu et al. [22] investigated flexible actuators fabricated using polymer ionogels, determining their electromechanical aspects theoretically. The results indicated that with changes in modulus, the mechanical displacement decreased while the blocking force increased, both of which are favorable for actuation properties.

Liu et al. [23] fabricated flexible films from electrically conducting PEDOT materials for thermoelectric energy harvesting. The final samples exhibited a high electrical conductivity of 62.91 S/cm, a Seebeck coefficient of 14.43 µV/K, and a power factor of 1.32 µW/m·K^2. Tameev et al. [24] reported the charge carrier mobility of poly-TPD-based composites with electron acceptor molecules. The electron and hole mobilities were lower for poly-TPD composites and higher for poly-TPD:PCBM. For example, the electronic mobility was approximately 4.2×10^{-6} cm^2/V·s for poly-TPD and as high as 8.3×10^{-6} cm^2/V·s for poly-TPD:PCBM. In another work, Cho et al. [25] prepared hydrogel electrolytes in rechargeable batteries based on Zn-MnO$_2$. They reported very interesting results, such as the battery's high durability, with negligible capacity loss from approximately 0.4 Ah/g at the 100th cycle to about 0.38 Ah/g at the 800th cycle. Similarly, Hrostea et al. [26] studied the influence of phase change materials as a third component on the optoelectric properties of organic blends. The best sample exhibited superior absorption of 1.07×10^5 cm^{-1} at a wavelength of 628 nm. Moreover, a carrier mobility of 1.41×10^{-4} cm^2/V·s was reported. Lastly, Wei et al. [27] reported a theoretical study of light actuation facilitated by a crystal elastomer fiber. The results were promising, with the elastic coefficient improving by 0.25% and light intensity showing a 20.88% increase. This study is useful for non-circular curved tracks, offering improved adaptability and versatility.

3. Summary and Future Outlook

Energy devices based on polymeric materials hold tremendous potential for the future of energy conversion and storage technologies. Continuous innovations in polymer chemistry and materials science are driving the development of new polymer-based systems with enhanced performance [28]. Moreover, the future of flexible and stretchable energy devices, sustainable polymers, and multifunctional materials is bright. For example, stretchable devices made from polymeric materials enable the development of energy systems that can be integrated into clothing, medical devices, and flexible electronics. Additionally, sustainable polymers are being prioritized for the development of biodegradable or recyclable polymeric materials for energy applications [29]. Finally, next-generation materials with diverse functionalities hold promise. For instance, these multifunctional materials can combine energy storage and structural properties within a single material, making them ideal for applications like electric vehicles and smart grids [30]. Overall, the versatility of polymeric materials can be enhanced by ongoing advancements in materials science. These novel materials position themselves as key enablers for next-generation energy technologies, leading to the evolution of new devices that offer more efficient, flexible, and sustainable energy solutions.

Acknowledgments: The authors thank to all the contributors and reviewers for their valuable contributions and support from section editors of this Special Issue.

Conflicts of Interest: The authors declare no conflicts of interest.

References

1. Rodrigues-Marinho, T.; Perinka, N.; Costa, P.; Lanceros-Mendez, S. Printable lightweight polymer-based energy harvesting systems: Materials, processes, and applications. *Mater. Today Sustain.* **2023**, *21*, 100292. [CrossRef]
2. Dissanayake, K.; Kularatna-Abeywardana, D. A review of supercapacitors: Materials, technology, challenges, and renewable energy applications. *J. Energy Storage* **2024**, *96*, 112563. [CrossRef]
3. Parin, F.N.; Demirci, F. Durability of polymer composite materials for high-temperature applications. In *Aging and Durability of FRP Composites and Nanocomposites*; Woodhead Publishing: Cambridge, UK, 2024; pp. 135–170.
4. Li, Z.; Fu, J.; Zhou, X.; Gui, S.; Wei, L.; Yang, H.; Guo, X. Ionic conduction in polymer-based solid electrolytes. *Adv. Sci.* **2023**, *10*, 2201718. [CrossRef] [PubMed]
5. Nandi, A.K.; Chatterjee, D.P. Hybrid polymer gels for energy applications. *J. Mater. Chem. A* **2023**, *11*, 12593–12642. [CrossRef]
6. Verma, S.K.; Dutt, S.; Jadhav, V.; Sabavath, G. Application of a polymer nanocomposite for energy harvesting. In *Recent Advances in Energy Harvesting Technologies*; River Publishers: Aalborg, Denmark, 2024; pp. 109–144.
7. Kausar, A.; Ahmad, I.; Zhao, T.; Aldaghri, O.; Eisa, M.H. Polymer/graphene nanocomposites via 3D and 4D printing—Design and technical potential. *Processes* **2023**, *11*, 868. [CrossRef]
8. Fu, X.; Wen, J.; Xia, C.; Liu, Q.; Zhang, R.; Hu, S. Nafion doped polyaniline/graphene oxide composites as electrode materials for high-performance flexible supercapacitors based on Nafion membrane. *Mater. Des.* **2023**, *236*, 112506. [CrossRef]
9. Shanbedi, M.; Ardebili, H.; Karim, A. Polymer-based triboelectric nanogenerators: Materials, characterization, and applications. *Prog. Polym. Sci.* **2023**, *144*, 101723. [CrossRef]
10. Lu, X.; Wang, Y.; Xu, X.; Yan, B.; Wu, T.; Lu, L. Polymer-based solid-state electrolytes for high-energy-density lithium-ion batteries—Review. *Adv. Energy Mater.* **2023**, *13*, 2301746. [CrossRef]

11. Pal, A.; Das, N.C. Polymeric materials for flexible batteries. In *Recent Advancements in Polymeric Materials for Electrochemical Energy Storage*; Springer Nature: Singapore, 2023; pp. 401–417.
12. Varghese, A.; Devi K R, S.; Kausar, F.; Pinheiro, D. Evaluative study on supercapacitance behavior of polyaniline/polypyrrole–metal oxide based composite electrodes: A review. *Mater. Today Chem.* **2023**, *29*, 101424. [CrossRef]
13. Alam, M.N.; Kumar, V.; Jeong, T.; Park, S.S. Nanocarbon black and molybdenum disulfide hybrid filler system for the enhancement of fracture toughness and electromechanical sensing properties in the silicone rubber-based energy harvester. *Polymers* **2023**, *15*, 2189. [CrossRef]
14. Jeżowski, P.; Kowalczewski, P.Ł. Isinglass as an alternative biopolymer membrane for green electrochemical devices: Initial studies of application in electric double-layer capacitors and future perspectives. *Polymers* **2023**, *15*, 3557. [CrossRef] [PubMed]
15. Guo, H.X.; Takemura, Y.; Tange, D.; Kurata, J.; Aota, H. Redox-active ferrocene polymer for electrode-active materials: Step-by-step synthesis on gold electrode using automatic sequential polymerization equipment. *Polymers* **2023**, *15*, 3517. [CrossRef] [PubMed]
16. Alam, M.N.; Kumar, V.; Jung, H.S.; Park, S.S. Fabrication of high-performance natural rubber composites with enhanced filler–rubber interactions by stearic acid-modified diatomaceous earth and carbon nanotubes for mechanical and energy harvesting applications. *Polymers* **2023**, *15*, 3612. [CrossRef] [PubMed]
17. Zappia, S.; Alloisio, M.; Valdivia, J.C.; Arias, E.; Moggio, I.; Scavia, G.; Destri, S. Silver Nanoparticle–PEDOT: PSS Composites as Water-Processable Anodes: Correlation between the Synthetic Parameters and the Optical/Morphological Properties. *Polymers* **2023**, *15*, 3675. [CrossRef]
18. Alshammari, K.; Alashgai, T.; Alshammari, A.H.; Abdelhamied, M.M.; Alotibi, S.; Atta, A. Effects of Nd_2O_3 nanoparticles on the structural characteristics and dielectric properties of PVA polymeric films. *Polymers* **2023**, *15*, 4084. [CrossRef] [PubMed]
19. Muñoz, B.K.; González-Banciella, A.; Ureña, D.; Sánchez, M.; Ureña, A. Electrochemical comparison of 2D-flexible solid-state supercapacitors based on a matrix of PVA/H_3PO_4. *Polymers* **2023**, *15*, 4036. [CrossRef]
20. Surisetty, J.; Sharifian, M.; Lucyshyn, T.; Holzer, C. Investigating the aging behavior of high-density polyethylene and polyketone in a liquid organic hydrogen carrier. *Polymers* **2023**, *15*, 4410. [CrossRef]
21. Yuan, Z.; Liu, J.; Qian, G.; Dai, Y.; Li, K. Self-rotation of electrothermally responsive liquid crystal elastomer-based turntable in steady-state circuits. *Polymers* **2023**, *15*, 4598. [CrossRef]
22. Xu, J.; Hu, H.; Zhang, S.; Cheng, G.; Ding, J. Flexible actuators based on conductive polymer ionogels and their electromechanical modeling. *Polymers* **2023**, *15*, 4482. [CrossRef]
23. Liu, F.; Gao, L.; Duan, J.; Li, F.; Li, J.; Ge, H.; Li, M. A novel and green method for preparing highly conductive PEDOT: PSS films for thermoelectric energy harvesting. *Polymers* **2024**, *16*, 266. [CrossRef]
24. Tameev, A.R.; Aleksandrov, A.E.; Sayarov, I.R.; Pozin, S.I.; Lypenko, D.A.; Dmitriev, A.V.; Nekrasova, N.V.; Chernyadyev, A.Y.; Tsivadze, A.Y. Charge carrier mobility in poly (N, N'-bis-4-butylphenyl-N, N'-bisphenyl) benzidine composites with electron acceptor molecules. *Polymers* **2024**, *16*, 570. [CrossRef] [PubMed]
25. Cho, J.; Turney, D.E.; Yadav, G.G.; Nyce, M.; Wygant, B.R.; Lambert, T.N.; Banerjee, S. Use of hydrogel electrolyte in Zn-MnO_2 rechargeable batteries: Characterization of safety, performance, and Cu^{2+} ion diffusion. *Polymers* **2024**, *16*, 658. [CrossRef] [PubMed]
26. Hrostea, L.; Oajdea, A.; Leontie, L. Impact of PCBM as a third component on optical and electrical properties in ternary organic blends. *Polymers* **2024**, *16*, 1324. [CrossRef]
27. Wei, L.; Hu, J.; Wang, J.; Wu, H.; Li, K. Theoretical analysis of light-actuated self-sliding mass on a circular track facilitated by a liquid crystal elastomer fiber. *Polymers* **2024**, *16*, 1696. [CrossRef] [PubMed]
28. Gaikwad, N.; Gadekar, P.; Kandasubramanian, B.; Kaka, F. Advanced polymer-based materials and mesoscale models to enhance the performance of multifunctional supercapacitors. *J. Energy Storage* **2023**, *58*, 106337. [CrossRef]
29. Agostinho, B.; Silvestre, A.J.; Coutinho, J.A.; Sousa, A.F. Synthetic (bio)degradable polymers—When does recycling fail? *Green Chem.* **2023**, *25*, 13–31. [CrossRef]
30. Yao, L.; Wang, Y.; Zhao, J.; Zhu, Y.; Cao, M. Multifunctional nanocrystalline-assembled porous hierarchical material and device for integrating microwave absorption, electromagnetic interference shielding, and energy storage. *Small* **2023**, *19*, 2208101. [CrossRef]

Disclaimer/Publisher's Note: The statements, opinions and data contained in all publications are solely those of the individual author(s) and contributor(s) and not of MDPI and/or the editor(s). MDPI and/or the editor(s) disclaim responsibility for any injury to people or property resulting from any ideas, methods, instructions or products referred to in the content.

Article

Theoretical Analysis of Light-Actuated Self-Sliding Mass on a Circular Track Facilitated by a Liquid Crystal Elastomer Fiber

Lu Wei, Junjie Hu, Jiale Wang, Haiyang Wu and Kai Li *

School of Civil Engineering, Anhui Jianzhu University, Hefei 230601, China; weilu@ahjzu.edu.cn (L.W.); 13966837296@163.com (J.H.); wjl1230504@163.com (J.W.); hywu@stu.ahjzu.edu.cn (H.W.)
* Correspondence: kli@ahjzu.edu.cn

Abstract: Self-vibrating systems obtaining energy from their surroundings to sustain motion can offer great potential in micro-robots, biomedicine, radar systems, and amusement equipment owing to their adaptability, efficiency, and sustainability. However, there is a growing need for simpler, faster-responding, and easier-to-control systems. In the study, we theoretically present an advanced light-actuated liquid crystal elastomer (LCE) fiber–mass system which can initiate self-sliding motion along a rigid circular track under constant light exposure. Based on an LCE dynamic model and the theorem of angular momentum, the equations for dynamic control of the system are deduced to investigate the dynamic behavior of self-sliding. Numerical analyses show that the theoretical LCE fiber–mass system operates in two distinct states: a static state and a self-sliding state. The impact of various dimensionless variables on the self-sliding amplitude and frequency is further investigated, specifically considering variables like light intensity, initial tangential velocity, the angle of the non-illuminated zone, and the inherent properties of the LCE material. For every increment of $\pi/180$ in the amplitude, the elastic coefficient increases by 0.25% and the angle of the non-illuminated zone by 1.63%, while the light intensity contributes to a 20.88% increase. Our findings reveal that, under constant light exposure, the mass element exhibits a robust self-sliding response, indicating its potential for use in energy harvesting and other applications that require sustained periodic motion. Additionally, this system can be extended to other non-circular curved tracks, highlighting its adaptability and versatility.

Keywords: self-sliding; liquid crystal elastomer; light-actuated; sliding mass; curved track

Citation: Wei, L.; Hu, J.; Wang, J.; Wu, H.; Li, K. Theoretical Analysis of Light-Actuated Self-Sliding Mass on a Circular Track Facilitated by a Liquid Crystal Elastomer Fiber. *Polymers* **2024**, *16*, 1696. https://doi.org/10.3390/polym16121696

Academic Editors: Vineet Kumar and Md Najib Alam

Received: 12 May 2024
Revised: 6 June 2024
Accepted: 11 June 2024
Published: 14 June 2024

Copyright: © 2024 by the authors. Licensee MDPI, Basel, Switzerland. This article is an open access article distributed under the terms and conditions of the Creative Commons Attribution (CC BY) license (https://creativecommons.org/licenses/by/4.0/).

1. Introduction

Self-vibration exists widely in natural phenomena and engineering applications [1–3]. It is a phenomenon wherein the system changes periodically under consistent external stimuli [4–6], equivalent to stimulation, controlling the phase of the movement. A self-vibrating system typically comprises vibrating elements, reliable energy sources and mechanisms for feedback control [7,8]. In particular, self-vibration is a steady-state cycle movement maintained by the system itself, which means that it can be sustained without the external force of periodic change [9]. This is why it is fundamentally different from forced vibration; hence, self-actuated vibration merits in-depth investigation [10–12]. Self-actuated vibration has varied positive aspects, namely, self-actuated vibration is capable of constantly maintaining periodic motion without periodic external stimuli and additional manual control components [13,14], which greatly reduces the requirements for system motion regulation, eliminating the necessity of designing intricate control systems [15–17]. Furthermore, this characteristic enables the control system to directly draw energy from a consistent external source to sustain periodic motion [3–5]. The self-vibration phenomenon exists in many fields, for instance, non-linear friction inspirational self-vibration [2], steering wheel vibration [11], rotor vortex [18], fluid incentive tremor [13], fluid electromagnetic vibration [19], chemical reaction system vibration [9], and similar instances.

In recent years, significant advancements have been achieved in the research on self-vibration systems, among which, active material-based self-vibration systems have attracted widespread research interest [20–24]. Active materials can respond to various external stimuli to undergo deformation and movement, such as light [2,3], heat [1], electricity [6], magnetism [19,23], etc. Light stands out among the various stimuli due to its distinct advantages, including environmental friendliness [25–27], precise controllability, contact-free nature, repeatability, and multifunctionality. In addition, with the intention of upsetting the system's equilibrium, advanced feedback systems have been established for energy compensation [6,9]. This disturbance prompts a steady and enduring response from the active material, subsequently leading to self-vibration. Examples of such systems include the self-adjusting shading device [28], the integration of large-scale deformation with chemical reactions [9], the connection of fluid evaporation and structural deformation [29], and gradients in surface tension due to photothermal heating [30,31]. The active materials that have the potential to produce self-actuated vibration phenomena include, but are not restricted to, hydrophilic polymer gel [32,33], ionic polymer gel [34], and liquid crystal elastomers (LCEs) [35–37].

Among these numerous active materials, LCEs are a Liquid Crystal Polymer material synthesized through the cross-linking of liquid crystal monomer molecules. When synthesizing liquid crystal elastomers, different intermediate molecules can be introduced to create different types of LCEs, providing them with the ability to react to multiple external triggers, including light [38], heat [39,40], electricity [41], magnetism [33], and so forth. LCE materials generally have numerous advantages, such as a rapid deformation response, recoverable deformation, and no noise [42–44]. Compared with other active materials categories, these materials have distinctive advantages, such as wireless non-contact driving, a lightweight structural design, and a reduced environmental impact [45,46]. Considering the advantageous characteristics of light, there is a wide range of self-vibrating systems enabled by light-actuated liquid crystal elastomers (LCEs), encompassing actions like bending [47,48], synchronization [49], rolling [50,51], shuttling [52], jumping [31], flying [53], floating [16], swimming [22], spinning [54], chaos [55,56], and various other self-vibration mechanisms. Light-responsive LCEs exhibit extensive applicability and broad prospects in the fields of micro robots [57–59], biomimetic soft robots [60–63], biomedicine [64], and energy harvesting [65], due to their reversible contraction [66] and relaxation properties [67,68] under light stimuli.

There have been many studies on self-vibration based on liquid crystal elastomers, especially the self-vibration mode. However, the diversity of self-vibration modes is not sufficient, and the construction of self-vibration systems is not systematic, which limits the application of self-vibration phenomena in many fields, such as energy harvesting, soft robots, medical equipment, amusement equipment and micro nano devices. In summary, it is imperative to fabricate additional LCE self-vibration structures which exhibit simplicity, controllability, and integrability. In light of this, the paper creatively proposes the LCE fiber–sliding mass system, which consists of a light-actuated LCE fiber, a sliding mass, and a rigid circular track, and investigates several critical aspects, including the conditions for obtaining self-sliding modes, the dynamic mechanisms underlying self-sliding, and the influence of system dimensionless variables on self-sliding modes, amplitudes, and periods. Compared with previous self-oscillating systems, the proposed system features a simplified structure, rapid response, high controllability, and multifunctionality. Importantly, the system's adaptability extends beyond circular tracks, accommodating a range of non-circular curved paths as well. The objective is to develop a novel light-actuated self-sliding system based on active materials.

The structure of this paper is outlined below. Firstly, based on the available dynamic behavior model of LCE materials and the theorem of angular momentum, the equations for dynamic control of the sliding mass enabled by LCE fibers are deduced in Section 2. Next, the two unique motion states of the LCE fiber–sliding mass system, namely, the static state and self-sliding state, are presented, and a detailed explanation of the operational principle of self-sliding is presented in Section 3. Then, a quantitative analysis is performed to explore the impact of diverse dimensionless variables on both the amplitude and frequency characteristics of self-sliding of the system, as explained in Section 4. Finally, this study's critical conclusions are precisely abstracted in Section 5.

2. Theoretical Model and Formulation

This section begins with a description of a light-actuated self-sliding system, which includes an LCE fiber, a sliding mass, and a rigid circular track. Subsequently, we derive the equations used for dynamic control of the sliding mass enabled by LCE fibers based on the dynamic model of LCE, the theorem of angular momentum, and vibrating theory. Finally, the nondimensionalization and the numerical solution method of the dynamic control equation are introduced.

2.1. Dynamics of Self-Sliding System

Figure 1 and Video S1 schematically present a dynamic simulation of a self-sliding actuated system via light, which is capable of sliding continuously and stably under designated initial tangential speed and illumination conditions. The self-sliding system comprises an azobenzene-based LCE fiber, a sliding mass, and a rigid circular track. The photosensitive molecules in LCE fibers, such as azobenzene molecules, are aligned along the fiber axis. The lower end of the LCE fiber is fastened to a horizontal fixed support, while the upper end is linked to a sliding mass with a weight of m. The sliding mass is mounted on the rigid circular track and can slide on it, while the upper end of the track is fixed to a horizontal rigid base. In addition, the weight of the LCE fiber is much lower than the weight of the sliding ball, so it is neglected in this study. In the reference state, the radius of the rigid circular track is denoted as r and the LCE fiber is stress-free with an original length of L_0, as shown in Figure 1a. We designate the starting position of the sliding mass as the center point of the polar coordinate system, establishing the polar axis, which extends in a radial direction from the core of the circular track. Since the sliding mass is set on the rigid circular track, it can only slide along the tangential direction of the track, with an angular displacement of θ.

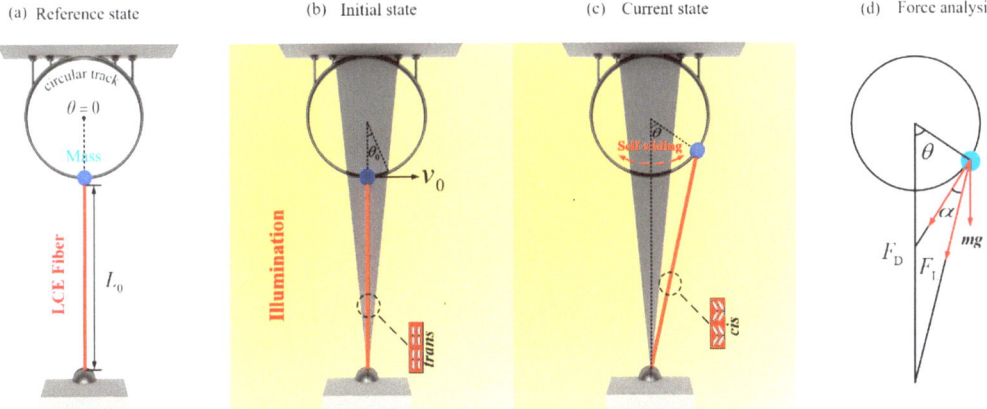

Figure 1. Diagram of the side view of a self-sliding system comprised of a light-actuated LCE fiber, a sliding mass, and a rigid circular track: (**a**) reference state; (**b**) initial state; (**c**) current state; and (**d**) force analysis. In the self-sliding model, the sliding mass with the LCE fiber can slide continuously and periodically along the rigid circular track under sustained illumination.

As depicted in Figure 1b,c, the bright section denotes the illuminated region, while the gray triangular section denotes the shadowed region, i.e., the non-illuminated region, with the angle from the origin to the first intersection point between the right side of the shade and the circular track being denoted as θ_0. Due to the initial tangential velocity in the tangential direction, the sliding mass continues its right counterclockwise motion until it enters the illuminated region. Within this region, the azobenzene molecules in the LCE fibers can absorb light energy and isomerize from straight trans to bent cis, so the LCE fibers contract. As the LCE fiber contracts and extends within the illuminated region, the system's elastic potential energy peaks when the sliding mass reaches maximum angular displacement. Subsequently, propelled by the tensile force in the LCE fiber, the sliding mass slides in the reverse direction. Upon entering the non-illuminated region, the light-actuated shrinkage of the LCE fiber is restored, resulting in a decrease in its tension until it reaches the lighted region on the opposite side. In this way, it stores potential energy from its elastic deformation and subsequently repeats the entire process. By appropriately selecting system variables and initial conditions, the light-actuated sliding mass structure can sustain consistent and steady self-sliding motion.

The sliding mass bears a tensile force F_L from the LCE fiber; the damping force F_D, and its gravity mg, as depicted in Figure 1d. Along the tangential direction, the dynamic control equation of the sliding mass can be expressed as follows [69]:

$$mr^2\ddot{\theta} = -mg\sin\theta \cdot r - F_L\cos\alpha \cdot r - F_D \cdot r \tag{1}$$

where $\ddot{\theta}$ represents the acceleration of the sliding mass; F_L denotes the tensile force of the LCE fiber; F_D refers to the damping force; θ is an angular displacement, measured relative to the vertical line, with a counterclockwise direction designated as positive; α is the angle between the force F_D and the force F_D; g is the gravitational acceleration.

According to the geometric relations in Figure 1d, $\cos\alpha = \dfrac{(L_0+r)\sin\theta}{\sqrt{r^2+(L_0+r)^2-2r(L_0+r)\cos\theta}}$, where L_0 is the initial length of the LCE fiber in a stress-free condition and r is the radius of the rigid circular track.

The tension of the LCE fiber is assumed to be proportional to elastic strain, and can be expressed as [70]:

$$F_L = KL_0\varepsilon_e(t) \tag{2}$$

where K represents the elastic coefficient of LCE; $\varepsilon_e(t)$ denotes the elastic strain in the LCE fiber. For simplicity, the elastic strain $\varepsilon_e(t)$ under small deformation can be assumed to be a linear combination of the total strain $\varepsilon_{tot}(t)$ and the light-actuated contraction strain $\varepsilon_L(t)$, i.e., $\varepsilon_{tot}(t) = \varepsilon_e(t) + \varepsilon_L(t)$. Therefore, the tension of LCE fiber in Equation (2) can be rewritten as:

$$F_L = KL_0(\varepsilon_{tot}(t) - \varepsilon_L(t)) \tag{3}$$

For simplicity, the total strain $\varepsilon_{tot}(t)$ is defined as $\varepsilon_{tot}(t) = \dfrac{L-L_0}{L_0}$. Thus, the tension of F_L in Equation (3) can be rewritten as:

$$F_L = K(L - L_0(1 + \varepsilon_L(t))) \tag{4}$$

where L is the length of the LCE fiber under stress, which can be expressed as $\sqrt{r^2 + (L_0+r)^2 - 2r(L_0+r)\cos\theta}$ based on the cosine theorem of a triangle.

For simplicity, when the velocity is low, the damping force is assumed to be approximately a quadratic function, which is always opposite to the direction of motion:

$$F_D = \beta_1 r\dot{\theta} + \beta_2 r^2\dot{\theta}^2 \tag{5}$$

where β_1 and β_2 represent the linear and quadratic damping coefficients, and $\dot{\theta}$ refers to the angular speed of the system.

By inserting Equations (4) and (5) into Equation (1), considering the counterclockwise and clockwise motion directions, we can obtain:

$$mr\ddot{\theta} = -sgn(\dot{\theta})mgsin\theta - sgn(\dot{\theta})Kr\left(1 - \frac{\frac{L_0}{r}(1+\varepsilon_L(t))}{\sqrt{1+\left(1+\frac{L_0}{r}\right)^2 - 2\left(1+\frac{L_0}{r}\right)cos\theta}}\right)(1+\frac{L_0}{r})\sin\theta - \beta_1 r\dot{\theta} - \beta_2 r^2 \dot{\theta}|\dot{\theta}| \quad (6)$$

2.2. Dynamic LCE Model

This part primarily illustrates the dynamic behavior of the light-actuated contraction in LCE fibers. For simplicity, the contraction of LCE fibers under small deformation is assumed to be a linearly related to the number fraction $\varphi(t)$ cis-isomer in LCE fibers, i.e.,

$$\varepsilon_L(t) = -C_0\varphi(t) \quad (7)$$

where C_0 denotes the contraction coefficient of LCE fibers.

The light-actuated strain of LCE fibers during contraction is dependent on the cis-isomer number fraction $\varphi(t)$ in the LCE. According to the research, the cis-isomer number fraction can be manipulated through exposure to UV light or laser irradiation at wavelengths specifically below 400 nm [71]. However, the precise wavelength within this range significantly impacts the efficiency and dynamics of the isomerization process, with 365 nm being a common choice for azobenzene-based systems due to its effectiveness in triggering the trans-to-cis transition [72]. The cis-isomer number fraction depends on thermally excited *trans*-to-*cis* isomerization, thermally driven relaxation from *cis* to *trans*, and light-driven *trans*-to-*cis* isomerization. Typically, thermally excited *trans*-to-*cis* isomerization can be neglected relative to light-powered *trans*-to-*cis* isomerization. Therefore, the number fraction of the cis-isomer in the LCE is governed by the following equation [73–75]:

$$\frac{\partial \varphi}{\partial t} = \eta_0 I(1-\varphi) - \frac{\varphi}{T_0} \quad (8)$$

where η_0 refers to the constant for light absorption, T_0 denotes the duration of thermally induced relaxation process from *cis* to *trans*, and I represents the intensity of light. By solving Equation (8), the number fraction of cis-isomer can be deduced as:

$$\varphi(t) = \frac{\eta_0 T_0 I}{\eta_0 T_0 I + 1} + \left(\varphi_0 - \frac{\eta_0 T_0 I}{\eta_0 T_0 I + 1}\right)exp\left[-\frac{t}{T_0}(\eta_0 T_0 I + 1)\right] \quad (9)$$

where φ_0 is the initial *cis* number fraction at $t = 0$.

In the region that is illuminated, the initial number fraction is $\varphi_0 = 0$. Then, Equation (9) can be simplified as:

$$\varphi(t) = \frac{\eta_0 T_0 I}{\eta_0 T_0 I + 1}\left\{1 - exp\left[-\frac{t}{T_0}(1+\eta_0 T_0 I)\right]\right\} \quad (10)$$

In the region that is non-illuminated, when the light intensity is set to $I = 0$, we can derive the following:

$$\varphi(t) = \varphi_0 exp\left(-\frac{t}{T_0}\right) \quad (11)$$

where φ_0 can be designated as the peak value of φ in Equation (9), i.e., $\varphi_0 = \frac{\eta_0 T_0 I}{\eta_0 T_0 I + 1}$. Then, Equation (11) can be simplified as:

$$\varphi(t) = \frac{\eta_0 T_0 I}{\eta_0 T_0 I + 1}exp\left(-\frac{t}{T_0}\right) \quad (12)$$

2.3. Nondimensionalization

To simplify calculations, enhance solution efficiency, and improve generality, the following quantities are subjected to nondimensionalization: $\bar{t} = t/T_0$, $\bar{\dot{\theta}} = \dot{\theta} T_0$, $\bar{\ddot{\theta}} = \ddot{\theta} T_0^2$, $\bar{g} = gT_0^2/r$, $\bar{K} = KT_0^2/m$, $\bar{I} = \eta_0 T_0 I$, $\bar{\beta_1} = \beta_1 T_0/m$, $\bar{\beta_2} = \beta_2 r/m$ and $\bar{\varphi} = \frac{\varphi(\eta_0 T_0 I + 1)}{\eta_0 T_0 I}$. Thus, Equation (6) can be written in a dimensionless form as:

$$\bar{\ddot{\theta}} = -sgn(\dot{\theta})\bar{g}\sin\theta - sgn(\dot{\theta})\bar{K}\left(1 - \frac{\frac{L_0}{r}(1+\varepsilon_L(t))}{\sqrt{1+\left(1+\frac{L_0}{r}\right)^2 - 2\left(1+\frac{L_0}{r}\right)\cos\theta}}\right)(1+\frac{L_0}{r})\sin\theta - \bar{\beta_1}\bar{\dot{\theta}} - \bar{\beta_2}\bar{\dot{\theta}}\left|\bar{\dot{\theta}}\right| \quad (13)$$

In the region that is illuminated, Equation (11) can be expressed in another way, as:

$$\bar{\varphi}(t) = 1 - \exp[-\bar{t}(\bar{I}+1)] \quad (14)$$

In the region that is non-illuminated, Equation (12) can be rewritten as:

$$\bar{\varphi}(t) = \exp(-\bar{t}) \quad (15)$$

Meanwhile, the tension of the LCE fiber in Equation (4) and the damping force in Equation (5) can be written in a dimensionless form as:

$$\bar{F_L} = \bar{K}\left(1 - \frac{\frac{L_0}{r}(1+\varepsilon_L(t))}{\sqrt{1+\left(1+\frac{L_0}{r}\right)^2 - 2\left(1+\frac{L_0}{r}\right)\cos\theta}}\right)(1+\frac{L_0}{r})\sin\theta \quad (16)$$

$$\bar{F_D} = \bar{\beta_1}\bar{\dot{\theta}} + \bar{\beta_2}\bar{\dot{\theta}}\left|\bar{\dot{\theta}}\right| \quad (17)$$

The dynamics of the light-actuated sliding mass system are governed by Equations (13)–(15), where the time-dependent number fraction of cis-isomer is intricately linked to the position of the sliding mass. Making use of dimensionless and given variables, including $\bar{I}$, C_0, $\bar{g}$, $\bar{K}$, $\bar{\beta_1}$, $\bar{\beta_2}$, and θ_0, we can derive the temporal evolution of the light-actuated contraction strain and the position of the sliding mass. Equations (13)–(15) are evidently nonlinear differential equations, and it is challenging to acquire the analytical calculation. In the study, the standard fourth-stage Runge–Kutta method and MATLAB R2021a software are applied to numerically calculate the steady-state responses, including the light-actuated contraction strain of the LCE fiber, and the angle and angular velocity of the sliding mass at any time. Moreover, we can further obtain the tension of the LCE fiber from Equation (16) and the damping force from Equation (17).

3. Two Motion States and Mechanism of Self-Sliding

Rooted in the control equations deduced in Section 2, in this section we discuss the dynamics of the light-actuated sliding mass system under a state of steady illumination. Two typical dynamic modes of self-sliding are first presented, which are categorized as the static state and self-sliding state. Subsequently, we describe the corresponding mechanism of self-sliding.

3.1. Two Motion States

To further investigate the self-sliding motion behavior of the LCE fiber–sliding mass system, it is first necessary to determine the typical values of the parameters in the dimensionless equations in Section 2. Based on previously studied outcomes and experimental results [76–78], the empirical values of the parameters required in a sliding mass system are collected and presented in Table 1. Using the variable data from Table 1, the corresponding dimensionless variable values can be derived and are listed in Table 2.

Table 1. Material properties and geometric parameters.

Parameter	Definition	Value	Unit
I	light intensity	0~80	kW/m^2
C_0	contraction coefficient of LCE fiber	0~0.5	/
K	elastic coefficient of LCE fiber	20~40	N/m
T_0	Cis to trans thermal relaxation time	0.02~0.45	s
η_0	light absorption constant	0.002	m^2/(s·W)
m	mass of sliding mass	0~0.02	kg
β_1	linear damping coefficient	0~0.3	kg/s
β_2	quadratic damping coefficient	0~0.15	kg/m
v_0	initial tangential velocity	0~2.5	m/s
θ_0	angle of non-illuminated zone	0~0.5	rad
r	radius of circular track	0.01~0.15	m
L_0	original length of LCE fiber	0.1~0.5	m

Table 2. Dimensionless parameters.

Parameter	$\bar{I}$	C_0	$\bar{K}$	$\bar{v_0}$	$\bar{\beta_1}$	$\bar{\beta_2}$	θ_0
Value	0~5	0~0.5	0~10	0~1	0~0.2	0~0.1	0~0.5

By numerically solving Equation (11), we can acquire the time–history curve of sliding as well as the phase trajectory diagram for the LCE fiber–sliding mass system, which are depicted in Figure 2. The results show the presence of two clearly distinguishable motion states, namely, the static state and self-sliding state, under different light intensities: $\bar{I} = 0.15$ and $\bar{I} = 0.6$. During the numerical computation, we set the other dimensionless variables of the sliding mass system, including $C_0 = 0.45$, $\bar{K} = 2.7$, $\bar{v_0} = 0.7$, $\bar{\beta_1} = 0.015$, $\bar{\beta_2} = 0.003$, $\theta_0 = 0.09$. For $\bar{I} = 0.15$, the system initially slides left and right but eventually comes to a stop due to the damping force, reaching a static state as depicted in Figure 2a. Corresponding to Figure 2a, the phase trajectory in Figure 2b stabilizes at a stationary point. In contrast, as illustrated in Figure 2c,d, where $\bar{I} = 0.6$, the sliding amplitude of the LCE fiber–sliding mass system progressively decreases until it reaches a steady value, which is termed the self-sliding mode. Similarly, in the phase trajectory diagram, the amplitude of the sliding mass gradually becomes constant, with its state eventually settling into a limit cycle, signifying a periodic and stable behavior.

3.2. Mechanism of Self-Sliding

This section is specifically intended to elaborate the self-sliding mechanism, focusing on the inherent energy-balancing mechanism in the system. To aid in a deeper understanding of this complex process, through the plotting of relationship curves, we depict the correlations among several vital variables that are involved in the self-sliding process, as depicted in Figure 3. Herein, the dimensionless variables of the system are selected as $\bar{I} = 0.6$, $C_0 = 0.45$, $\bar{K} = 2.7$, $\bar{v_0} = 0.7$, $\bar{\beta_1} = 0.015$, $\bar{\beta_2} = 0.003$, and $\theta_0 = 0.09$. Figure 3a depicts the change in the angular displacement of the LCE fiber–sliding mass system over time. The yellow highlighted area denotes the illumination region where the LCE fiber is illuminated. It is apparent that the LCE fiber–sliding mass system maintains a steady amplitude and cycle, with the sliding mass moving to and fro within the illuminated regions on the right and left sides. Figure 3b illustrates how the number fraction in the LCE fiber varies with time due to light exposure. When the angular displacement of the sliding mass is more than the angle of non-illuminated zone θ_0, the LCE fiber enters the illuminated regions, and the LCE fiber's number fraction progressively goes up, approaching a specific maximum. When the sliding mass moves from the illuminated regions into the non-illuminated regions, the LCE fiber's number fraction sharply decreases to zero. With the system consistently traversing in and out of the illuminated regions, the LCE fiber's number fraction experiences recurring variations.

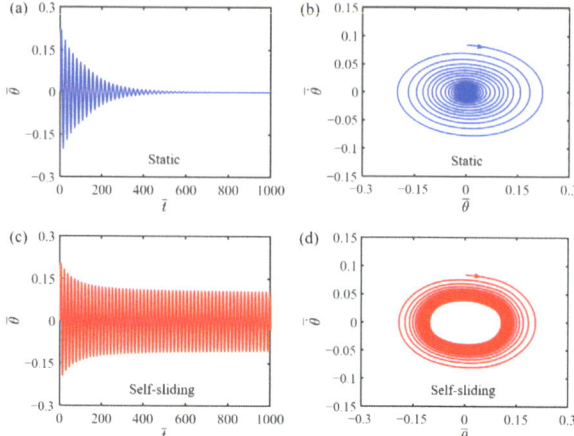

Figure 2. Two typical dynamic modes of the self-sliding structure: static state and self-sliding state. (**a**) Chronological graph of the angular displacement when $\bar{I} = 0.15$; (**b**) phase space plot when $\bar{I} = 0.15$; (**c**) chronological graph of the angular displacement when $\bar{I} = 0.6$; (**d**) the limit cycle in phase space when $\bar{I} = 0.6$. If the other dimensionless variables are the same, two distinct dynamic states can be attained under different light intensities.

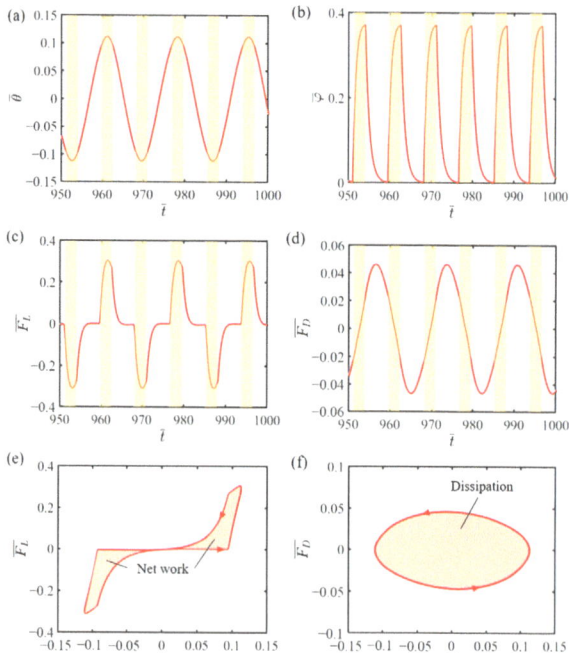

Figure 3. Self-sliding mechanism of the LCE-fiber-based mass system. (**a**) Time–history curve of the angular displacement of the mass. (**b**) Time–history curve of the number fraction of the LCE fiber. (**c**) Variation in the tension of the LCE fiber with time. (**d**) Variation in the damping force with time. (**e**) Dependence of the tension of the LCE fiber on the angular displacement. (**f**) Dependence of the damping force on the angular displacement. Stable self-sliding is maintained in the system due to the light-actuated elastic force, compensating for damping dissipation.

Figure 3c demonstrates the time-dependent tension variations in LCE fiber. The tension changes due to the cyclical self-sliding of the system, showing a periodic trend. As the LCE fiber enters the illuminated regions, the tension in the LCE fiber increases as a consequence of the light-actuated shrinkage. However, when the system departs from the illuminated regions, the tension decreases as the light-actuated contraction reverses, as clearly shown in Figure 3c. As is evident from Figure 3d, the damping force displays a periodic variation over time, similar to the tension variations observed in the LCE fiber.

From Figure 3e, it is evident that the hysteresis loop formed by the tension of the LCE fiber illustrates the net work carried out the tension in one complete sliding cycle, which is numerically evaluated to be 0.085. Similarly, Figure 3f reveals the linkage of the damping force with the angular displacement, with the enclosed hysteresis curve signifying the amount of work carried out by the damping force in a full sliding cycle, representing the system's damping dissipation. Calculations reveal that the area encompassed by the hysteresis loop in Figure 3f also equals 0.085, indicating that the energy dissipated due to the damping force during self-sliding is balanced by the work generated by the tensile force of the LCE fiber. Consequently, the self-sliding of the LCE fiber–sliding mass system remains sustainable.

Additionally, Figure 4 displays several defining snapshots representing the self-sliding motion of the LCE-fiber-based mass system during a complete sliding cycle under constant light exposure. As the LCE-fiber-based mass system moves from a non-illuminated region to an illuminated region, the increase in the number fraction $\varphi(t)$ of the LCE fiber results in a corresponding increase in the contraction strain, reaching a maximum value, as shown in Figure 4a,c. During this process, the system decelerates due to the combined effects of damping forces and the negative work performed via the tension of the LCE fiber, ultimately reducing the velocity to zero. Conversely, when the sliding mass moves from the illuminated regions into the non-illuminated regions, the sharp decrease in the LCE fiber's number fraction leads to a gradual reduction in the contraction strain, as depicted in Figure 4b,d. Despite this decrease in contraction strain, the tension of the LCE fiber has a positive effect, enabling the system to re-enter the illuminated region and absorb light energy to compensate for the negative effects of damping. This cyclic process allows for continuous self-sliding motion. The variation in the contraction strain of the LCE fiber, as described in Figure 4, aligns with the theoretical framework outlined in the study and specifically with Equation (7).

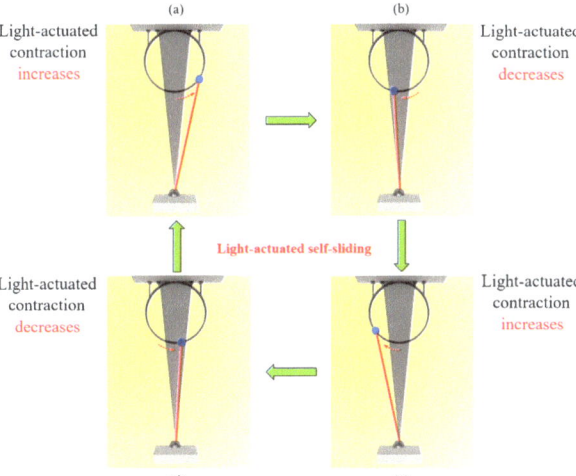

Figure 4. Snapshots of the self-sliding motion in a cycle. (**a**) Move into right illuminated region; (**b**) Sliding in right illuminated region; (**c**) Move into the left illuminated region; (**d**). Sliding in left illuminated region. Under constant light exposure, the periodic changes in contraction actuated by light result in a continual and recurring self-sliding behavior of the system.

4. Parameter Study

When dynamically modeling the self-sliding behavior of the LCE fiber–sliding mass system, we consider six dimensionless system variables, namely, $\bar{I}$, C_0, $\bar{K}$, $\bar{v}_0$, $\bar{\beta}_1$, $\bar{\beta}_2$, and θ_0. In this section, we undertake a quantitative analysis to examine the impact of these variables on the self-sliding characteristics of the LCE fiber–sliding mass system, particularly highlighting their influence on the amplitude and frequency. Specifically, the dimensionless amplitude and frequency of self-sliding are denoted as A and F, respectively.

4.1. Influence of the Light Intensity

Figure 5 demonstrates how light intensity influences the self-sliding, considering specific values of the additional dimensionless variables where $C_0 = 0.45$, $\bar{K} = 2.7$, $\bar{v}_0 = 0.7$, $\bar{\beta}_1 = 0.015$, $\bar{\beta}_2 = 0.003$, and $\theta_0 = 0.09$. The corresponding limit cycles of self-sliding for $\bar{I}$ values 0.6, 0.9, and 1.2 are presented in Figure 5a. The amplitude of self-sliding is expressed through the horizontally measured width of the limit cycle, with the vertical height representing the angular velocity of the self-sliding. It can be seen from Figure 5a that the critical light intensity separating the static state from self-sliding state is $\bar{I} = 0.596$. At light intensities lower than 0.596, the LCE fiber lacks the necessary light energy absorption to overcome the damping dissipation, resulting in a transition to a static state due to an inability to sustain motion. In contrast, light intensities exceeding 0.596 allow the LCE fiber to absorb sufficient energy to overcome damping dissipation, thereby sustaining continuous and steady self-sliding, which characterizes the self-sliding state. The relationship between light intensity and its influence on amplitude and frequency is shown in Figure 5b. It can be seen from Figure 5b that the amplitude and frequency exhibit a direct correlation as the light intensity rises. The reason for this is stronger light intensities facilitate the LCE fiber's ability to obtain a greater amount of energy and convert it into kinetic energy, enabling the system to achieve a higher amplitude. The findings indicate that increasing the light intensity plays an important role in enhancing the efficiency of energy utilization within the LCE fiber–sliding mass system.

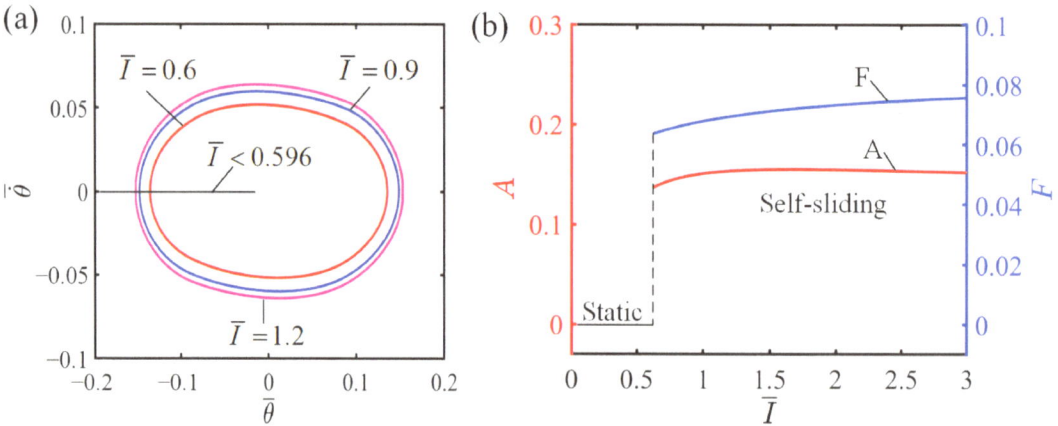

Figure 5. Influence of light intensity on the self-sliding system. (a) Limit cycles with $\bar{I} = 0.6$, $\bar{I} = 0.9$, and $\bar{I} = 1.2$. (b) Changes in amplitude and frequency with varying light intensities.

4.2. Influence of the Contraction Coefficient of LCE

Figure 6 demonstrates how the contraction coefficient influences the self-sliding considering specific values of the additional dimensionless variables, where $\bar{I} = 0.6$, $\bar{K} = 2.7$, $\bar{v}_0 = 0.7$, $\bar{\beta}_1 = 0.015$, $\bar{\beta}_2 = 0.003$, and $\theta_0 = 0.09$. The corresponding limit cycles of self-sliding for C_0 values 0.4, 0.45, and 0.5 are presented in Figure 6a. As observed in Figure 6a, the limit cycle associated with a higher contraction coefficient completely envelopes the one with a lower coefficient, showing that a decrease in the contraction coefficient weakens the LCE fiber–sliding mass system's ability to absorb light, leading to a decreased amplitude and kinetic energy. Figure 6b illustrates how the amplitude and frequency of self-sliding are influenced by the contraction coefficient. As observed in Figure 6b, a distinct threshold value exists for the contraction coefficient, which is mathematically determined to be 0.398, and serves as the tipping point for inducing self-sliding. Below a contraction coefficient of 0.398, the sliding mass maintains a stationary condition. However, once the contraction coefficient rises above 0.398, the system enters a self-sliding state. The finding suggests that the efficient conversion of light energy to mechanical energy can be improved by increasing the contraction coefficient of the LCE fiber.

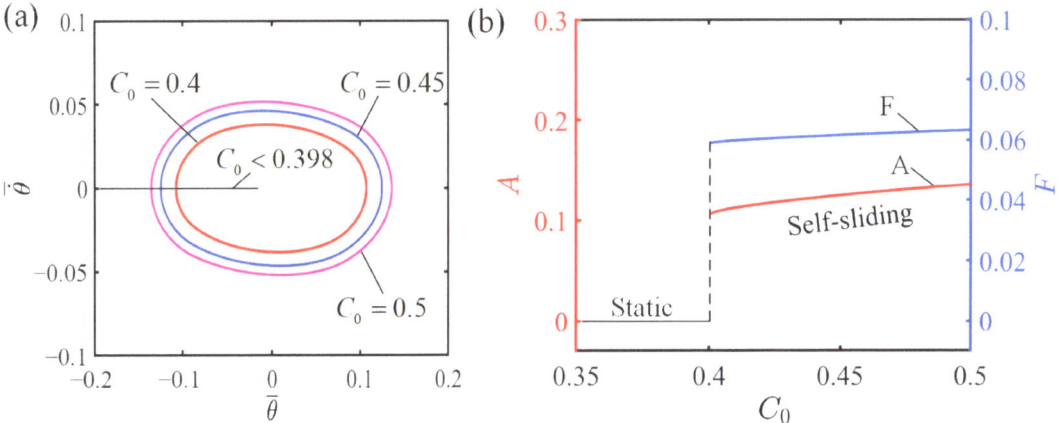

Figure 6. Influence of contraction coefficient on the self-sliding system. (**a**) Limit cycles with $C_0 = 0.4$, $C_0 = 0.45$, and $C_0 = 0.5$. (**b**) Changes in amplitude and frequency with varying contraction coefficients.

4.3. Influence of the Elastic Coefficient of LCE

Figure 7 demonstrates how the elastic coefficient influences the self-sliding, considering the specific values of the additional dimensionless variables, where $\bar{I} = 0.6$, $C_0 = 0.45$, $\bar{v}_0 = 0.7$, $\bar{\beta}_1 = 0.015$, $\bar{\beta}_2 = 0.003$, and $\theta_0 = 0.09$. For $\bar{K} = 2.7$, $\bar{K} = 3.7$, and $\bar{K} = 4.7$, the corresponding limit cycles of self-sliding are presented in Figure 7a. From Figure 7a, it is evident that $\bar{K} = 2.257$ represents the threshold value of the elastic coefficient that determines whether the system remains in static mode or self-sliding mode. Under steady illumination, if the elastic coefficient is smaller than 2.257, the LCE fiber is unable to absorb adequate light energy. Consequently, the system lacks sufficient energy to counteract damping dissipation and eventually settles into a static state. Conversely, when the elastic coefficient is more than 2.257, the LCE fiber has the ability to absorb an ample amount of energy, counteracting the system's damping dissipation and thereby sustaining self-sliding. As depicted in Figure 7b, the elastic coefficient plays a vital role in determining the magnitude and periodicity of the self-sliding. As the elastic coefficient increases, there is a corresponding increase in the magnitude and periodicity of self-sliding. This is because a greater elastic coefficient generates a stronger elastic force from the LCE fiber. As a result, the system gains more elastic potential energy that can be converted into kinetic energy, thus leading to a higher amplitude of self-sliding. Therefore, selecting the appropriate elastic coefficient is essential in achieving superior performance when designing an LCE-based tension system.

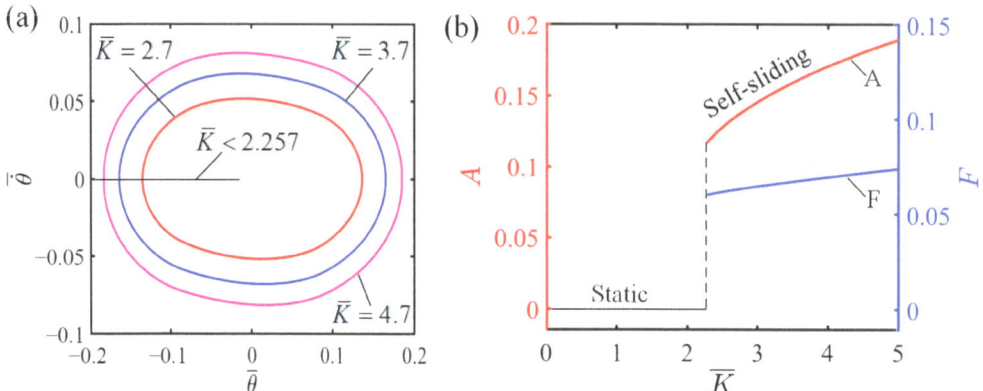

Figure 7. Influence of elastic coefficient on the self-sliding system. (**a**) Limit cycles with $\overline{K} = 2.7$, $\overline{K} = 3.7$, and $\overline{K} = 4.7$ (**b**) Changes in amplitude and frequency with varying elastic coefficients.

4.4. Influence of the Initial Tangential Velocity

Figure 8 demonstrates how initial tangential velocity influences the self-sliding, considering specific values of the additional dimensionless variables where $\overline{I} = 0.6$, $C_0 = 0.45$, $\overline{K} = 2.7$, $\overline{\beta_1} = 0.015$, $\overline{\beta_2} = 0.003$, and $\theta_0 = 0.09$. The numerical simulations reveal that initiating self-sliding in the LCE fiber–sliding mass system is possible with initial tangential velocities of $\overline{v_0} = 0.3$, $\overline{v_0} = 0.5$, and $\overline{v_0} = 0.7$. The limit cycles resulting from these velocities are presented in Figure 8a and it is worth mentioning that all three limit cycles overlap.

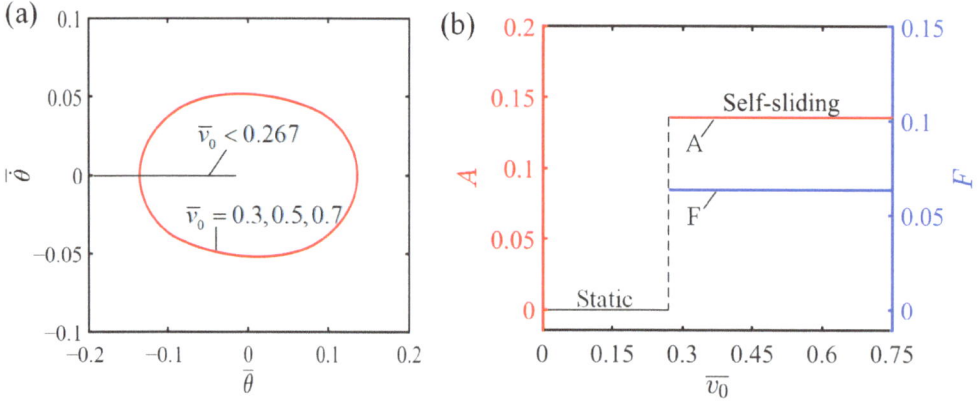

Figure 8. Influence of initial tangential velocity on the self-sliding system. (**a**) Limit cycles with $\overline{v_0} = 0.3$, $\overline{v_0} = 0.5$, and $\overline{v_0} = 0.7$. (**b**) Changes in amplitude and frequency with varying initial tangential velocities.

Figure 8b represents the dependency of the characteristics of self-sliding on the initial tangential velocity, namely its amplitude and frequency. Based on Figure 8b, it is evident that the system enters a static state when the initial tangential velocity is below 0.267. This occurs because the insufficient initial tangential velocity prevents the LCE fiber from accessing the illumination zone, thereby hindering the absorption of the light energy necessary to maintain dynamic motion, ultimately resulting in a static state. When the initial tangential velocity exceeds 0.267, given $\overline{v_0} = 0.3$, $\overline{v_0} = 0.5$, and $\overline{v_0} = 0.7$, the system enters a self-sliding state, during which the ultimate amplitude and frequency are unaffected by the variable initial tangential velocities. The reason for this is that the amplitude of self-sliding is determined by the energy conversion process between damping dissipation and the net work carried out by the LCE fiber, which form the system's internal properties. Since the initial tangential velocity has no impact on the system's energy conversion mechanism, the amplitude remains unaffected.

4.5. Influence of the Damping Coefficients

Figure 9 demonstrates how damping coefficients influence the self-sliding, considering specific values of the additional dimensionless variables where $\bar{I} = 0.6$, $C_0 = 0.45$, $\bar{K} = 2.7$, $\bar{v}_0 = 0.7$, and $\theta_0 = 0.09$.

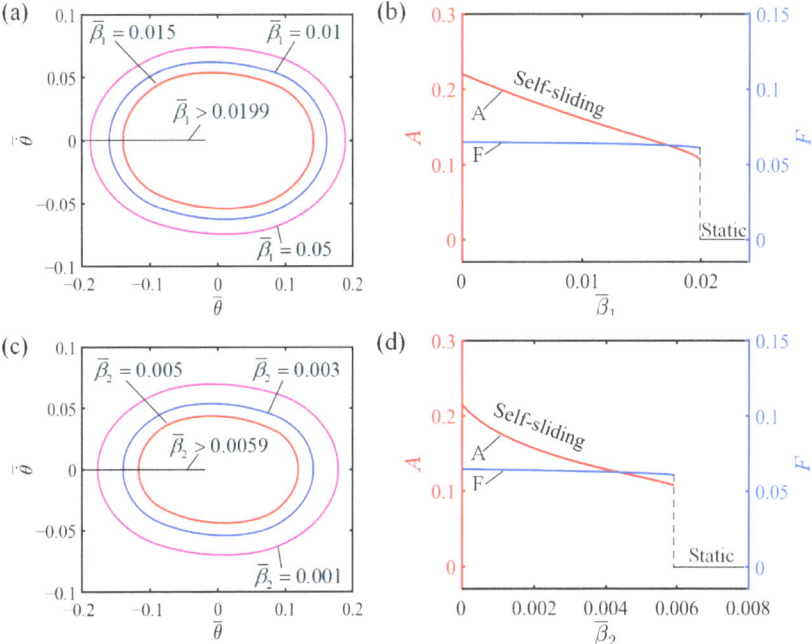

Figure 9. Influence of damping coefficients on the self-sliding system. (a) Limit cycles with $\bar{\beta}_1 = 0.005$, $\bar{\beta}_1 = 0.01$, and $\bar{\beta}_1 = 0.015$. (c) Limit cycles with $\bar{\beta}_2 = 0.001$, $\bar{\beta}_2 = 0.003$, and $\bar{\beta}_2 = 0.005$. (b,d) Changes in amplitude and frequency with varying damping coefficients.

Figure 9a,c further illustrate that as the damping coefficient increases, the limit cycle decreases in size. When the damping coefficient surpasses a certain critical value, the system transitions from a self-sliding state to a static state. As the damping coefficient rises, a greater amount of energy becomes necessary to overcome the energy dissipation that occurs due to the damping forces, thus marking this transitional change. The energy harvested by the LCE in the illuminated zone becomes insufficient to overcome the damping forces, ultimately leading the system to settle into a static state. In practical system design, it is imperative to devise strategies that minimize energy losses due to resistance, enabling the energy generated by the LCE fiber in the illuminated zone to be more efficiently utilized for the system's intrinsic self-sliding motion. From Figure 9b,d, it is evident that whether it is $\bar{\beta}_1$ or $\bar{\beta}_2$, a smaller damping coefficient results in a larger amplitude of self-sliding, and vice versa. The amplitude is significantly affected by the damping coefficient, whereas the impact on the frequency of self-sliding is less significant.

4.6. Influence of the Angle of Non-Illuminated Zone

Figure 10 demonstrates how non-illuminated zone angle influences the self-sliding, considering specific values of the additional dimensionless variables where $\bar{I} = 0.6$, $C_0 = 0.45$, $\bar{K} = 2.7$, $\bar{v_0} = 0.7$, $\bar{\beta_1} = 0.015$, and $\bar{\beta_2} = 0.003$. For $\theta_0 = 0.03$, $\theta_0 = 0.09$, and $\theta_0 = 0.15$, the corresponding limit cycles of self-sliding are presented in Figure 10a. It can be seen from Figure 10a that the critical angle of the non-illuminated zone separating the static state from the self-sliding state is $\theta_0 = 0.154$. Once the angle of the non-illuminated zone exceeds 0.154, the LCE fiber is incapable of entering the illuminated zone to gather sufficient light energy to overcome the damping-induced energy loss. Consequently, due to the exhaustion of the kinetic energy, the system inevitably transitions into a static state. In contrast, when the angle of the non-illuminated zone is less than 0.154, the LCE fiber is able to reach the illuminated zone to absorb sufficient energy to overcome damping dissipation, thereby sustaining continuous and steady self-sliding, which characterizes the self-sliding state. The relationship between the angle of the non-illuminated zone and its influence on amplitude and frequency is shown in Figure 10b. It can be seen from Figure 10b that the amplitude exhibits a direct correlation with the increase in the angle of the non-illuminated zone. The explanation lies in the fact that, with a small angular non-illumination, the LCE fiber promptly enters the illuminated zone, causing its increasing tension to restrict the sliding mass from continuing its forward movement and thereby limiting the amplitude of self-sliding. In contrast, a larger angular non-illumination allows for greater angular displacement of the LCE fiber–sliding mass system before it reaches the illuminated zone, subsequently enhancing the total amplitude of self-sliding.

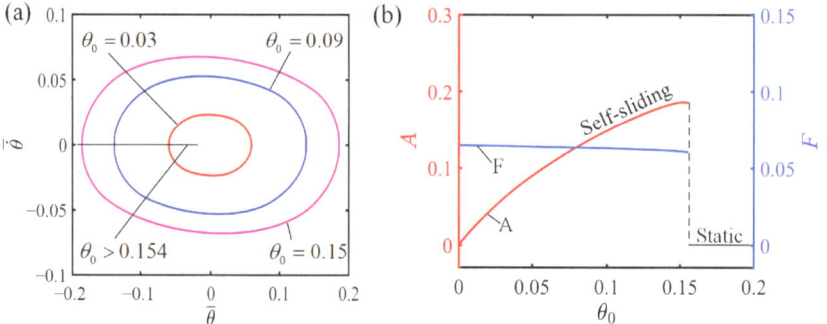

Figure 10. Influence of the angle of the non-illuminated zone on the self-sliding system. (**a**) Limit cycles with $\theta_0 = 0.03$, $\theta_0 = 0.09$, and $\theta_0 = 0.15$. (**b**) Changes in amplitude and frequency with varying angles of non-illuminated zone.

5. Conclusions

Existing self-vibrating systems, despite offering adaptability, efficiency, and sustainability, are complex, hard to produce, and challenging to control. This highlights the urgency of developing more advanced and practical self-oscillation structures. In the study, we propose a light-actuated LCE fiber–sliding mass system consisting of a light-actuated LCE fiber, a sliding mass, and a rigid circular track, which can sustain periodic and continuous sliding under constant light exposure. Derived from the available dynamic behavior model of LCEs, and the theorem of angular momentum and vibration theory, the equations for dynamic control of the sliding mass enabled by the light-actuated LCE fiber are deduced. Using the standard fourth-stage Runge–Kutta method and MATLAB R2021a software, the numerical calculation of the dynamic control equations is acquired. According to the results, two motion states of the self-sliding mass system are explained in detail, which are categorized as the static state and self-sliding state. In particular, the self-sliding process, along with its energy balancing system, is clarified. Herein, the energy drawn from a constant external source serves to balance out the dissipation caused by system damping, preserving its dynamic equilibrium.

Moreover, the influences of light intensity, contraction coefficient, elastic coefficient, initial tangential velocity, linear and quadratic damping coefficients, and angle of the non-illuminated zone are quantitatively analyzed. The numerical calculation results reveal an increase in light intensity, contraction coefficient, elastic coefficient, and angle of the non-illuminated zone, which leads to an augmentation in both the amplitude and frequency of the system. Specifically, the elastic coefficient and the angle of the non-illuminated zone exert a profound influence on the amplitude, with each $\pi/180$ increment in amplitude leading to a 0.25% increase in the elastic coefficient and a 1.63% expansion in the angle of the non-illuminated zone. Additionally, light intensity plays a crucial role, contributing to a 20.88% augmentation for every increment of $\pi/180$ in the amplitude. In contrast, while an increase in the damping coefficient leads to a notable reduction in both the amplitude and frequency, the initial tangential velocity has no discernible effect on either the amplitude or the frequency of the system.

While the simplicity, controllability, and rapid response to the light of the proposed LCE fiber–sliding mass system render it a promising candidate for widespread adoption, it is important to acknowledge the limitations of the small deformation assumption, the simplified damping consideration, and the neglect of viscoelastic effects in LCE fibers. To further enhance the system's potential for broad applications, future research will aim to incorporate viscoelasticity into the model and investigate the system's performance on non-circular curved tracks.

Supplementary Materials: The following supporting information can be downloaded at: https://www.mdpi.com/article/10.3390/polym16121696/s1, Video S1: The process of the self-sliding motion of the system in a cycle.

Author Contributions: Conceptualization, K.L.; methodology, L.W.; software, H.W.; validation, K.L.; investigation, J.H. and J.W.; data curation, H.W., J.H., and J.W.; drafting the original manuscript, L.W.; reviewing and revising the manuscript, L.W. and K.L.; supervision, K.L. All authors have read and agreed to the published version of the manuscript.

Funding: This research is supported by University Natural Science Research Project of Anhui Province (Grant Nos. KJ2021A0609 and 2022AH020029), National Natural Science Foundation of China (Grant No. 12172001), Anhui Provincial Natural Science Foundation (No. 2208085Y01), and Housing and Urban-Rural Development Science and Technology Project of Anhui Province (Grant No. 2023-YF036).

Institutional Review Board Statement: Not applicable.

Data Availability Statement: The original contributions presented in the study are included in the article/Supplementary Material, further inquiries can be directed to the corresponding authors.

Conflicts of Interest: The authors declare no conflicts of interest.

References

1. Ding, W. *Self-Excited Vibration*; Tsing-Hua University Press: Beijing, China, 2009.
2. Wu, H.; Lou, J.; Dai, Y.; Zhang, B.; Li, K. Bifurcation analysis in liquid crystal elastomer spring self-oscillators under linear light fields. *Chaos Solitons Fractals* **2024**, *181*, 114587. [CrossRef]
3. Bai, C.; Kang, J.; Wang, Y.Q. Light-induced motion of three-dimensional pendulum with liquid crystal elastomeric fiber. *Int. J. Mech. Sci.* **2024**, *266*, 108911. [CrossRef]
4. Yang, H.; Zhang, C.; Chen, B.; Wang, Z.; Xu, Y.; Xiao, R. Bioinspired design of stimuli-responsive artificial muscles with multiple actuation modes. *Smart Mater. Struct.* **2023**, *32*, 085023. [CrossRef]
5. Wang, Y.; Liu, J.; Yang, S. Multi-functional liquid crystal elastomer composites. *Appl. Phys. Rev.* **2022**, *9*, 011301. [CrossRef]
6. Wang, X.-Q.; Tan, C.F.; Chan, K.H.; Lu, X.; Zhu, L.; Kim, S.-W.; Ho, G.W. In-built thermo-mechanical cooperative feedback mechanism for self-propelled multimodal locomotion and electricity generation. *Nat. Commun.* **2018**, *9*, 1–10. [CrossRef]
7. Ge, F.; Yang, R.; Tong, X.; Camerel, F.; Zhao, Y. A multifunctional dye-doped liquid crystal polymer actuator: Light-guided transportation, turning in locomotion, and autonomous motion. *Angew. Chem.* **2018**, *130*, 11932–11937. [CrossRef]

8. Papangelo, A.; Putignano, C.; Hoffmann, N. Self-excited vibrations due to viscoelastic interactions. *Mech. Syst. Signal Process.* **2020**, *144*, 106894. [CrossRef]
9. Fu, M.; Burkart, T.; Maryshev, I.; Franquelim, H.G.; Merino-Salomón, A.; Reverte-López, M.; Frey, E.; Schwille, P. Mechanochemical feedback loop drives persistent motion of liposomes. *Nat. Phys.* **2023**, *19*, 1211–1218. [CrossRef]
10. Preston, D.J.; Jiang, H.J.; Sanchez, V.; Rothemund, P.; Rawson, J.; Nemitz, M.P.; Lee, W.-K.; Suo, Z.; Walsh, C.J.; Whitesides, G.M. A soft ring oscillator. *Sci. Robot.* **2019**, *4*, eaaw5496. [CrossRef]
11. Liu, J.; Shi, F.; Song, W.; Dai, Y.; Li, K. Modeling of self-oscillating flexible circuits based on liquid crystal elastomers. *Int. J. Mech. Sci.* **2024**, *270*, 109099. [CrossRef]
12. Wu, H.; Zhao, C.; Dai, Y.; Li, K. Modeling of a light-fueled self-paddling boat with a liquid crystal elastomer-based motor. *Phys. Rev. E* **2024**, *109*, 044705. [CrossRef]
13. Korner, K.; Kuenstler, A.S.; Hayward, R.C.; Audoly, B.; Bhattacharya, K. A nonlinear beam model of photomotile structures. *Proc. Natl. Acad. Sci. USA* **2020**, *117*, 9762–9770. [CrossRef]
14. Brighenti, R.; Artoni, F.; Cosma, M.P. Mechanics of materials with embedded unstable molecules. *Int. J. Solids Struct.* **2019**, *162*, 21–35. [CrossRef]
15. Lee, V.; Bhattacharya, K. Actuation of cylindrical nematic elastomer balloons. *J. Appl. Phys.* **2021**, *129*, 114701. [CrossRef]
16. Parrany, A.M. Nonlinear light-induced vibration behavior of liquid crystal elastomer beam. *Int. J. Mech. Sci.* **2018**, *136*, 179–187. [CrossRef]
17. Bai, R.; Bhattacharya, K. Photomechanical coupling in photoactive nematic elastomers. *J. Mech. Phys. Solids* **2020**, *144*, 104115. [CrossRef]
18. Lee, Y.; Koehler, F.; Dillon, T.; Loke, G.; Kim, Y.; Marion, J.; Antonini, M.J.; Garwood, I.C.; Sahasrabudhe, A.; Nagao, K. Magnetically Actuated Fiber-Based Soft Robots. *Adv. Mater.* **2023**, *35*, 2301916. [CrossRef]
19. Zhao, D.; Liu, Y.; Liu, C. Transverse vibration of nematic elastomer Timoshenko beams. *Phys. Rev. E* **2017**, *95*, 012703. [CrossRef]
20. Zheng, R.; Ma, L.; Feng, W.; Pan, J.; Wang, Z.; Chen, Z.; Zhang, Y.; Li, C.; Chen, P.; Bisoyi, H.K. Autonomous Self-Sustained Liquid Crystal Actuators Enabling Active Photonic Applications. *Adv. Funct. Mater.* **2023**, *33*, 2301142. [CrossRef]
21. Serak, S.; Tabiryan, N.; Vergara, R.; White, T.J.; Vaia, R.A.; Bunning, T.J. Liquid crystalline polymer cantilever oscillators fueled by light. *Soft Matter* **2010**, *6*, 779–783. [CrossRef]
22. Haberl, J.M.; Sánchez-Ferrer, A.; Mihut, A.M.; Dietsch, H.; Hirt, A.M.; Mezzenga, R. Liquid-crystalline elastomer-nanoparticle hybrids with reversible switch of magnetic memory. *Adv. Funct. Mater.* **2013**, *25*, 1787–1791. [CrossRef]
23. Tang, R.; Liu, Z.; Xu, D.; Liu, J.; Yu, L.; Yu, H. Optical pendulum generator based on photomechanical liquid-crystalline actuators. *ACS Appl. Mater. Interfaces* **2015**, *7*, 8393–8397. [CrossRef]
24. Hou, W.; Wang, J.; Lv, J. Bioinspired liquid crystalline spinning enables scalable fabrication of high-performing fibrous artificial muscles. *Adv. Mater.* **2023**, *35*, 2211800. [CrossRef]
25. Cheng, Y.C.; Lu, H.C.; Lee, X.; Zeng, H.; Priimagi, A. Kirigami-based light-induced shape-morphing and locomotion. *Adv. Mater.* **2020**, *32*, 1906233. [CrossRef]
26. Vantomme, G.; Gelebart, A.; Broer, D.; Meijer, E. A four-blade light-driven plastic mill based on hydrazone liquid-crystal networks. *Tetrahedron* **2017**, *73*, 4963–4967. [CrossRef]
27. Hauser, A.W.; Sundaram, S.; Hayward, R.C. Photothermocapillary oscillators. *Phys. Rev. Lett.* **2018**, *121*, 158001. [CrossRef]
28. Shin, B.; Ha, J.; Lee, M.; Park, K.; Park, G.H.; Choi, T.H.; Cho, K.-J.; Kim, H.-Y. Hygrobot: A self-locomotive ratcheted actuator powered by environmental humidity. *Sci. Robot.* **2018**, *3*, eaar2629. [CrossRef]
29. Kuenstler, A.S.; Chen, Y.; Bui, P.; Kim, H.; DeSimone, A.; Jin, L.; Hayward, R.C. Blueprinting photothermal shape-morphing of liquid crystal elastomers. *Adv. Mater.* **2020**, *32*, 2000609. [CrossRef]
30. Hua, M.; Kim, C.; Du, Y.; Wu, D.; Bai, R.; He, X. Swaying gel: Chemo-mechanical self-oscillation based on dynamic buckling. *Matter* **2021**, *4*, 1029–1041. [CrossRef]
31. Kim, Y.; van den Berg, J.; Crosby, A.J. Autonomous snapping and jumping polymer gels. *Nat. Mater.* **2021**, *20*, 1695–1701. [CrossRef]
32. Pathak, S.; Jain, K.; Pant, R. Improved magneto-viscoelasticity of cross-linked PVA hydrogels using magnetic nanoparticles. *Colloids Surf. A Physicochem. Eng. Asp.* **2018**, *539*, 273–279.
33. Jin, B.; Liu, J.; Shi, Y.; Chen, G.; Zhao, Q.; Yang, S. Solvent-assisted 4D programming and reprogramming of liquid crystalline organogels. *Adv. Mater.* **2022**, *34*, 2107855. [CrossRef] [PubMed]
34. Wang, Y.; Yin, R.; Jin, L.; Liu, M.; Gao, Y.; Raney, J.; Yang, S. 3D-Printed Photoresponsive Liquid Crystal Elastomer Composites for Free-Form Actuation. *Adv. Funct. Mater.* **2023**, *33*, 2210614. [CrossRef]
35. Wang, L.; Wei, Z.; Xu, Z.; Yu, Q.; Wu, Z.L.; Wang, Z.; Qian, J.; Xiao, R. Shape Morphing of 3D Printed Liquid Crystal Elastomer Structures with Precuts. *ACS Appl. Polym. Mater.* **2023**, *5*, 7477–7484. [CrossRef]
36. Dai, L.; Wang, L.; Chen, B.; Xu, Z.; Wang, Z.; Xiao, R. Shape memory behaviors of 3D printed liquid crystal elastomers. *Soft Sci.* **2023**, *3*, 4.
37. Chakrabarti, A.; Choi, G.P.; Mahadevan, L. Self-excited motions of volatile drops on swellable sheets. *Phys. Rev. Lett.* **2020**, *124*, 258002. [CrossRef]
38. Guo, Y.; Liu, N.; Cao, Q.; Cheng, X.; Zhang, P.; Guan, Q.; Zheng, W.; He, G.; Chen, J. Photothermal diol for NIR-responsive liquid crystal elastomers. *ACS Appl. Polym. Mater.* **2022**, *4*, 6202–6210. [CrossRef]

39. Cui, Y.; Yin, Y.; Wang, C.; Sim, K.; Li, Y.; Yu, C.; Song, J. Transient thermo-mechanical analysis for bimorph soft robot based on thermally responsive liquid crystal elastomers. *Appl. Math. Mech.* **2019**, *40*, 943–952. [CrossRef]
40. Zhao, D.; Liu, Y. A prototype for light-electric harvester based on light sensitive liquid crystal elastomer cantilever. *Energy* **2020**, *198*, 117351. [CrossRef]
41. Lu, X.; Zhang, H.; Fei, G.; Yu, B.; Tong, X.; Xia, H.; Zhao, Y. Liquid-crystalline dynamic networks doped with gold nanorods showing enhanced photocontrol of actuation. *Adv. Mater.* **2018**, *30*, 1706597. [CrossRef]
42. Kageyama, Y.; Ikegami, T.; Satonaga, S.; Obara, K.; Sato, H.; Takeda, S. Light-Driven Flipping of Azobenzene Assemblies—Sparse Crystal Structures and Responsive Behaviour to Polarised Light. *Chem. A Eur. J.* **2020**, *26*, 10759–10768. [CrossRef] [PubMed]
43. Gelebart, A.H.; Jan Mulder, D.; Varga, M.; Konya, A.; Vantomme, G.; Meijer, E.; Selinger, R.L.; Broer, D.J. Making waves in a photoactive polymer film. *Nature* **2017**, *546*, 632–636. [CrossRef] [PubMed]
44. Wang, Y.; Dang, A.; Zhang, Z.; Yin, R.; Gao, Y.; Feng, L.; Yang, S. Repeatable and reprogrammable shape morphing from photoresponsive gold nanorod/liquid crystal elastomers. *Adv. Mater.* **2020**, *32*, 2004270. [CrossRef] [PubMed]
45. Pilz da Cunha, M.; Peeketi, A.R.; Ramgopal, A.; Annabattula, R.K.; Schenning, A.P. Light-Driven Continual Oscillatory Rocking of a Polymer Film. *ChemistryOpen* **2020**, *9*, 1149–1152. [CrossRef] [PubMed]
46. Manna, R.K.; Shklyaev, O.E.; Balazs, A.C. Chemical pumps and flexible sheets spontaneously form self-regulating oscillators in solution. *Proc. Natl. Acad. Sci. USA* **2021**, *118*, e2022987118. [CrossRef] [PubMed]
47. Li, Z.; Myung, N.V.; Yin, Y. Light-powered soft steam engines for self-adaptive oscillation and biomimetic swimming. *Sci. Robot.* **2021**, *6*, eabi4523. [CrossRef] [PubMed]
48. Wu, H.; Zhang, B.; Li, K. Synchronous behaviors of three coupled liquid crystal elastomer-based spring oscillators under linear temperature fields. *Phys. Rev. E* **2024**, *109*, 024701. [CrossRef] [PubMed]
49. Bazir, A.; Baumann, A.; Ziebert, F.; Kulić, I.M. Dynamics of fiberboids. *Soft Matter* **2020**, *16*, 5210–5223. [CrossRef] [PubMed]
50. Qiu, Y.; Wu, H.; Dai, Y.; Li, K. Behavior prediction and inverse design for self-rotating skipping ropes based on random forest and neural network. *Mathematics* **2024**, *12*, 1019. [CrossRef]
51. Liu, X.; Liu, Y. Spontaneous photo-buckling of a liquid crystal elastomer membrane. *Int. J. Mech. Sci.* **2021**, *201*, 106473. [CrossRef]
52. Wu, H.; Zhao, C.; Dai, Y.; Li, K. Light-fueled self-fluttering aircraft with a liquid crystal elastomer-based engine. *Commun. Nonlinear Sci. Numer. Simul.* **2024**, *132*, 107942. [CrossRef]
53. Kumar, K.; Knie, C.; Bléger, D.; Peletier, M.A.; Friedrich, H.; Hecht, S.; Broer, D.J.; Debije, M.G.; Schenning, A.P. A chaotic self-oscillating sunlight-driven polymer actuator. *Nat. Commun.* **2016**, *7*, 11975. [CrossRef] [PubMed]
54. Wu, H.; Dai, Y.; Li, K.; Xu, P. Theoretical study of chaotic jumping of liquid crystal elastomer ball under periodic illumination. *Nonlinear Dyn.* **2024**, *112*, 7799–7815. [CrossRef]
55. Xu, P.; Chen, Y.; Wu, H.; Dai, Y.; Li, K. Chaotic motion behaviors of liquid crystal elastomer pendulum under periodic illumination. *Results Phys.* **2024**, *56*, 107332. [CrossRef]
56. Liao, B.; Zang, H.; Chen, M.; Wang, Y.; Lang, X.; Zhu, N.; Yang, Z.; Yi, Y. Soft rod-climbing robot inspired by winding locomotion of snake. *Soft Robot.* **2020**, *7*, 500–511. [PubMed]
57. Wehner, M.; Truby, R.L.; Fitzgerald, D.J.; Mosadegh, B.; Whitesides, G.M.; Lewis, J.A.; Wood, R.J. An integrated design and fabrication strategy for entirely soft, autonomous robots. *Nature* **2016**, *536*, 451–455. [CrossRef] [PubMed]
58. Nocentini, S.; Parmeggiani, C.; Martella, D.; Wiersma, D.S. Optically driven soft micro robotics. *Adv. Opt. Mater.* **2018**, *6*, 1800207. [CrossRef]
59. Liao, W.; Yang, Z. The integration of sensing and actuating based on a simple design fiber actuator towards intelligent soft robots. *Adv. Mater. Technol.* **2022**, *7*, 2101260. [CrossRef]
60. Bartlett, N.W.; Tolley, M.T.; Overvelde, J.T.; Weaver, J.C.; Mosadegh, B.; Bertoldi, K.; Whitesides, G.M.; Wood, R.J. A 3D-printed, functionally graded soft robot powered by combustion. *Science* **2015**, *349*, 161–165. [CrossRef]
61. He, Q.; Yin, R.; Hua, Y.; Jiao, W.; Mo, C.; Shu, H.; Raney, J.R. A modular strategy for distributed, embodied control of electronics-free soft robots. *Sci. Adv.* **2023**, *9*, eade9247. [CrossRef]
62. Herbert, K.M.; Fowler, H.E.; McCracken, J.M.; Schlafmann, K.R.; Koch, J.A.; White, T.J. Synthesis and alignment of liquid crystalline elastomers. *Nat. Rev. Mater.* **2022**, *7*, 23–38. [CrossRef]
63. He, Q.; Wang, Z.; Wang, Y.; Wang, Z.; Li, C.; Annapooranan, R.; Zeng, J.; Chen, R.; Cai, S. Electrospun liquid crystal elastomer microffber actuator. *Sci. Robot* **2021**, *6*, eabi9704. [CrossRef] [PubMed]
64. Chun, S.; Pang, C.; Cho, S.B. A micropillar-assisted versatile strategy for highly sensitive and efficient triboelectric energy generation under in-plane stimuli. *Adv. Mater.* **2020**, *32*, 1905539. [CrossRef] [PubMed]
65. Sun, J.; Wang, Y.; Liao, W.; Yang, Z. Ultrafast, high-contractile electrothermal-driven liquid crystal elastomer fibers towards artificial muscles. *Small* **2021**, *17*, 2103700. [CrossRef] [PubMed]
66. Zhao, D.; Liu, Y. Effects of director rotation relaxation on viscoelastic wave dispersion in nematic elastomer beams. *Math. Mech. Solids* **2019**, *24*, 1103–1115. [CrossRef]
67. Wu, H.; Lou, J.; Zhang, B.; Dai, Y.; Li, K. Stability analysis of a liquid crystal elastomer self-oscillator under a linear temperature field. *Appl. Math. Mech.* **2024**, *45*, 337–354. [CrossRef]
68. Yong, Y.; Hu, H.; Dai, Y.; Li, K. Modeling the light-powered self-rotation of a liquid crystal elastomer fiber-based engine. *Phys. Rev. E* **2024**, *109*, 034701.
69. Menard, K.P.; Menard, N. *Dynamic Mechanical Analysis*; CRC Press: Boca Raton, FL, USA, 2020.

70. Warner, M.; Terentjev, E.M. *Liquid Crystal Elastomers*; Oxford University Press: Oxford, UK, 2007.
71. Merritt, I.C.; Jacquemin, D.; Vacher, M. Cis→ trans photoisomerisation of azobenzene: A fresh theoretical look. *Phys. Chem. Chem. Phys.* **2021**, *23*, 19155–19165. [CrossRef] [PubMed]
72. Bandara, H.M.; Burdette, S.C.; Burdette, S.C. Photoisomerization in different classes of azobenzene. *Chem. Soc. Rev.* **2012**, *41*, 1809–1825. [CrossRef]
73. Finkelmann, H.; Nishikawa, E.; Pereira, G.; Warner, M. A new opto-mechanical effect in solids. *Phys. Rev. Lett.* **2001**, *87*, 015501. [CrossRef]
74. Chen, B.; Liu, C.; Xu, Z.; Wang, Z.; Xiao, R. Modeling the thermo-responsive behaviors of polydomain and monodomain nematic liquid crystal elastomers. *Mech. Mater.* **2024**, *188*, 104838. [CrossRef]
75. Nägele, T.; Hoche, R.; Zinth, W.; Wachtveitl, J. Femtosecond photoisomerization of cis-azobenzene. *Chem. Phys. Lett.* **1997**, *272*, 489–495. [CrossRef]
76. Braun, L.B.; Hessberger, T.; Pütz, E.; Müller, C.; Giesselmann, F.; Serra, C.A.; Zentel, R. Actuating thermo-and photo-responsive tubes from liquid crystalline elastomers. *J. Mater. Chem. C* **2018**, *6*, 9093–9101. [CrossRef]
77. Zhao, T.; Zhang, Y.; Fan, Y.; Wang, J.; Jiang, H.; Lv, J.-a. Light-modulated liquid crystal elastomer actuator with multimodal shape morphing and multifunction. *J. Mater. Chem. C* **2022**, *10*, 3796–3803. [CrossRef]
78. Jampani, V.; Volpe, R.; Reguengo de Sousa, K.; Ferreira Machado, J.; Yakacki, C.; Lagerwall, J. Liquid crystal elastomer shell actuators with negative order parameter. *Sci. Adv.* **2019**, *5*, eaaw2476. [CrossRef]

Disclaimer/Publisher's Note: The statements, opinions and data contained in all publications are solely those of the individual author(s) and contributor(s) and not of MDPI and/or the editor(s). MDPI and/or the editor(s) disclaim responsibility for any injury to people or property resulting from any ideas, methods, instructions or products referred to in the content.

Article

Impact of PCBM as a Third Component on Optical and Electrical Properties in Ternary Organic Blends

Laura Hrostea [1,*], Anda Oajdea [2] and Liviu Leontie [2,*]

[1] Research Center on Advanced Materials and Technologies (RAMTECH), Department of Exact and Natural Sciences, Institute of Interdisciplinary Research, Alexandru Ioan Cuza University of Iasi, 11 bd. Carol I, 700506 Iasi, Romania
[2] Faculty of Physics, Alexandru Ioan Cuza University of Iasi, 11 bd. Carol I, 700506 Iasi, Romania; anda.oajdea@gmail.com
* Correspondence: laura.hrostea@uaic.ro (L.H.); lleontie@uaic.ro (L.L.)

Abstract: This paper investigates the influence of constituent weight ratios on optical and electrical properties, with a particular focus on the intrinsic properties (such as electrical mobility) of ternary organic blends, highlighting the role of a third component. The study explores novel donor:acceptor1:acceptor2 (D:A_1:A_2) matrix blends with photovoltaic potential, systematically adjusting the ratio of the two acceptors in the mixtures, while keeping constant the donor:acceptor weight ratio (D:A = 1:1.4). Herein, depending on this adjustment, six different samples of 100–400 nm thickness are methodically characterized. Optical analysis demonstrates the spectral complementarity of the component materials and exposes the optimal weight ratio (D:A_1:A_2 = 1:1:0.4) for the highest optical absorption coefficient. Atomic force microscopy (AFM) analysis reveals improved and superior morphological attributes with the addition of the third component (fullerene). In terms of the electrical mobility of charge carriers, this study finds that the sample in which A_1 = A_2 has the greatest recorded value [$\mu_{max} = 1.41 \times 10^{-4} cm^2/(Vs)$]. This thorough study on ternary organic blends reveals the crucial relationship between acceptor ratios and the properties of the final blend, highlighting the critical function of the third component in influencing the intrinsic factors such as electrical mobility, offering valuable insights for the optimization of ternary organic solar cells.

Keywords: ternary organic films; fullerenes; constituents' compatibility; electrical mobility; CELIV method

Citation: Hrostea, L.; Oajdea, A.; Leontie, L. Impact of PCBM as a Third Component on Optical and Electrical Properties in Ternary Organic Blends. *Polymers* **2024**, *16*, 1324. https:// doi.org/10.3390/polym16101324

Academic Editor: Md Najib Alam

Received: 15 April 2024
Revised: 25 April 2024
Accepted: 30 April 2024
Published: 8 May 2024

Copyright: © 2024 by the authors. Licensee MDPI, Basel, Switzerland. This article is an open access article distributed under the terms and conditions of the Creative Commons Attribution (CC BY) license (https:// creativecommons.org/licenses/by/ 4.0/).

1. Introduction

In the context of the worldwide energy crisis and the issues arising from the greenhouse effect, renewable energy sources are gradually overtaking traditional sources of energy [1–4]. The sun is the most significant and cleanest energy resource since it is a limitless source of free fuel that has the enormous potential to supply much more energy than the planet needs. Nevertheless, there are certain restrictions on how photovoltaic devices (solar cells) can absorb solar energy and convert it into electricity [5]. The most crucial requirements for solar cells are a high photon-to-electron conversion efficiency and affordable production costs, adding to these the lifetime, stability, availability, and performance of the materials on which they are based [6]. Although first- and second-generation silicon solar cells have shown themselves to be a successful evolving photovoltaic technology, their current high costs restrict them from being used widely around the world. As a result of this, the development of photovoltaics has progressed along two paths: either silicon is substituted with other materials, or silicon is used in various solar cell architectures (such as silicon-based tandem solar cells and perovskite solar cells) [7]. In contrast, third-generation photovoltaics (OSCs—organic solar cells), based on organic materials, have shown notable scientific and technological progress [8].

In organic solar cells, the development of acceptor materials can be distinguished by comparing fullerenes with the more modern non-fullerene acceptors (NFAs) [8,9]. Although fullerenes have benefits, such as strong electron withdrawal and good miscibility with donor polymers, they also have disadvantages, including poor light absorption and challenging chemical tuning. However, because of their wide absorption range, simplicity in synthesis, and chemical tunability, NFAs show great promise. Furthermore, this field's research recognizes commensurate developments in donor polymers, emphasizing the application of chlorinated conjugated polymers for accurate molecular-level modifications [10,11].

An approach to improving the photovoltaic performance of solar cell devices consists of the development of ternary organic solar cells (TOSCs) [12], which exploit the synergistic effects of tandem solar cells with those of classical bulk-heterojunction solar cells—all-in-one. TOSCs are based on an active layer consisting of the classical donor/acceptor (D/A) matrix that incorporates, as a third element, an electron donor or an electron acceptor material [13,14]. This approach gives the opportunity of choosing materials with the most promising photovoltaic potential and complementary features [15,16], while the role of the third component is crucial, as it supports an expanded optical absorption spectrum and facilitates the charge carrier transport, demonstrated in previous papers [17]. More than that, improvements to the polymers' chemical structures, which serve as donor materials, can be carried out concomitantly [18,19]. Another important advantage, besides the photovoltaic potential improvement, is the simplicity and the ease of fabricating this type of cell, requiring only a combination of straightforward steps similar to those used in bi-layered cells fabrication. The latest (2024) record efficiency of a ternary organic solar cell is 19.13% [20].

Knowledge about the transport mechanism in ternary organic blends helps in understanding and developing further photovoltaic devices, given that charge carrier transit may limit solar energy conversion. Experimentally, a precise charge transport model in multi-component bulk heterojunction films has not yet been fully developed, despite the existence of some empirical criteria for mobility evaluation [21]. Thus, four different models are proposed to suitably describe the transport mechanism. Based on the energy properties and matrix type, they can be used to explain the basic mechanism of ternary cells. The charge transfer model, the parallel model, the alloy model, and the energy transfer model are the corresponding models [14].

Since charge carrier mobility is one of the inherent difficulties, assessing it might help identify any constraints or potential challenges to raising the efficiency of photon-to-electron conversion. Due to these key details, a number of methods for measuring charge carrier mobility have been created. The well-known method for measuring electrical mobility, presented by G. Juska in 2000 [22], called Charge Extraction by Linearly Increasing Voltage (CELIV), still has a number of advantages over other methods, the including Space Charge Limited Current (SCLC) and Time-of-Flight (TOF) techniques. One significant benefit is the ability to investigate multiple factors at once, including charge carrier mobility, relaxation time, and charge density. A triangle voltage pulse is applied to the blocking electrode (anode) of a sample structure in order to evaluate the equilibrium charge carrier current transients. As a result, the applied voltage pulse will move the equilibrium charge carriers to the opposite electrode (cathode, which is an ITO electrode connected to the negative terminal), where they will be extracted [23–29].

The novelty of this study is in how the adjustment of the constituents' weight ratio in a ternary organic blend, especially the presence of the third component, impacts the electrical mobility and other significant intrinsic parameters. New emergent donor:acceptor1:acceptor2 matrix blends for ternary solar cell applications, based on a polymer (PBDB-T-2Cl), a non-fullerene (ITIC-F), and a fullerene (PCBM), are a main focus of this study.

2. Materials and Methods

The samples include blended constituent materials, separately bought from Sigma Aldrich (Steinheim, Germany): a conjugated polymer, PBDB-T-2Cl (Poly[(2,6-(4,8-bis(5-(2-ethylhexyl-3-chloro) thiophen-2-yl)-benzo[1,2-b:4,5-b']dithiophene))-alt-(5,5-(1',3'-di-2-thienyl-5',7'-is(2ethyl hexyl)benzo[1',2'-c:4',5'-c'] dithiophene-4,8-dione)]), a non-fullerene ITIC–F (3,9-bis(2-methylene-((3-(1,1-dicyanomethylene)-6,7-difluoro)-indanone))-5,5,11,11tetrakis(4-exyl phenyl)-dithieno[2,3d:2',3'-d']-sindaceno[1,2-b:5,6-b']dithiophene), and a fullerene, ([6,6]-Phenyl-C61-butyric acid methyl ester).

Both Figure 1a,b show the schematic device construction (not to scale) and the optical characteristics of the neat constituents (polymer, non-fullerene, and fullerene as individual thin films). Furthermore, Figure 1c describes the molecular orbital energy levels (HOMO and LUMO) [30].

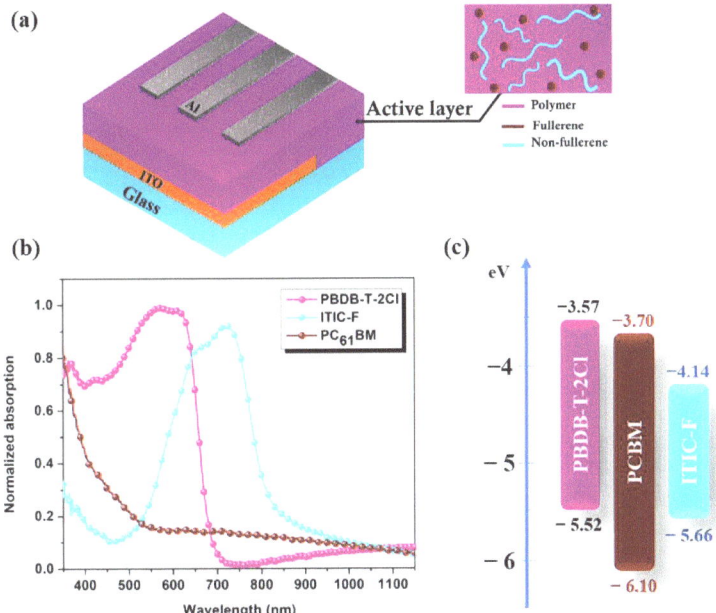

Figure 1. Schematic device structure (**a**); normalized absorption spectra of neat PBDB-T-2Cl, ITIC-F and PCBM thin films (**b**); and energy levels diagram of constituent materials (**c**).

Chlorobenzene was used as a solvent. In advance, a mixture containing chlorobenzene (CB) and 1,8-diiodooctane (DIO) with a concentration of 3% was prepared. The first step was to dissolve the total amount of polymer in the prepared solvent (CB:DIO). This solution was magnetically stirred for 1.5 h at a temperature of 50 °C. Subsequently, five mixtures of 20 mg/mL concentration were prepared using the initial polymer solution, varying the acceptors' weight ratio, labelled as [sample name = polymer:non-fullerene:fullerene]: #S1 = 1:0:0; #S2 = 1:1.4:0; #S3 = 1:1:0.4; #S4 = 1:0.7:0.7; #S5 = 1:0.4:1; #S6 = 1:0:1.4. The substrates were cleaned for 10 min in an ultrasonic bath with acetone, ethanol and detergent and the solutions were spun coated as thin films (100–400 nm thickness as presented in Table 1) at 1500 rpm spinning speed. Samples were dried at 100 °C for 10 min after deposition. The entire preparation process was carried out in a clean room under a lab hood that provided a normal/constant environment, and samples were kept in the dark. The aluminum digital cathode was deposited by thermal evaporation in a vacuum.

Table 1. Thickness and roughness values of investigated samples.

Sample	D:A$_1$:A$_2$ Weight Ratio	Thickness (nm)	Absorption Edge (nm)	R$_{RMS}$ (nm)
S1	1:0:0	100	656	4.97
S2	1:1.4:0	400	685	8.24
S3	1:1:0.4	250	780	4.65
S4	1:0.7:0.7	285	775	3.91
S5	1:0.4:1	350	775	3.37
S6	1:0:1.4	210	656	2.42

A DektakXT Stylus profilometer (Bruker France S.A.S., Wissembourg, France) was utilized to determine the thickness of the samples. A TEC5 spectrophotometer was used for the analysis of the optical absorption. Using an NT-MDT Solver Pro-M system (from NT-MDT, Moscow, Russia), atomic force microscopy (AFM) pictures were acquired. The measurements were performed using a SiN cantilever (NSC21 from Mikromasch, Tallinn, Estonia) in non-contact mode at room temperature. Using Nova software (version 1.0.26.1443) from NT-MDT, the samples' root mean square roughness (RRMS) was calculated for a 1.5 × 1.5 µm^2 scanned area. The XPS analysis was performed on a Physical Electronics PHI 5000 VersaProbe instrument (Ulvac-PHI, Inc., Chikasaki, Japan), equipped with a monochromatic AlKα X-ray source (hν = 1486.6 eV). The take-off angle of the photoelectrons was equal to 45°. All the XPS peak positions in the survey spectra were calibrated with respect to the C 1 s peak—binding energy (BE) = 284.6 eV. The CELIV (charge extraction by linearly increasing voltage) method was used to determinate the electrical mobility, wherein a triangular-shaped bias voltage pulse was applied from a AFG31022 function generator and the extracted current transient's signal was recorded by a digital oscilloscope. The analysis involved changing the applied voltage from 1 V to 10 V while maintaining a consistent signal period of 30 µs. Under normal lighting conditions, several heating and cooling cycles were applied during the electrical measurements, which were performed in a perpendicular geometry configuration. The temperatures ranged from 30 °C to 120 °C.

3. Results and Discussions

Figure 1b shows the optical absorption spectra (normalized to 1) of the components in the blends. The preferred photon-harvesting wavelength ranges of each of the constituent materials, deposited as individual material thin films, are as follows: PBDB-T-2Cl thin film reveals an absorption band within the visible range between 450 and 700 nm, ITIC-F thin film displays a red-shifted extended band ranging between 500 and 850 nm with a maximum peak at 760 nm, and, in comparison to the non-fullerene and polymer components' properties, PCBM thin film shows a notable absorption near the ultraviolet wavelengths, up to 400 nm, with a smaller absorption peak. Furthermore, the absorption of photons with wavelengths longer than 500 nm by PCBM continues to decrease towards the infrared region [31]. Because of the optical absorption complementarity of these materials, the photon harvesting of ternary blend thin films is increased and extended, as seen in Figure 2. As a result, NFA-based (binary and ternary) blends have broadened absorption spectra ranging from 500 to 800 nm, with two peaks indicating polymer and NFA fingerprints' involvement. Small adjustments in the quantity of the third element can result in significant changes to the dielectric constant. The dielectric environment changes as acceptor elements are introduced into the polymeric matrix, enhancing its character, as previously reported in [32]. Considering solely ternary blends, the absorption coefficient decreases as the amount of NFA in the blends diminishes. Sample #S3 (1:1:0.4) has the highest absorption coefficient and the longest wavelength as absorption edge (Table 1), considering the fact that the optical absorption edge shifts to lower energies (longer wavelengths) when the non-fullerene is present in the matrix.

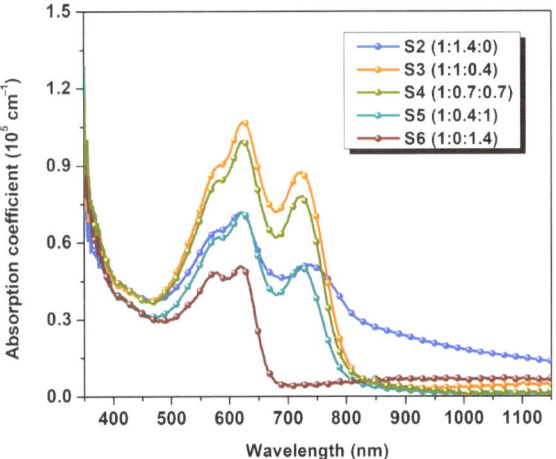

Figure 2. Absorption coefficient spectra of binary and ternary blends thin films.

Thus, the AFM height images (2D and 3D—Figure 3a) and the AFM phase images (Figure 3b) illustrate a key characteristic of the samples. The AFM height images describe the samples by determining the roughness. According to Table 1, the nanoscale smooth topography of the samples is observed, where the specific root mean square roughness (R_{RMS}) of the polymer thin film (#S1) is 4.97 nm. By adding the non-fullerenes, this value increases to 8.24 nm and then gradually moderates with the addition of fullerene. This is due to a higher miscibility of the fullerenes and non-fullerenes in the polymer matrix. Good miscibility between materials reduces the driving force for phase separation, which results in smaller impurity domains, Figure 3b, that support efficient electron-hole dissociation [31,33]. Higher-performance OSCs require compatible and favorable film morphology in the active layer. The ternary blend thin films exhibit smoother topography compared to polymer (#S1) or fullerene-free thin films (#S2), confirming the enhanced miscibility of the three constituent materials and remarking on the crucial role of the third component—PCBM, which can be considered a morphology regulator in adjusting the molecular arrangement of the polymer:non-fullerene host matrix [34]. When the amount of fullerene in the polymer:non-fullerene host matrix is raised, it is evident, from comparing the AFM phase images of all the samples, that highly well-defined and shaped nanodomains are revealed. This fact is based on PCBM's higher tendency to form clusters, which favors the development of nanodomains. In contrast to fullerene-free blends, the composition differences between fullerene domains, non-fullerene domains, and the polymer matrix become more prominent at increasing PCBM concentrations, providing a stronger contrast in AFM phase images of more easily discernible nanodomains [35].

A well-mixed material surface is revealed by the XPS survey spectra, Figure 4, which displays the surface compositional profiles down to around 10 nm in depth. Even though the third component, PCBM, is hydrophilic, this surface composition is consistent with [17], which emphasizes the hydrophobic nature of the layer surface even though it may not accurately reflect the distribution of elements within the complete layers. According to this information, PCBM is anticipated to settle in the region between the polymer and non-fullerene host matrix. In addition, certain non-fullerene's atoms, such the fluor atom, are likely to be highlighted at the surface of thin films based on binary and ternary blends. As a result, this does suggest that the polymer matrix contains the two integrated acceptor materials, fullerene and non-fullerene. Together with the XPS data, [34] also confirms the compatibility and good miscibility of the PCBM and ITIC-F materials, which are incorporated in the polymeric matrix to form an effective electron transport network.

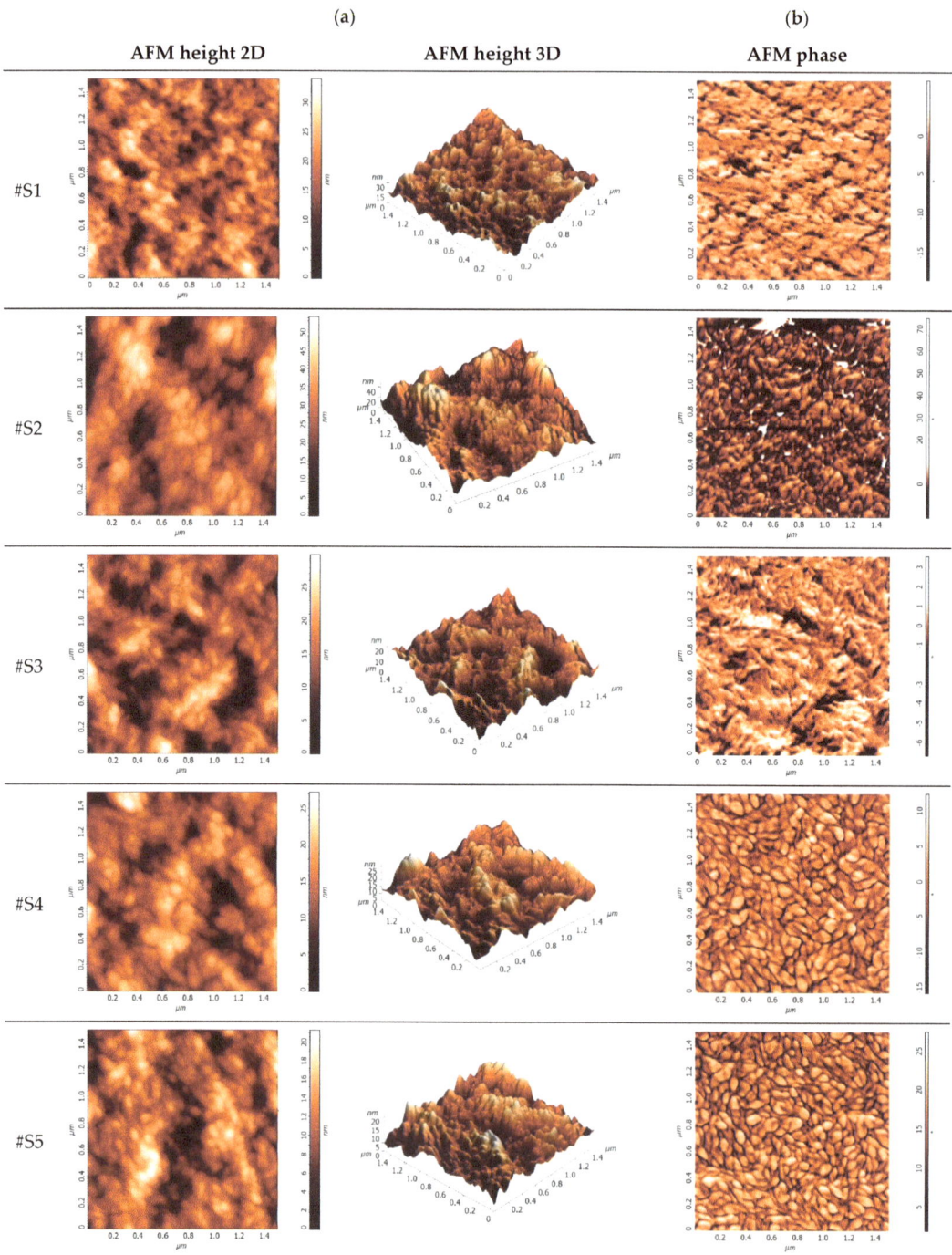

Figure 3. *Cont.*

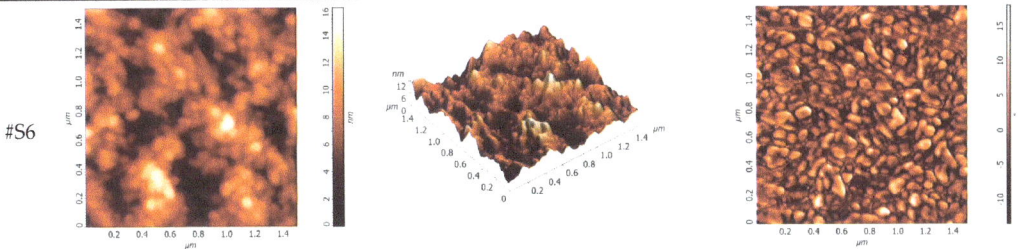

Figure 3. AFM analysis: height images in 2D and 3D representation (**a**) and phase images (**b**).

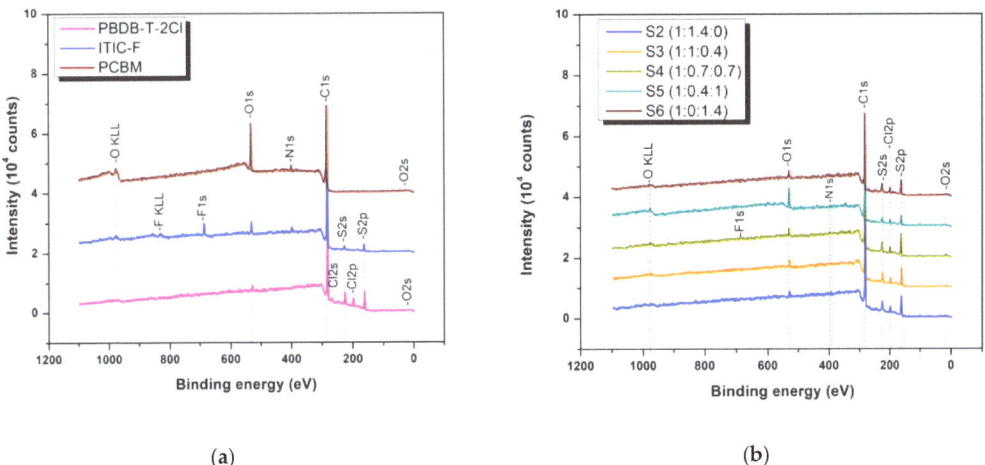

Figure 4. XPS survey spectra of individual material thin films coated on glass (**a**) and binary and ternary organic thin films (**b**).

To analyze the intrinsic parameters, such as charge carrier mobility, the CELIV method is widely used for determining the bulk transport properties of materials in sandwich geometry, especially in disordered materials such as organics [36–38]. Regarding the transport mechanism, the parallel model provides the best description of the transport process in the case of the investigated D:A$_1$:A$_2$ matrix, where the active layer functions as two intercalated bilayers (D:A$_1$ and D:A$_2$) [17]. Due to the third element's function as an electron transfer channel, the compounds' compatibility and miscibility are crucial. In this case, holes are created in the D domains as electrons move to the nanoscale domains of A$_1$ and A$_2$. Therefore, knowing that the investigated samples are based on donor:acceptor blends of a 1:1.4 weight ratio, this suggests that since the acceptor concentration is higher, the positive charge carriers (holes) are the majority carriers. The method is based on investigation of the extraction of equilibrium carries, in which a linearly increasing voltage is applied. Hence, by applying the voltage [$U(t) = A \cdot t$, where A is the voltage ramp], a current transient signal is obtained, as shown in Figure 5a. Charge injection is avoided due to the presence of a blocking contact from the sandwich-like structure of the device. The total current transient signal is composed from $j(0)$, the initial current step, caused by the geometrical capacity of the sample, giving information about the material permittivity or about the interelectrode distance [$j(0) = \frac{\varepsilon \varepsilon_0 A}{d}$], and Δj, caused by the electrical conductivity of the sample [$\sigma = \frac{3\varepsilon \varepsilon_0 \Delta j}{2 t_{max} j(0)}$], which is the supplementary current formed by the extracted charge carrier. The total current transient, j_{max}, reaches a maximum value at t_{max}, as is exemplified in Figure 5a inset. By applying Formula (1), knowing the thickness d, the mobility of the faster charges can be determined:

$$\mu = \frac{2}{3} \frac{2d^2}{At_{max}^2 \left[1 + 0.36\left(\frac{\Delta j}{j(0)}\right)\right]} \quad (1)$$

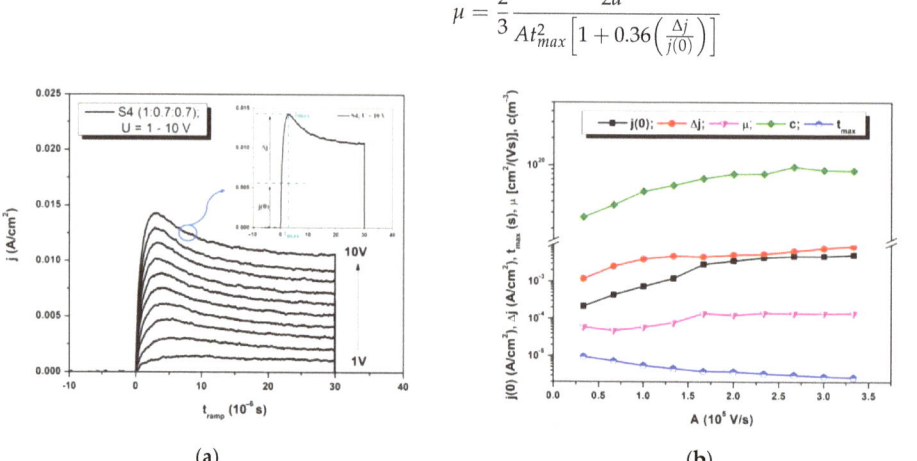

Figure 5. Analyze on sample S4: CELIV signal variation with the applied voltage (a); intrinsic parameters determined from the current transients (b).

The electric-field dependence of the intrinsic behavior of the charge carrier was determined by varying the maximum of a triangular voltage pulse between 1 and 10 V, and the current transient signals can be seen in Figure 5a for one sample, #S4.

Herein, the current transient increases rapidly until the majority of charge carriers have been extracted at t_{max}, at which point it sharply declines [28]. Compared to other methods, CELIV is based on the extraction of equilibrium charge carriers (shallow trapped) and, increasing the voltage, the extracted carriers are located even on deeper states, leading to a prominent CELIV current [39]. More than that, the evolution of some parameters is presented in Figure 5b.

Using the analyses of sample #S4 as an example, it is noticeable that the increase of the voltage exerts an influence on the transport of charge carriers within the thin film (Table 2). In this context, the mobility exhibits a proportional increase with the applied voltage, attaining its maximum value at $\mu_{10V} = 1.41 \times 10^{-4} \text{cm}^2/\text{Vs}$, corresponding to the values reported in [31] or even higher [40]. This fact is based on Poole–Frenkel model, consisting of an increased probability of electric-field-stimulated charge carrier release from localized states [41]. Conversely, there is an anticipated inverse correlation observed in the time required for the current to reach its peak value [42], denoted as $t_{max} = 2.66 \times 10^{-6}$ s. T_{max} increases as the speed of the voltage rise A decreases, indicating the electrical field dependence of mobility [42]. The concentration of extracted charge carriers ($c \sim 10^{20}$ m^{-3}) is estimated, calculated from the area under the extraction peak. As inferred from the current transient components [$j(0)$ and Δj], it also manifests an elevation concomitant with the incremental voltage ramp.

Table 2. Determined parameters compared to the literature.

Parameters	Maximum Determined Values	Literature Reported Values	
μ_{max}	1.41×10^{-4} cm^2/Vs	$1 \div 6 \times 10^{-5}$ cm^2/Vs	[40]
c	10^{20} m^{-3}	4.73×10^{20} m^{-3}	[43]
t_{tr}	0.57 µs	0.55 ÷ 0.70 µs	[44]
α	10^5 cm^{-1}	3.5×10^4 cm^{-1}	[45]

Acceptors may affect the morphology of the blend inside the layer, as the AFM images demonstrate. For hole transfer, a well-defined interconnected network is essential. The ideal ratio of constituents can be found by adjusting the weight ratio of the acceptors, which will improve the packing of the donor polymer while making hole hopping between donor molecules easier. Furthermore, the effectiveness of charge hopping depends on the energy levels of the donor polymer and both acceptors. The right weight ratio minimizes the energy barrier for charge transport and facilitates effective transfer from the donor to the acceptors. By adding more intermolecular interactions to the blend, the PCBM acceptor acting as third component has an impact on mobility and the electronic coupling between donor molecules.

Despite maintaining a constant donor:acceptor ratio across all the blends of thin films (D:A = 1:1.4), the variation in the weight ratio of the acceptors (A_1:A_2) is observed to impact charge carrier transport, as illustrated in Figure 6. Notably, the electrical conductivity experiences an increase with the augmentation of fullerene content in the blend, a fact confirmed by four-probe method assessment, wherein the highest conductivity value is $\sigma_{S4} = 4.4 \times 10^{-7} \Omega^{-1} \text{cm}^{-1}$, corresponding to sample #S4. Additionally, in terms of mobility, sample #S4 (wherein A_1:A_2 = 1) attains the highest value [$\mu_{max} = 1.41 \times 10^{-4}$ cm^2/(Vs)], at $U = 10$ V, followed by sample #S5, wherein the non-fullerene component is absent. Once the fullerene amount decreases, the electrical mobility registers lower values. Consequently, both t_{max} and t_{tr} (carrier transit time of interelectrode distance) reach their minimal values for the same blend ($t_{max}^{S4} = 2.66$ μs and $t_{tr}^{S4} = 0.57$ μs).

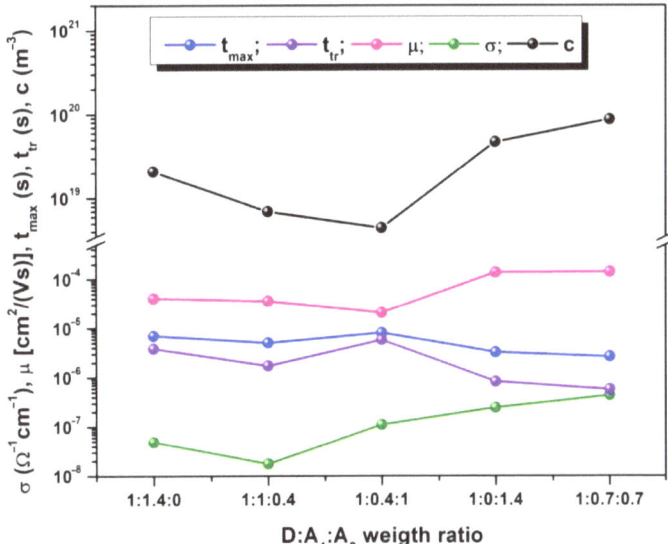

Figure 6. CELIV parameters of samples depending on the weight ratio variation.

4. Conclusions

The present study into donor:acceptor1:acceptor2 (D:A_1:A_2 = PBDB-T-2Cl:ITIC-F:PCBM) ternary organic blends demonstrates the significant impact of varying the acceptors' weight ratio on their optical, morphological, and electrical properties. Sample #S3, with a weight ratio $A_1 > A_2$ and an absorption maximum $\alpha = 1.07 \times 10^5 \text{cm}^{-1}$ at 628 nm wavelength, has superior optical properties. AFM characterization shows a very smooth nanoscale topography, remarking that a decrease in roughness is correlated with an increase in fullerene content in the blend. This fact indicates that fullerenes function as binders in the polymer-fullerene matrix. The electrical properties are also improved as the amount of fullerene in the blends increases; sample #S4 ($A_1 = A_2$) exhibits the highest electrical conductivity

($\sigma_{S4} = 4.4 \times 10^{-7} \Omega^{-1} \text{cm}^{-1}$) and charge carrier mobility [$\mu_{S4} = 1.41 \times 10^{-4}$ cm^2/(Vs)]. This study also reveals the electric-field dependence of intrinsic parameters, determined using the CELIV method, and the variability of the applied voltage shows significant changes in the transient current density, highlighted by #S4 analysis. These results highlight how important it is to take into consideration the trade-off between optical, morphological, and electrical properties when designing ternary blends suitable for photovoltaic applications. This work shows that electrical mobility and other intrinsic properties are significantly influenced by varying the weight ratio of acceptors, specifically the impact of the third component. This study offers perspectives for developing highly efficient organic solar cells with an optimized structure.

Author Contributions: Conceptualization, L.H.; methodology, L.H. and A.O.; software, L.H. and A.O.; validation, L.H., A.O. and L.L.; formal analysis, L.H.; investigation, L.H. and A.O.; resources, L.H.; data curation, L.H. and L.L.; writing—original draft preparation, L.H.; writing—review and editing, L.H., A.O. and L.L.; visualization, L.H.; supervision, L.H.; project administration, L.H.; funding acquisition, L.H. All authors have read and agreed to the published version of the manuscript.

Funding: This work was supported by a grant of the "Alexandru Ioan Cuza" University of Iasi, within the Research Grants program, Grant UAIC, code GI-UAIC-2021-07.

Institutional Review Board Statement: Not applicable.

Data Availability Statement: Data are contained within the article.

Acknowledgments: The authors acknowledge the support to MSc George Jitariu for providing the schematic device structure representation, Dr. Vasile Tiron for the AFM analysis and Dr. Marius Dobromir for XPS analysis.

Conflicts of Interest: The authors declare no conflicts of interest.

References

1. Hassan, Q.; Viktor, P.; Al-Musawi, T.J.; Ali, B.M.; Algburi, S.; Alzoubi, H.M.; Al-Jiboory, A.K.; Sameen, A.Z.; Salman, H.M.; Jaszczur, M. The Renewable Energy Role in the Global Energy Transformations. *Renew. Energy Focus* **2024**, *48*, 100545. [CrossRef]
2. Moosavian, S.F.; Noorollahi, Y.; Shoaei, M. Renewable Energy Resources Utilization Planning for Sustainable Energy System Development on a Stand-Alone Island. *J. Clean. Prod.* **2024**, *439*, 140892. [CrossRef]
3. Babazadeh Dizaj, R.; Sabahi, N. Optimizing LSM-LSF Composite Cathodes for Enhanced Solid Oxide Fuel Cell Performance: Material Engineering and Electrochemical Insights. *World J. Adv. Res. Rev.* **2023**, *20*, 1284–1291. [CrossRef]
4. Amirjani, A.; Amlashi, N.B.; Ahmadiani, Z.S. Plasmon-Enhanced Photocatalysis Based on Plasmonic Nanoparticles for Energy and Environmental Solutions: A Review. *ACS Appl. Nano Mater.* **2023**, *6*, 9085–9123. [CrossRef]
5. Scharber, M.C. Efficiency of Emerging Photovoltaic Devices under Indoor Conditions. *Sol. RRL* **2024**, *8*, 2300811. [CrossRef]
6. Paci, B.; Righi Riva, F.; Generosi, A.; Guaragno, M.; Mangiacapre, E.; Brutti, S.; Wagner, M.; Distler, A.; Egelhaaf, H.-J. Semitransparent Organic Photovoltaic Devices: Interface/Bulk Properties and Stability Issues. *Nanomaterials* **2024**, *14*, 269. [CrossRef]
7. Li, Y.; Ru, X.; Yang, M.; Zheng, Y.; Yin, S.; Hong, C.; Peng, F.; Qu, M.; Xue, C.; Lu, J. Flexible Silicon Solar Cells with High Power-to-Weight Ratios. *Nature* **2024**, *626*, 105–110. [CrossRef]
8. Ding, P.; Yang, D.; Yang, S.; Ge, Z. Stability of Organic Solar Cells: Toward Commercial Applications. *Chem. Soc. Rev.* **2024**. [CrossRef]
9. Yi, J.; Zhang, G.; Yu, H.; Yan, H. Advantages, Challenges and Molecular Design of Different Material Types Used in Organic Solar Cells. *Nat. Rev. Mater.* **2024**, *9*, 46–62. [CrossRef]
10. Wang, H.; He, F. Recent Advances of Chlorination in Organic Solar Cells. *Synlett* **2021**, *32*, 1297–1302. [CrossRef]
11. Cao, J.; Wang, H.; Yang, L.; Du, F.; Yu, J.; Tang, W. Chlorinated Unfused Acceptor Enabling 13.57% Efficiency and 73.39% Fill Factor Organic Solar Cells via Fine-Tuning Alkoxyl Chains on Benzene Core. *Chem. Eng. J.* **2022**, *427*, 131828. [CrossRef]
12. Li, M.; He, F. Organic Solar Cells Developments: What's next? *Next Energy* **2024**, *2*, 100085. [CrossRef]
13. Cai, Y.; Xie, C.; Li, Q.; Liu, C.; Gao, J.; Jee, M.H.; Qiao, J.; Li, Y.; Song, J.; Hao, X. Improved Molecular Ordering in a Ternary Blend Enables All-Polymer Solar Cells over 18% Efficiency. *Adv. Mater.* **2023**, *35*, 2208165. [CrossRef]
14. Doumon, N.Y.; Yang, L.; Rosei, F. Ternary Organic Solar Cells: A Review of The Role of the Third Element. *Nano Energy* **2022**, *94*, 106915. [CrossRef]
15. Yang, X.; Sun, R.; Wang, Y.; Chen, M.; Xia, X.; Lu, X.; Lu, G.; Min, J. Ternary All-Polymer Solar Cells with Efficiency up to 18.14% Employing a Two-Step Sequential Deposition. *Adv. Mater.* **2023**, *35*, 2209350. [CrossRef]
16. An, Q.; Zhang, F.; Zhang, J.; Tang, W.; Deng, Z.; Hu, B. Versatile Ternary Organic Solar Cells: A Critical Review. *Energy Environ. Sci.* **2016**, *9*, 281–322. [CrossRef]

17. Hrostea, L.; Dumitras, M.; Leontie, L. Study of Electrical Properties of PBDB-T-2Cl Based Ternary Thin Films for Photovoltaic Applications. *Mater. Sci. Semicond. Process.* **2023**, *166*, 107743. [CrossRef]
18. Hrostea, L.; Leontie, L.; Girtan, M. Chemical Sensitization for Electric Properties Improvement of PBDB-T-SF Polymer for Solar Cells Application. *IOP Conf. Ser. Mater. Sci. Eng.* **2020**, *877*, 012002. [CrossRef]
19. Hrostea, L.; Leontie, L.; Girtan, M. Characterization of PBDB-T-SF: Fullerene Blend Thin Films for Solar Cell Applications. *Rom. Rep. Phys.* **2020**, *72*, 504.
20. Huang, T.; Zhang, Y.; Wang, J.; Cao, Z.; Geng, S.; Guan, H.; Wang, D.; Zhang, Z.; Liao, Q.; Zhang, J. Dual-Donor Organic Solar Cells with 19.13% Efficiency through Optimized Active Layer Crystallization Behavior. *Nano Energy* **2024**, *121*, 109226. [CrossRef]
21. Hu, H.; Mu, X.; Qin, W.; Gao, K.; Hao, X.; Yin, H. Rationalizing Charge Carrier Transport in Ternary Organic Solar Cells. *Appl. Phys. Lett.* **2022**, *120*, 023302. [CrossRef]
22. Juska, G.; Arlauskas, K.; Viliunas, M.; Genevicius, K.; Osterbacka, R.; Stubb, H. Charge Transport in Pi-Conjugated Polymers from Extraction Current Transients. *Phys. Rev. B* **2000**, *62*, R16235–R16238. [CrossRef]
23. Juska, G.; Arlauskas, K.; Genevicius, K. Charge carrier transport and recombination in disordered materials. *Lith. J. Phys.* **2016**, *56*, 182–189. [CrossRef]
24. Grynko, O.; Juška, G.; Reznik, A. Charge Extraction by Linearly Increasing Voltage (CELIV) Method for Investigation of Charge Carrier Transport and Recombination in Disordered Materials. *Photocond. Photocond. Mater. Fundam. Tech. Appl.* **2022**, *1*, 339–368.
25. Khan, M.D.; Nikitenko, V.R. On the Charge Mobility in Disordered Organics from Photo-CELIV Measurements. *Chem. Phys.* **2020**, *539*, 110954. [CrossRef]
26. Aukstuolis, A.; Girtan, M.; Mousdis, G.A.; Mallet, R.; Socol, M.; Rasheed, M.; Stanculescu, A. measurement of charge carrier mobility in perovskite nanowire films by photo-celiv method. *Proc. Rom. Acad. Ser. A-Math. Phys. Tech. Sci. Inf. Sci.* **2017**, *18*, 34–41.
27. Stephen, M.; Genevičius, K.; Juška, G.; Arlauskas, K.; Hiorns, R.C. Charge Transport and Its Characterization Using Photo-CELIV in Bulk Heterojunction Solar Cells: Photo-CELIV to Probe Charge Transport in Solar Cells. *Polym. Int.* **2017**, *66*, 13–25. [CrossRef]
28. Semeniuk, O.; Juska, G.; Oelerich, J.-O.; Wiemer, M.; Baranovskii, S.D.; Reznik, A. Charge Transport Mechanism in Lead Oxide Revealed by CELIV Technique. *Sci. Rep.* **2016**, *6*, 33359. [CrossRef]
29. Zhang, Y.; Li, X.; Dai, T.; Ha, W.; Du, H.; Li, S.; Wang, K.; Meng, F.; Xu, D.; Geng, A. Charge Transport and Extraction of Bilayer Interdiffusion Heterojunction Organic Solar Cells. *J. Phys. Chem. C* **2019**, *123*, 24446–24452. [CrossRef]
30. Chang, L.; Sheng, M.; Duan, L.; Uddin, A. Ternary Organic Solar Cells Based on Non-Fullerene Acceptors: A Review. *Org. Electron.* **2021**, *90*, 106063. [CrossRef]
31. Yin, P.; Yin, Z.; Ma, Y.; Zheng, Q. Improving the Charge Transport of the Ternary Blend Active Layer for Efficient Semitransparent Organic Solar Cells. *Energy Environ. Sci.* **2020**, *13*, 5177–5185. [CrossRef]
32. Hrostea, L.; Bulai, G.-A.; Tiron, V.; Leontie, L. Study of Tunable Dielectric Permittivity of PBDB-T-2CL Polymer in Ternary Organic Blend Thin Films Using Spectroscopic Ellipsometry. *Polymers* **2023**, *15*, 3771. [CrossRef]
33. An, K.; Zhong, W.; Peng, F.; Deng, W.; Shang, Y.; Quan, H.; Qiu, H.; Wang, C.; Liu, F.; Wu, H. Mastering Morphology of Non-Fullerene Acceptors towards Long-Term Stable Organic Solar Cells. *Nat. Commun.* **2023**, *14*, 2688. [CrossRef]
34. Gao, J.; Wang, J.; An, Q.; Ma, X.; Hu, Z.; Xu, C.; Zhang, X.; Zhang, F. Over 16.7% Efficiency of Ternary Organic Photovoltaics by Employing Extra PC 71 BM as Morphology Regulator. *Sci. China Chem.* **2020**, *63*, 83–91. [CrossRef]
35. Liu, X.; Du, S.; Fu, Z.; Chen, C.; Tong, J.; Li, J.; Zheng, N.; Zhang, R.; Xia, Y. Ternary Solar Cells via Ternary Polymer Donors and Third Component PC71BM to Optimize Morphology with 13.15% Efficiency. *Sol. Energy* **2021**, *222*, 18–26. [CrossRef]
36. Juska, G.; Genevicius, K.; Viliunas, M.; Arlauskas, K.; Osterbacka, R.; Stubb, H. Transport Features of Photogenerated and Equilibrium Charge Carriers in Thin PPV Polymer Layers. In Proceedings of the Optical Organic and Inorganic Materials; Asmontas, S.P., Gradauskas, J., Eds.; Semicond Phys Inst; SPIE Balt Chapter; Lithuanian Minist & Educ & Res; Lithuanian State Sci & Studies Fdn; SPIE. European Commiss: Vilnius, Lituania, 2001; Volume 4415, pp. 145–149.
37. Aukstuolis, A.; Nekrasas, N.; Genevicius, K.; Juska, G. Investigation of Charge Carrier Mobility and Recombination in PBDTTPD Thin Layer Structures. *Org. Electron.* **2021**, *90*, 106066. [CrossRef]
38. Zhang, L.; Yang, F.; Meng, X.; Yang, S.; Ke, L.; Zhou, C.; Yan, H.; Hu, X.; Zhang, S.; Ma, W. Regulating Crystallization to Maintain Balanced Carrier Mobility via Ternary Strategy in Blade-Coated Flexible Organic Solar Cells. *Org. Electron.* **2021**, *89*, 106027. [CrossRef]
39. Juška, G.; Arlauskas, K.; Viliūnas, M.; Kočka, J. Extraction Current Transients: New Method of Study of Charge Transport in Microcrystalline Silicon. *Phys. Rev. Lett.* **2000**, *84*, 4946. [CrossRef]
40. Dahlstrom, S.; Liu, X.; Yan, Y.; Sandberg, O.J.; Nyman, M.; Liang, Z.; Osterbacka, R. Extraction Current Transients for Selective Charge-Carrier Mobility Determination in Non-Fullerene and Ternary Bulk Heterojunction Organic Solar Cells. *ACS Appl. Energy Mater.* **2020**, *3*, 9190–9197. [CrossRef]
41. Juška, G.; Genevičius, K.; Arlauskas, K.; Österbacka, R.; Stubb, H. Charge Transport at Low Electric Fields in π-Conjugated Polymers. *Phys. Rev. B* **2002**, *65*, 233208. [CrossRef]
42. Mozer, A.J.; Sariciftci, N.S.; Pivrikas, A.; Österbacka, R.; Juška, G.; Brassat, L.; Bässler, H. Charge Carrier Mobility in Regioregular Poly (3-Hexylthiophene) Probed by Transient Conductivity Techniques: A Comparative Study. *Phys. Rev. B* **2005**, *71*, 035214. [CrossRef]

43. Rao, A.D.; Murali, M.G.; Kesavan, A.V.; Ramamurthy, P.C. Experimental Investigation of Charge Transfer, Charge Extraction, and Charge Carrier Concentration in P3HT: PBD-DT-DPP: PC70BM Ternary Blend Photovoltaics. *Sol. Energy* **2018**, *174*, 1078–1084. [CrossRef]
44. Dai, T.; Li, X.; Zhang, Y.; Xu, D.; Geng, A.; Zhao, J.; Chen, X. Performance Improvement of Polymer Solar Cells with Binary Additives Induced Morphology Optimization and Interface Modification Simultaneously. *Sol. Energy* **2020**, *201*, 330–338. [CrossRef]
45. Liu, X.; Yan, Y.; Yao, Y.; Liang, Z. Ternary Blend Strategy for Achieving High-efficiency Organic Solar Cells with Nonfullerene Acceptors Involved. *Adv. Funct. Mater.* **2018**, *28*, 1802004. [CrossRef]

Disclaimer/Publisher's Note: The statements, opinions and data contained in all publications are solely those of the individual author(s) and contributor(s) and not of MDPI and/or the editor(s). MDPI and/or the editor(s) disclaim responsibility for any injury to people or property resulting from any ideas, methods, instructions or products referred to in the content.

Article

Use of Hydrogel Electrolyte in Zn-MnO$_2$ Rechargeable Batteries: Characterization of Safety, Performance, and Cu^{2+} Ion Diffusion

Jungsang Cho [1], Damon E. Turney [1,*], Gautam Ganapati Yadav [2], Michael Nyce [1], Bryan R. Wygant [3], Timothy N. Lambert [3,4] and Sanjoy Banerjee [1]

1. The CUNY Energy Institute, City University of New York, 160 Convent Ave, New York, NY 10031, USA; chojs0114@gmail.com (J.C.); mnyce1957@gmail.com (M.N.); sanjoy@urbanelectricpower.com (S.B.)
2. Urban Electric Power, Pearl River, NY 10965, USA; gautam@urbanelectricpower.com
3. Sandia National Laboratories, Department of Photovoltaics and Materials Technology, Albuquerque, NM 87185, USA; bwygant@sandia.gov (B.R.W.); tnlambe@sandia.gov (T.N.L.)
4. Sandia National Laboratories, Center of Integrated Nanotechnologies, Albuquerque, NM 87185, USA
* Correspondence: dturney@ccny.cuny.edu

Citation: Cho, J.; Turney, D.E.; Yadav, G.G.; Nyce, M.; Wygant, B.R.; Lambert, T.N.; Banerjee, S. Use of Hydrogel Electrolyte in Zn-MnO$_2$ Rechargeable Batteries: Characterization of Safety, Performance, and Cu^{2+} Ion Diffusion. *Polymers* **2024**, *16*, 658. https://doi.org/10.3390/polym16050658

Academic Editors: Vineet Kumar and Md Najib Alam

Received: 1 January 2024
Revised: 19 February 2024
Accepted: 21 February 2024
Published: 28 February 2024

Copyright: © 2024 by the authors. Licensee MDPI, Basel, Switzerland. This article is an open access article distributed under the terms and conditions of the Creative Commons Attribution (CC BY) license (https://creativecommons.org/licenses/by/4.0/).

Abstract: Achieving commercially acceptable Zn-MnO$_2$ rechargeable batteries depends on the reversibility of active zinc and manganese materials, and avoiding side reactions during the second electron reaction of MnO$_2$. Typically, liquid electrolytes such as potassium hydroxide (KOH) are used for Zn-MnO$_2$ rechargeable batteries. However, it is known that using liquid electrolytes causes the formation of electrochemically inactive materials, such as precipitation Mn$_3$O$_4$ or ZnMn$_2$O$_4$ resulting from the uncontrollable reaction of Mn^{3+} dissolved species with zincate ions. In this paper, hydrogel electrolytes are tested for MnO$_2$ electrodes undergoing two-electron cycling. Improved cell safety is achieved because the hydrogel electrolyte is non-spillable, according to standards from the US Department of Transportation (DOT). The cycling of "half cells" with advanced-formulation MnO$_2$ cathodes paired with commercial NiOOH electrodes is tested with hydrogel and a normal electrolyte, to detect changes to the zincate crossover and reaction from anode to cathode. These half cells achieved ≥700 cycles with 99% coulombic efficiency and 63% energy efficiency at C/3 rates based on the second electron capacity of MnO$_2$. Other cycling tests with "full cells" of Zn anodes with the same MnO$_2$ cathodes achieved ~300 cycles until reaching 50% capacity fade, a comparable performance to cells using liquid electrolyte. Electrodes dissected after cycling showed that the liquid electrolyte allowed Cu ions to migrate more than the hydrogel electrolyte. However, measurements of the Cu diffusion coefficient showed no difference between liquid and gel electrolytes; thus, it was hypothesized that the gel electrolytes reduced the occurrence of Cu short circuits by either (a) reducing electrode physical contact to the separator or (b) reducing electro-convective electrolyte transport that may be as important as diffusive transport.

Keywords: hydrogel; zinc; manganese dioxide; rechargeable; diffusion; energy storage

1. Introduction

Zinc (Zn)–manganese dioxide (MnO$_2$) rechargeable batteries have drawn research interest because of their safe, affordable, and environmentally friendly properties. Further, both Zn and MnO$_2$ have high theoretical specific capacities of 820 mAh/g and 617 mAh/g, respectively, which creates an opportunity for a commercially feasible battery. The high capacity of MnO$_2$ is predicated on it hosting two-electron reaction chemistry. Aqueous potassium hydroxide (KOH) has facilitated these electrochemical reactions in MnO$_2$ rechargeable batteries [1–4], shown to be a first electron reaction involving a proton insertion reaction and a second electron reaction involving an MnO$_2$ dissolution—precipitation reaction. For the anode, Zn undergoes dissolution on discharge and precipitation on charge, as described in Equations (1)–(4):

Zn anode:

$$Zn + 4OH^- \leftrightarrow Zn(OH)_4^{2-} + 2e^- \quad (1)$$

$$Zn(OH)_4^{2-} \leftrightarrow ZnO + 2OH^- + H_2O \quad \text{direct dissolution–precipitation} \quad (2)$$

MnO$_2$ cathode:

$$MnO_2 + H_2O + e^- \rightarrow MnOOH + OH^- \quad \text{proton insertion reaction} \quad (3)$$

$$MnOOH + H_2O + e^- \rightarrow Mn(OH)_2 + OH^- \quad \text{dissolution–precipitation reaction} \quad (4)$$

It is hypothesized that using liquid electrolytes can exacerbate battery failure. Mn ions dissolve during the second electron reaction process, which can lead to Mn loss due to migration and diffusion to distant sites, hypothetically accelerating the formation of inactive phases like spinel hausmannite (Mn$_3$O$_4$) when [Mn-OH] complexes react with each other or re-precipitate at non-local sites, all resulting in the loss of active Mn ions. At the Zn electrode, liquid electrolytes amplify Zn redistribution, which can lead to Zn electrode shape change and passivation, in turn leading to pore plugging and uncontrollable redeposition of Zn during charge at high current densities and dendrite formation [5,6]. Also, liquid electrolytes used for Zn-MnO$_2$ rechargeable batteries have lower viscosity (relative to gelled electrolytes) and therefore leak easily through cracks in the battery housing, if damaged, thereby creating safety issues for battery transportation. The previous literature [7–12] found that copper (Cu) and bismuth (Bi) helped Mn reversibility, accessing the second electron reaction region with long battery cycle life. Compared to the recent research [13,14], our group's method of electrode fabrication [9,10] is easier than electrodeposition, and the cycling performance paired with Zn achieved a longer cycle life. We thus denote this electrode fabrication as "advanced" compared to previous methods. However, failure mechanisms during the second electron reaction region were also reported, specifically, that if zincate ions diffuse across the separator into the cathode region, they can react with dissolved Mn^{3+} ions to form hetaerolite (ZnMn$_2$O$_4$), which is electrochemically inert. This results in the loss of active Mn^{3+} ions, and battery performance is directly affected by the reversibility of Zn and MnO$_2$ during the dissolution–precipitation reaction, as shown in Equations (1)–(4).

Our group previously reported that gel electrolytes can mitigate failure mechanisms for Zn-MnO$_2$ batteries constrained to just the first electron reaction of MnO$_2$ [15]. Gel electrolytes have been under research for beneficial properties between solid and liquid electrolytes, with improved safety, yet retaining the self-healing property of liquid electrolyte [16–19]. We showed that the gel electrolyte reduced the migration of zincate ions (Zn(OH)$_4{}^{2-}$), suggesting that the formation of the electrochemically inactive material, hetaerolite (ZnMn$_2$O$_4$), was mitigated. Moreover, we found that gel electrolyte reduced dissolution of each active material. Recent studies containing hydrogels, which are chemically similar to the hydrogel reported in this manuscript, have demonstrated their ability to enhance the performance of zinc-based batteries when compared to the use of liquid KOH electrolytes [20–22]. However, the battery chemistry referenced in Ref. [20] is based on zinc–nickel, which typically yields an energy density of approximately 140 Wh L^{-1}. This is lower than that of Zn-MnO$_2$ batteries, which can reach >400 Wh L^{-1}. In the cases of Refs. [21,22], Zn-MnO$_2$ battery chemistry was utilized. However, the results obtained were cycled in the first electron reaction of MnO$_2$, suggesting that higher energy achievement can be made by accessing the second electron reaction of MnO$_2$.

Here, we expand those experiments into the second electron reaction of MnO$_2$ where higher cathode capacity is accessed. We also report on our safety analysis of the liquid vs. hydrogel electrolytes, via U.S. Department of Transportation (DOT) standards that define a battery to be non-spillable if the electrolyte does not flow from a rupture or crack in the battery case [23]. To develop highly energy-dense batteries for the second electron reaction technology of Zn-MnO$_2$ batteries with gel electrolytes, we present battery cycling results under second electron reaction chemistry, along with ion diffusion properties in hydrogel electrolytes.

2. Materials and Methods

2.1. Hydrogel Synthesis

Hydrogel electrolytes, specifically potassium polyacrylate gels, were synthesized with liquid KOH electrolytes, acrylic acid (AA, Sigma Aldrich, St. Louis, MO, USA), N,N'-methylenebisacrylamide (MBA, cross-linker, Sigma Aldrich), and potassium persulfate (K$_2$S$_2$O$_8$, initiator, Sigma Aldrich). The final pH of the hydrogel electrolyte was ensured to be ~25 wt.% KOH. The synthesis process is the same as reported in Ref. [15]. Liquid KOH electrolytes were made with KOH pellets purchased from Fisher Scientific, and the AA, MBA, and K$_2$S$_2$O$_8$ were purchased from Sigma Aldrich. All components were used without further treatment. The mole fraction composition of the hydrogel was 1:0.156:0.0484:4.096 $\times$ 10^{-6} in terms of H$_2$O:KOH:Acrylic acid:Initiator, and the MBA addition to this was varied and optimized as described below.

2.2. Battery Preparation

The cathode comprised electrolytic manganese dioxide (EMD, γ-MnO$_2$) at 55 wt.%, carbon nanotubes (CNTs) at 35 wt.%, and bismuth oxide at 10 wt.%, the same as proposed by Ref. [9]. Each component was ball-milled together for 1 h. EMD was purchased from Borman (Henderson, NV, USA). CNTs were purchased from Cnano Technology Ltd., (Santa Clara, CA, USA), and bismuth oxide was purchased from Sigma Aldrich. The ball-milled mix was wetted with deionized (DI) water and then hand-cast onto a Cu-Ni current collector; then, the cathode was sealed with pellon and cellophane and pressed until the desired thickness of ~0.035 inches. Each current collector held ~23 mg of copper per cm^2. Commercial sintered nickel (NiOOH) electrodes were used and purchased from Jiangsu Highstar Battery Manufacturing (Qidong, China). The size of the anodes and cathodes was 2.54 cm $\times$ 2.54 cm for the cyclic voltammetry (CV) and galvanostatic experiments, and was 5.08 cm $\times$ 7.62 cm for Zn-MnO$_2$ full-cell cycling. Anodes and cathodes were assembled into a polysulfone box (8.255 cm width $\times$ 5.3975 cm depth $\times$ 15.875 cm height) and compressed with polypropylene shims. Then, the box was filled with 75 mL electrolyte on average. All cells were under vacuum for 30 min to soak the porous Zn and MnO$_2$ electrodes. Mercury mercuric oxide (Hg-HgO) reference electrodes were used in each cell box to track half-cell voltages.

2.3. Electrolyte Spillability Safety Measurements

Battery "spillability" was measured according to U.S. DOT rules that declare the electrolyte is non-spillable if it does not flow through cracks or rupture in a battery case. Since commercial battery cases are packed tight with electrode stacks, creating typical flow gaps of ~1 mm, we measured spillability as the flow from 2.4 mm I.D. glass capillary tubes. Once 30 mm of hydrogel electrolyte was set up inside the end of these capillary tubes; each tube was dropped 10 mm end-first, under their own gravity, onto a hard surface to provide force promoting the electrolyte to flow out. Electrolyte flow out was recorded for different hydrogel cross-linking formulations. Additional hydrogel flow measurements were recorded when 100 mL of hydrogel fabricated in the bottom of a 250 mL beaker (diameter 75 mm) was tipped over to provide a force promoting flow; however, these measurements were not important for defining spillability according to US DOT regulations, because ruptured batteries have flow from capillary length scales much smaller than 75 mm.

2.4. Electrochemical Measurements

Cyclic voltammetry (CV) and electrochemical impedance spectroscopy (EIS) were performed through Biologic potentiostat/galvanostat (VSP Modular 5-channel). A multichannel Arbin BT 2000 was used for galvanostatic experiments. After cycling, electrodes were removed, washed, and then soaked in DI water for 6 h and dried overnight in the air dryer at 50 °C. Scanning electron microscopy (SEM) and energy-dispersive X-ray spectrometry (EDX, EDS) were performed. SEM and EDX were performed by a FEI Helios Nanolab 660 Dualbeam FIB-SEM Operation fitted with an EDX (Thermo Fisher Scientific, Waltham, MA, USA).

2.5. Cu Diffusion Coefficient

Glass cuvettes (3.5 cm width × 1 cm depth × 4.5 cm height) were filled with either liquid or gel electrolytes that contained a known concentration of copper hydroxide ($Cu(OH)_2$, from Alfa Aesar, Haverhill, MA, USA); see Figure S1. At the start of an experiment, a cuvette containing 1 molar $Cu(OH)_2$ was fixed as the bottom cuvette. Next, in the case of hydrogel experiments, a cuvette with zero molar $Cu(OH)_2$ hydrogel electrolyte was quickly fixed upside down on top of the first cuvette, creating a "step function" in Cu^{2+} concentration at the initiation of the experiment, as clearly seen by the sharp step in blue color in Figure S1. In the case of a liquid electrolyte experiment, an empty cuvette was fixed upside down on top of the bottom cuvette, and the liquid electrolyte with zero molar $Cu(OH)_2$ was filled into the top cuvette via a hole drilled through the cuvette at its top. Photographs were taken on an hourly basis. The concentration profile of Cu^{2+} ions was calculated "colorimetrically" by use of the Beer–Lambert law, as explained in the Supplemental Information near Figure S1. Due to the precision shape of the glass cuvettes, there was no leakage between the cuvettes over the experiment. The experiments were conducted for 120 h.

3. Results and Discussion

3.1. Non-Spillable Hydrogel Experiment

To find the optimal hydrogel cross-linker formulation, the US DOT spillability methods described above were repeated on hydrogels with varying amounts of cross-linker. More cross-linker causes a higher overvoltage during cycling, so less cross-linker was preferred. We settled on a hydrogel formulation that had the smallest amount of cross-linker but kept zero flow from the 2.4 mm glass capillary experiments. In other words, we kept the electrolyte "non-spillable" but otherwise minimized hydrogel cross-linking (data shown in Table 1). As the data of Table 1 show, a 3.92×10^{-5} mole fraction of MBA was determined as the optimal cross-linked hydrogel. All the tested hydrogels in Table 1 retained their initial rheology over the time studied.

Table 1. Hydrogel flow measurement used to optimize hydrogel formulation. Mole fraction of MBA to H_2O is the first column. Cross-linking was allowed to proceed for at least 16 h prior to experiments.

Mole Fraction MBA:H_2O	Flow from ~1 mm Gap	Flow from ~75 mm Gap
2.61×10^{-5}	Flow	Flow
3.40×10^{-5}	Flow	Flow
3.92×10^{-5}	No Flow	Flow
4.70×10^{-5}	No Flow	Flow
5.20×10^{-5}	No Flow	Flow
6.00×10^{-5}	No Flow	Flow
6.50×10^{-5}	No Flow	No Flow
7.30×10^{-5}	No Flow	No Flow
7.80×10^{-5}	No Flow	No Flow

Using the optimized hydrogel formulation from Table 1, the optimized hydrogel electrolyte was tested for spillability from a real battery cell manufactured by Urban

Electric Power (UEP), as described in Figure 1a. Figure S2 shows the electrode stack inside the cell box, which confirms the ~2 mm capillary length scales of our spillability methods. The UEP manufactured batteries were filled with liquid and gel electrolyte separately up to the top of the electrodes. Cuts in the cell box were intentionally made using a razor blade, at locations 2 cm height from the bottom of the cells. The width of each cut was 2 cm, as shown in Figure 1a. Due to gravitational force, the electrolyte attempted to flow from the cracks. Any electrolyte emerging from the cracks was wiped away with paper towels, and the total weight of the batteries was measured every 30 min for several hours. Mass loss is due to leakage of electrolyte through the cut to the case. Figure 1b shows the resulting data, wherein the cell with liquid electrolyte had 49.69 g of mass loss, while the cell with gel electrolyte had only 0.36 g of mass loss. Therefore, the hydrogel electrolyte was determined to satisfy the DOT regulation for non-spillable batteries.

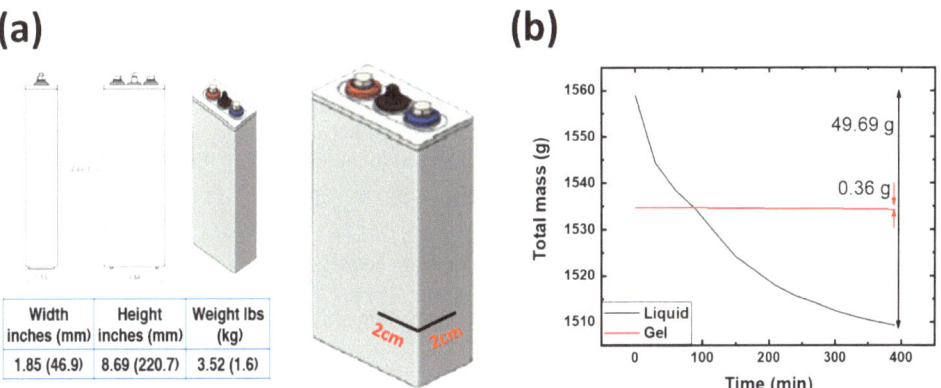

Figure 1. (a) A manufactured prismatic cell information and (b) time vs. total mass change after making cracks.

3.2. Electrochemical Performance

The non-spillable optimized hydrogel electrolyte was then used in measurements of Zn^{2+} and Cu^{2+} ion diffusion. To understand Zn^{2+} diffusion behavior, two Zn foil symmetric electrodes were used (as shown in Figure S3) for EIS measurements, which were repeated three times with liquid and gel electrolyte. The EIS scanning frequency range was 100 kHz to 0.01 Hz. As shown in Figure 2a, due to the viscous properties and polymer chain of the gel electrolyte, the solution resistance of the gel electrolyte was ~0.4 Ohm higher than the liquid electrolyte, and the curve had a deeper sigmoidal shape. In the low-frequency range, the plots of the liquid electrolyte had a slope close to 45 degrees, whereas the gel electrolyte plots showed a lower slope. This suggests that the zinc transfer rate in the gel electrolyte is slower than in the liquid electrolyte but not negligible, as evidenced by the plotted slope. The ionic conductivity was calculated by Equation (5):

$$\sigma = \left(\frac{1}{R}\right) * \left(\frac{l}{A}\right) \tag{5}$$

where R is charge-transfer resistance, l is thickness, and A is area [24]. From Figure 2a, the value of charge-transfer resistance was obtained, and the ionic conductivity of Zn with the gel electrolyte was ~5.45 mS/cm. One of the recent studies reported that their hydrogel electrolyte had 83 mS/cm [25]. Even though their components are similar to our gel electrolytes, the difference could be due to soaking their hydrogel electrolyte in $ZnSO_4$ solutions overnight, so that additional species such as sulfate ions helped obtain higher ionic conductivity than ours. Moreover, Figure S12 of Ref. [25] showed that their gel electrolyte was solid-like. The phase difference also made the difference.

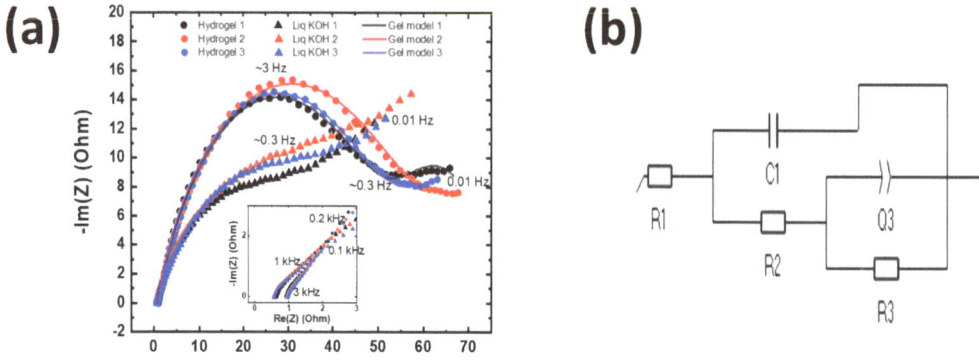

Figure 2. (a) Cell impedance measurement of Zn plate symmetric cell with the gel electrolyte and a liquid electrolyte. The experiment was conducted three times for each electrolyte. Transparent data points are the fit. (b) The equivalent circuit model from the fits in (a). R1 is the solution resistance; R2 is the charge-transfer resistance at Zn and electrolyte interface; R3 is the resistance associated with OH^- transport; C1 is the capacitance; Q3 is the constant phase element for non-faradaic charging of the double layer.

Figure 3a,b present CV plots with liquid and gel electrolyte, respectively. With liquid electrolyte, the Mn^{3+} to Mn^{2+} peaks at -0.7 V vs. Hg-HgO were sharper than with gel. As this voltage range is for the dissolution–precipitation reaction, dissolved Mn ions were intercalated with dissolved Cu ions, leading to Cu^{2+}-intercalated Bi-birnessite [9]. However, as the Cu^{1+} to Cu^0 peak at -0.6 V vs. Hg-HgO started fading at the 12th cycle, it suggested that dissolved Cu ions were not able to react with Mn ions during the cycle. Cu and Mn peaks at the same voltage were stronger than with hydrogel electrolyte until the 20th cycle. In the oxidation portion, the gel electrolyte cell showed Mn peaks at -0.2 and 0.2 vs. Hg-HgO for all cycles, stronger than for liquid electrolyte, especially after the first few cycles. This is because hydrogel electrolyte limited the dissolution of active materials, and thereby mitigated active ion loss.

To investigate the reason for the CV measurement, with liquid electrolyte showing the Cu peak reduced after the 12th cycle, two identical cells with MnO_2 vs. NiOOH electrodes at C/20 (capacity of MnO_2) were built, one with liquid KOH and the other with hydrogel KOH, and both were cycled until the 1st, 5th, or 12th cycle and dissected. The blue color seen in the dissected materials (see Figure 3c,d and Table 2) is due to Cu^{2+} ions. This increased migration of blue-colored Cu^{2+} in cells with liquid KOH correlates with more metallic-colored Cu deposited on the separators of cells with liquid electrolyte; see Table 2 and Figure S4. But with hydrogel electrolyte, this deposition of Cu ions was limited and more localized in the electrode area. After the first cycle with liquid electrolyte, the deposition of Cu ions was observed on the second layer of the separator. After the fifth cycle, the separator was observed to be degraded and the blue color was significantly denser than in the experiment with hydrogel electrolyte. To quantify Cu ions from the separators, identical experiments conducted as in Table 2 were carried out. The amount of cathode materials and the dimension of the electrodes were proportionally scaled down to fit 1.27 cm × 1.27 cm and then dissected after their cycling. The cells were cycled until the fifth cycle so that the separator would retain structural integrity, as shown in Table 2, enabling the analysis of the blue color in separators for both electrolytes. As specifically described in Figure 4a, SEM/EDX analysis was conducted at the four corners, and the EDX mapping results are shown in Figures S5–S8. The mapping analysis detected six species corresponding to the initial battery species, and elemental Cu, one of the six, was diffused in the separators of both liquid and gel electrolyte. However, it is noted that, in Figure 4b, the corners of the gel electrolyte separator detected less than 0.05 atomic % Cu, while the separator of cycled liquid electrolyte had detected 6–15 atomic % times higher Cu. This supports the results from Figure 3 and Figure S4 and Table 2 and agrees that hydrogel electrolytes mitigated Cu ion diffusion while cycling.

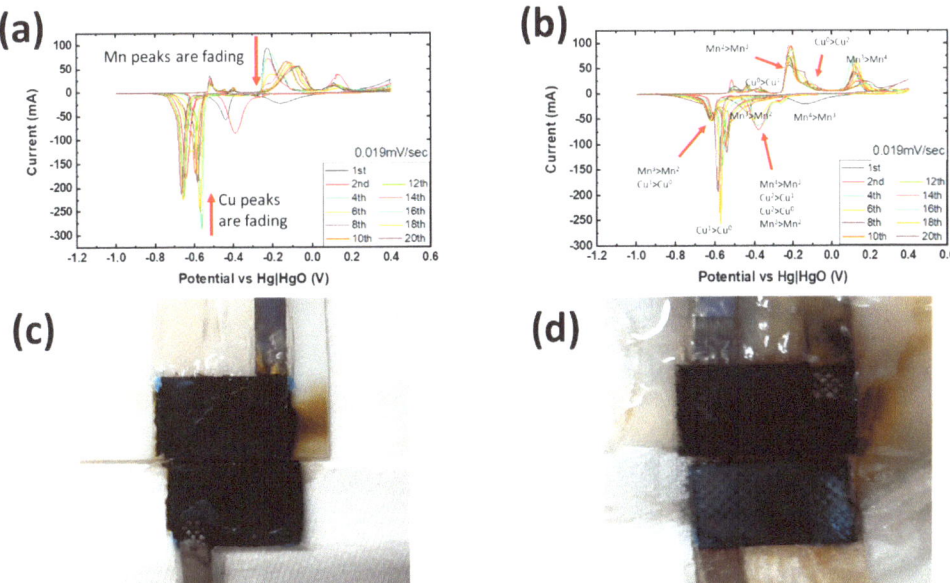

Figure 3. The CV experiments were conducted at 0.019 mV/s, which means it takes 20 h to charge and discharge. The CV results with (**a**) liquid electrolyte and (**b**) gel electrolyte. The picture of the dissected cell 12th cycle after galvanostatic with (**c**) liquid and (**d**) gel electrolyte.

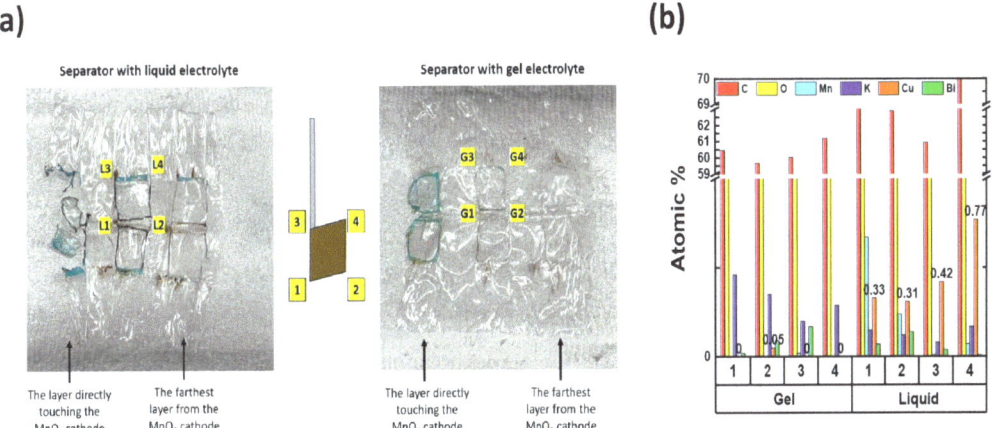

Figure 4. (**a**) SEM/EDX analysis was conducted on the separator at the four corners where the electrode was as shown in the electrode scheme in the middle. 1 and 2 represent the bottom two corners, and 3 and 4 represent the top two corners. L and G mean liquid and gel electrolyte, respectively. (**b**) The atomic % of six species from the four corners of the separators in (**a**).

Table 2. Separators from dissected cells after C/20 galvanostatic experiments.

	1st Cycle	5th Cycle	12th Cycle
Liquid KOH Electrolyte			
Gel Electrolyte			

Figure 5a,b demonstrate galvanostatic cycling performance from two identical MnO$_2$ half cells (vs. NiOOH) with gel electrolyte, which achieved ≥700 cycles with 99% coulombic efficiency and 63% energy efficiency. Our group reported [9] that the cycle life with liquid electrolyte is over 1000 cycles at the same cycling rate. Even though our present study of gelled KOH did not replicate the cycling performance reported in Ref. [9], our present results indicate optimistic outcomes, as we have observed that the gel electrolyte mitigated Cu diffusion so that it keeps MnO$_2$ reversibility. In Figure 5c,d, the two liquid-containing cells achieved ~500 cycles, while the cell with gel electrolyte performed 300 cycles until it showed 50% theoretical capacity fade. This is because the gel-containing cell had a twice greater Zn utilization than the two liquid-containing cells. If the gel-containing cell had the same Zn utilization, it is hypothesized that its cycle life would be at least equal to the liquid-containing cell's performance due to the mitigation of failure mechanisms caused by zinc. These electrochemical results indicated that the gel electrolyte helps reduce Cu ion loss so that active ions, such as Mn, Cu, and Bi, were able to react with each other, leading to the stable electrochemical reaction reversibility of the [(Cu-Bi)Mn] complex. In this way, using gel electrolytes will deliver a longer battery cycle life than cells with liquid electrolyte.

A reduction of Cu^{2+} diffusion was hypothesized to explain why copper migrates less in a cell with hydrogel electrolyte. To test this hypothesis, we measured the Cu^{2+} diffusion coefficient in liquid and gel electrolyte by fitting analytical solutions of Fick's law to the data we collected on time-varying concentration of Cu^{2+} in our cuvette experiments; see Methods Section and Supplemental Information near Figure S1. Fick's law holds

$$\frac{\partial C}{\partial t} = D \frac{\partial^2 C}{\partial y^2} \tag{6}$$

where C is the concentration of Cu, t is time, D is the diffusion coefficient, y is position. The cuvette experiments were homogenous in all directions except the y-direction. The boundary conditions (B.C.) and the initial conditions (I.C.) were experimentally, and analytically, as follows:

$$\text{B.C.} \quad \frac{\partial C}{\partial y} = 0 \quad \text{at } y = 0 \text{ and } L \tag{7}$$

$$\text{I.C.} \quad C(y,0) = \begin{cases} 0, & H < y < L \\ 1, & 0 < y < H \end{cases} \tag{8}$$

The analytical solution C(y,t) satisfies B.C., I.C., and the governing equation, and was determined by separation of variables, then Fourier series reconstruction. Using the eigenfunctions of this system, the entire solution is described as

$$C_n(y,t) = A_0 + \sum_{n=1}^{\infty} A_n \cos\frac{n\pi y}{L} e^{-\lambda_n^2 t} \text{ and } \lambda_n = \sqrt{D}\frac{n\pi}{7} \tag{9}$$

The coefficients, A_0 and A_n, were determined by Fourier cosine series. In our experiments, L is 7 cm and H is 3.5 cm, so the final analytical solution is

$$C(y,t) = \frac{1}{2} + \sum_{n=1}^{\infty} \frac{2}{n\pi} \sin\frac{n\pi}{2} \cos\frac{n\pi y}{7} e^{-\lambda_n^2 t} \text{ and } \lambda_n = \sqrt{D}\frac{n\pi}{7} \quad (10)$$

Using the first 30 terms of this Fourier series, the experimental data and analytical solution are overplotted in Figure 6. The diffusion coefficient was determined by fitting the experimental and theoretical data, with $D = 1.9 \times 10^{-6} \frac{cm^2}{s}$ being the best fit. We find the same diffusion coefficient in liquid and hydrogel electrolytes to within and experimental error of ~$0.3 \times 10^{-6} \frac{cm^2}{s}$, which in retrospect is less surprising when considering the volume fraction of water in the hydrogel is 95%. The spike in data points near the middle (near y = 3.5 cm) of the plots in Figure 6 have a high experimental error due to this location being where the two cuvettes touched, which caused optical refraction due to the glass edges.

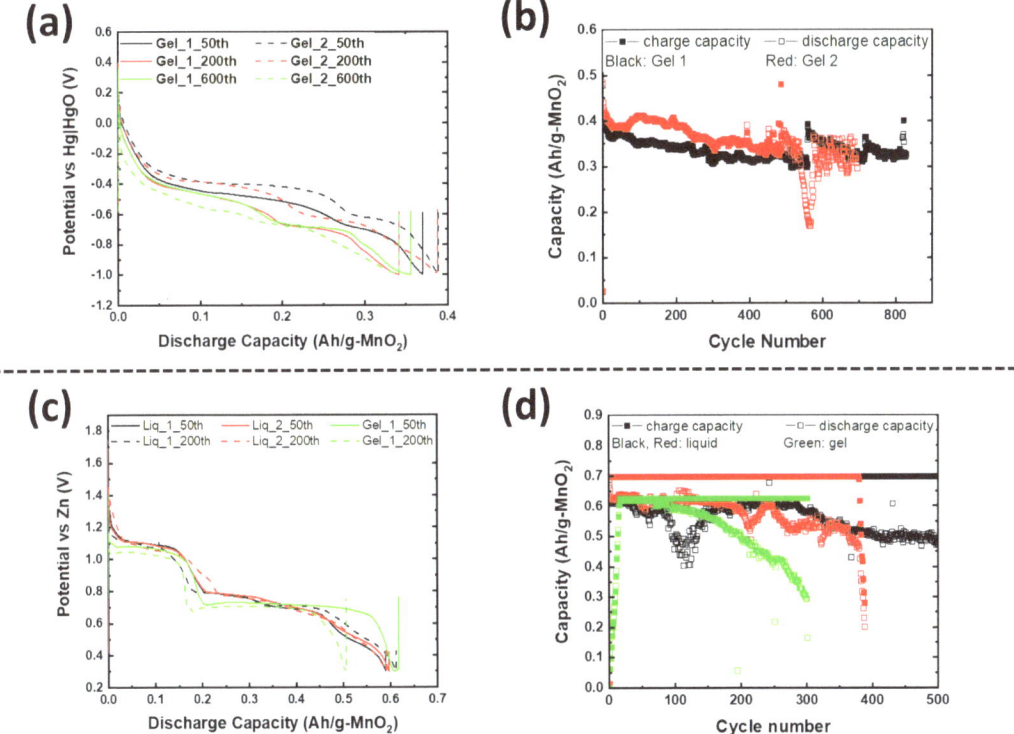

Figure 5. (a) The galvanostatic data of identical MnO_2 half cells against NiOOH with gel electrolytes at C/3, where C is based on 2-electron MnO_2 capacity. (b) Capacity retention of the MnO_2 cells from (a). (c) The galvanostatic data of full cells with liquid and gel electrolytes at C/20, where C is based on 2-electron MnO_2 capacity. (d) Capacity retention of the full cells from (c).

Since Cu^{2+} ion diffusion on the molecular level appears to be the same in gel and liquid KOH electrolyte, we must hypothesize a different explanation for the dissected battery cells showing reduced Cu migration. We speculate that gel electrolyte reduced convection of the electrolyte, which is forced by several factors (electrical, expansion/shrinkage cycles, bubble growth) [26,27]. This can be supported with Figure S9. The pictures in Figure S9 were taken while charging. As shown, bubbles went upwards and were removed from the same spot where they generated. They did not go anywhere in the electrolyte, supporting the speculation of reduced convection by gel electrolyte. Further, the separators with gel electrolyte in Table 2 showed light reflection on each layer of separators. This means that the gel electrolyte was evenly applied during the vacuum process. With this, we hypothesize that gel electrolyte leaves a thin film between the separators that can reduce the direct contact of conductive Cu depositions on the separator, thus reducing short circuit severity or frequency. Further studies will be needed to confirm these hypotheses.

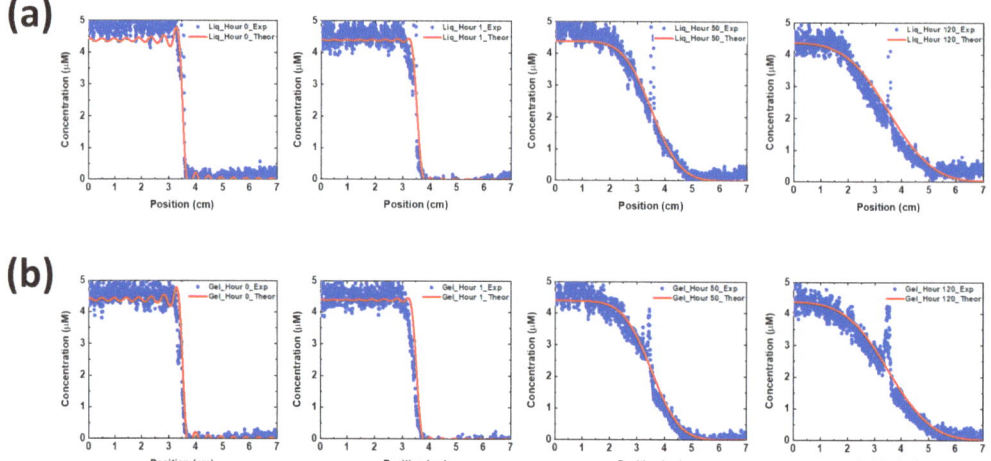

Figure 6. Cu diffusion quantification and plots with (**a**) liquid (top four figures in a row) and (**b**) gel electrolyte (bottom four figures in a row). The experiments were conducted for 120 h. Blue dots represent the experimental data and solid line represents the theoretical solution.

4. Conclusions

Cross-linked hydrogel electrolytes were optimized to pass the US Department of Transportation guidelines for non-spillable batteries to allow transportability. This was achieved by investigating electrolyte flow from cracks or ruptures to the battery cell box. The optimized gel electrolyte was then studied for the second electron Mn cathode reaction. A series of electrochemical experiments with the hydrogel electrolyte showed a successful electrochemical reaction of each active material under the second electron reaction chemistry. The CV experiments suggested that the gel electrolyte helped stabilize the Cu-Bi Mn complex, helping the reversibility of Bi-birnessite and therefore the long-term cycle life of the battery. Dissected cells showed less Cu^{2+} migration in gel electrolyte as compared to liquid electrolyte. The Cu^{2+} diffusion coefficient was measured to be the same in liquid and gel electrolyte, and so we suggest the reason for the reduced Cu^{2+} migration is that hydrogel electrolyte reduces convection and also provides a non-conductive physical barrier between conductive materials and the separator itself. This study supports the optimism that gel electrolytes could be applied to long-duration energy storage applications, continuously providing a second electron Mn cathode reaction and a two-electron Zn anode reaction.

Supplementary Materials: The following supporting information can be downloaded at: https://www.mdpi.com/article/10.3390/polym16050658/s1, Figure S1: Cu diffusion coefficient measured by colorimetry applied to diffusion of $Cu(OH)_2$ from bottom cuvette into top inverted cuvette, photos taken at (a) the beginning of an experiment, and (b) 120 hours afterwards. The four cuvettes sitting on the table on the left hold reference concentrations of 0, 0.3, 0.6 and 1 molarity; Figure S2: The top view of a prismatic cell and the information of the gap between the electrodes and left and right side of the cell box; Figure S3: The view of Zn plate symmetric experiment; Figure S4: Identical cell construction after the 60th cycle with liquid and gel electrolyte cycled at C/20 where C is 2-electron MnO_2 capacity; Figure S5: SEM images and EDX mapping results from the four corners of the separator with liquid electrolyte. All scale bars are 500 µm; Figure S6: SEM images and EDX mapping results from the four corners of the separator with gel electrolyte. All scale bars are 500 µm; Figure S7: The peak intensity results for the 6 species with liquid electrolyte; Figure S8: The peak intensity results for the 6 species with gel electrolyte; Figure S9: Bubble behavior in the gel electrolyte while charging; Table S1: Molar absorptivity of reference cuvettes.

Author Contributions: S.B. and D.E.T. originated the project. D.E.T. and J.C. conceptualized and executed the experiments. J.C. and D.E.T. wrote the manuscript with other authors' suggestions. J.C. and D.E.T. synthesized the gels. M.N. and J.C. fabricated electrodes and cycled cells. G.G.Y., T.N.L., B.R.W. and S.B. gave input and helped with the overall project. All authors have read and agreed to the published version of the manuscript.

Funding: This work was supported by the U.S. Department of Energy Office of Electricity (grant number DE-NA0003525) and the U.S. Department of Energy Office of Science (grant number FWP: 1601331).

Institutional Review Board Statement: Not applicable.

Data Availability Statement: The data presented in this study are available on request from the corresponding author.

Acknowledgments: This work was performed, in part, at the Center for Integrated Nanotechnologies, an Office of Science User Facility operated for the U.S. Department of Energy (DOE) Office of Science. The views expressed in this article do not necessarily represent the views of the U.S. Department of Energy or the United States Government. This article has been authored by an employee of National Technology & Engineering Solutions of Sandia, LLC under Contract No. DE-NA0003525 with the U.S. Department of Energy (DOE). The National Technology & Engineering Solutions of Sandia, LLC employee owns all right, title and interest to their contribution to the article and is responsible for its contents. The United States Government retains and the publisher, by accepting the article for publication, acknowledges that the United States Government retains a non-exclusive, paid-up, irrevocable, world-wide license to publish or reproduce the published form of this article or allow others to do so, for United States Government purposes. The DOE will provide public access to these results of federally sponsored research in accordance with the DOE Public Access Plan https://www.energy.gov/downloads/doe-public-access-plan (accessed on 31 December 2023).

Conflicts of Interest: Author Gautam Ganapati Yadav was employed by the Urban Electric Power. The remaining authors declare that the research was conducted in the absence of any commercial or financial relationships that could be construed as a potential conflict of interest.

References

1. Gallaway, J.W.; Erdonmez, C.K.; Zhong, Z.; Croft, M.; Sviridov, L.A.; Sholklapper, T.Z.; Turney, D.E.; Banerjee, S.; Steingart, D.A. Real-time materials evolution visualized within intact cycling alkaline batteries. *J. Mater. Chem. A* **2014**, *2*, 2757–2764. [CrossRef]
2. Turney, D.E.; Gallaway, J.W.; Yadav, G.G.; Ramirez, R.; Nyce, M.; Banerjee, S.; Chen-Weigart, Y.K.; Wang, J.; D'Ambrose, M.J.; Kolhekar, S.; et al. Rechargeable zinc alkaline anodes for long-cycle energy storage. *Chem. Mater.* **2017**, *29*, 4819–4832. [CrossRef]
3. Ingale, N.D.; Gallaway, J.W.; Nyce, M.; Couzis, A.; Banerjee, S. Rechargeability and economic aspects of alkaline zinc-manganese dioxide cells for electrical storage and load leveling. *J. Power Sources* **2015**, *276*, 7–18. [CrossRef]
4. Lim, M.B.; Lambert, T.N.; Chalamala, B.R. Rechargeable alkaline zinc-manganese oxide batteries for grid storage: Mechanisms, challenges and developments. *Mater. Sci. Eng.* **2021**, *143*, 100593. [CrossRef]
5. D'Ambrose, M.J.; Turney, D.E.; Yadav, G.G.; Nyce, M.; Banerjee, S. Material failure mechanisms of alkaline Zn rechargeable conversion electrodes. *ACS Appl. Energy Mater.* **2021**, *4*, 3381–3392. [CrossRef]
6. Hawkins, B.E.; Turney, D.E.; Messinger, R.J.; Kiss, A.M.; Yadav, G.G.; Banerjee, S.; Lambert, T.N. Electroactive ZnO: Mechanisms, Conductivity, and Advances in Zn Alkaline Battery Cycling. *Adv. Energy Mater.* **2022**, *12*, 2103294. [CrossRef]
7. Gallaway, J.W.; Yadav, G.G.; Turney, D.E.; Nyce, M.; Huang, J.; Chen-Weigart, Y.-C.K.; Williams, G.; Thieme, J.; Okasinksi, J.S.; Wei, X. An operando study of the initial discharge of Bi and Bi/Cu modified MnO_2. *J. Electrochem. Soc.* **2018**, *165*, A2935. [CrossRef]
8. Seo, J.K.; Shin, J.; Chung, H.; Meng, P.Y.; Wang, X.; Meng, Y.S. Intercalation and Conversion Reactions of Nanosized β-MnO_2 Cathode in the Secondary Zn/MnO_2 Alkaline Battery. *J. Phys. Chem. C* **2018**, *122*, 11177–11185. [CrossRef]
9. Yadav, G.G.; Gallaway, J.W.; Turney, D.E.; Nyce, M.; Huang, J.; Wei, X.; Banerjee, S. Regenerable Cu-intercalated MnO_2 layered cathode for highly cyclable energy dense batteries. *Nat. Commun.* **2017**, *8*, 14424. [CrossRef]
10. Yadav, G.G.; Wei, X.; Huang, J.; Gallaway, J.W.; Turney, D.E.; Nyce, M.; Secor, J.; Banerjee, S. A conversion-based highly energy dense Cu^{2+} intercalated Bi-birnessite/Zn alkaline battery. *J. Mater. Chem. A* **2017**, *5*, 15845–15854. [CrossRef]

11. Bruck, A.M.; Kim, M.A.; Ma, L.; Ehrlich, S.N.; Okasinksi, J.S.; Gallaway, J.W. Bismuth enables the formation of disordered birnessite in rechargeable alkaline batteries. *J. Electrochem. Soc.* **2020**, *167*, 110514. [CrossRef]
12. Schorr, N.B.; Arnot, D.J.; Bruck, A.M.; Duay, J.; Kelly, M.; Having, R.L.; Ricketts, L.S.; Vigil, J.A.; Gallaway, J.W.; Lambert, T.N. Rechargeable alkaline Zinc/Copper oxide batteries. *ACS Appl. Energy Mater.* **2021**, *4*, 7073–7082. [CrossRef]
13. Chen, Y.; Gu, S.; Wu, S.; Ma, X.; Hussain, I.; Sun, Z.; Lu, Z.; Zhang, K. Copper activated near-full two-electron Mn^{4+}/Mn^{2+} redox for mild aqueous Zn/MnO_2 battery. *Chem. Eng. J.* **2022**, *450*, 137923. [CrossRef]
14. Sun, Y.; Zhuang, S.; Ren, Y.; Jiang, S.; Pan, X.; Sun, G.; Zhu, B.; Wen, Y.; Li, X.; Tu, F.; et al. Promoting cycle stability and rate performance of birnessite-type MnO_2 cathode via copper and bismuth dual ions pre-intercalation for aqueous zinc-ion batteries. *J. Energy Storage* **2023**, *74*, 109589. [CrossRef]
15. Cho, J.; Yadav, G.G.; Weiner, M.; Huang, J.; Upreti, A.; Wei, X.; Yakobov, R.; Hawkins, B.E.; Nyce, M.; Lambert, T.N.; et al. Hydroxyl conducting hydrogels enable low-maintenance commercially sized rechargeable $Zn–MnO_2$ Batteries for Use in Solar Microgrids. *Polymers* **2022**, *14*, 417. [CrossRef]
16. Zhu, X.; Yang, H.; Cao, Y.; Ai, X. Preparation and electrochemical characterization of the alkaline polymer gel electrolyte polymerized from acrylic acid and KOH solution. *Electrochim. Acta* **2004**, *49*, 2533–2539. [CrossRef]
17. Mohamad, A.A. $Zn/gelled\ 6M\ KOH/O_2$ zinc-air battery. *J. Power Sources* **2006**, *159*, 752–757. [CrossRef]
18. Choudhury, N.A.; Sampath, S.; Shukla, A.K. Hydrogel-polymer electrolytes for electrochemical capacitors: An overview. *Energy Environ. Sci.* **2009**, *2*, 55–67. [CrossRef]
19. Maitra, J.; Shukla, V.K. Cross-linking in hydrogels—A review. *Am. J. Polym. Sci.* **2014**, *4*, 25–31.
20. Li, S.; Fan, X.; Liu, X.; Zhao, Z.; Xu, W.; Wu, Z.; Feng, Z.; Zhong, C.; Hu, W. Potassium Polyacrylate-Based Gel Polymer Electrolyte for Practical Zn−Ni Batteries. *ACS Appl. Mater. Interfaces* **2022**, *14*, 22847–22857. [CrossRef] [PubMed]
21. Hu, F.; Li, M.; Gao, G.; Fan, H.; Ma, L. The Gel-State Electrolytes in Zinc-Ion Batteries. *Batteries* **2022**, *8*, 214. [CrossRef]
22. Zhang, J.; Huang, Y.; Li, Z.; Gao, C.; Jin, S.; Zhang, S.; Wang, X.; Zhou, H. Polyacrylic acid assisted synthesis of free-standing $MnO_2/CNTs$ cathode for Zinc-ion batteries. *Nanotechnology* **2020**, *31*, 375–401. [CrossRef] [PubMed]
23. The Code of Federal Regulations (CFR), 49 CFR 173. Available online: https://www.ecfr.gov/current/title-49/subtitle-B/chapter-I/subchapter-C/part-173/subpart-E/section-173.159a?msclkid=7e67d3fdb43911ecb601bcd46a0d0bb0 (accessed on 31 December 2023).
24. Vadhva, P.; Hu, J.; Johnson, M.J.; Stocker, R.; Braglia, M.; Brett, D.J.L.; Rettie, A.J.E. Electrochemical impedance spectroscopy for all-solid-state batteries: Theory, methods and future outlook. *ChemElectroChem* **2021**, *8*, 1930–1947. [CrossRef]
25. Sun, M.; Ji, G.; Zheng, J. A hydrogel electrolyte with ultrahigh ionic conductivity and transference number benefit from Zn^{2+} "highways" for dendrite-free $Zn-MnO_2$ battery. *Chem. Eng. J.* **2023**, *463*, 142535. [CrossRef]
26. Tél, A.; Bauer, R.A.; Varga, Z.; Zrínyi, M. Heat conduction in poly(N-isopropylacrylamide) hydrogels. *Int. J. Therm. Sci.* **2014**, *85*, 47–53. [CrossRef]
27. Han, J.; Jang, S.; Kim, B.-K.; Park, K. Electrochemical study of agarose hydrogels for natural convection on macroelectrodes and ultramicroelectrodes. *J. Anal. Sci. Technol.* **2023**, *14*, 10. [CrossRef]

Disclaimer/Publisher's Note: The statements, opinions and data contained in all publications are solely those of the individual author(s) and contributor(s) and not of MDPI and/or the editor(s). MDPI and/or the editor(s) disclaim responsibility for any injury to people or property resulting from any ideas, methods, instructions or products referred to in the content.

Charge Carrier Mobility in Poly(N,N′-bis-4-butylphenyl-N,N′-bisphenyl)benzidine Composites with Electron Acceptor Molecules

Alexey R. Tameev *, Alexey E. Aleksandrov, Ildar R. Sayarov, Sergey I. Pozin, Dmitry A. Lypenko, Artem V. Dmitriev, Natalia V. Nekrasova, Andrey Yu. Chernyadyev and Aslan Yu. Tsivadze

Frumkin Institute of Physical Chemistry and Electrochemistry of the Russian Academy of Sciences, Leninsky Prosp., 31, Bld. 4, 119071 Moscow, Russia
* Correspondence: tameev@elchem.ac.ru

Abstract: Polymer composites based on poly(N,N′-bis-4-butylphenyl-N,N′-bisphenyl)benzidine (poly-TPD) with PCBM and copper(II) pyropheophorbide derivative (Cu-PP) were developed. In thin films of the poly-TPD and Cu-PP composites, the charge carrier mobility was investigated for the first time. In the ternary poly-TPD:PCBM:Cu-PP composite, the electron and hole mobilities are the most balanced compared to binary composites and the photoconductivity is enhanced due to the sensitization by Cu-PP in blue and red spectral ranges. The new composites are promising for use in the development of photodetectors.

Keywords: poly-TPD; copper (II) pyropheophorbide; polymer composite; charge carrier mobility; CELIV; SCLC; photoconductivity

1. Introduction

In materials design, polymers are important because their electronic properties can be controlled during both the synthesis and the creation of solid-state materials for various applications, particularly those related to organic electronics [1]. Charge carrier mobility is a key parameter characterizing the electrical properties of polymer and organic semiconductors being developed for use in electronic devices. [2]. Interest in poly(N,N′-bis-4-butylphenyl-N,N′-bisphenyl)benzidine, also known as poly-TPD (Figure 1), is caused by its ability to form thin films of high quality with excellent hole transport, i.e., monopolar electrical conductivity. The hole mobility in thin polymer layers was reported in the order of $(1 \div 2) \times 10^{-3}$ cm^{2}V^{-1}s^{-1} [3,4], 1.7×10^{-4} cm^{2}V^{-1}s^{-1} [5] and 4×10^{-7} cm^{2}V^{-1}s^{-1} [6] as extracted from the space charge limited current (SCLC) mode. In the organic field-effect transistor architecture, the hole mobility of poly-TPD was found to be 1×10^{-4} cm^{2}V^{-1}s^{-1} [7]. Possessing reasonably extended hole mobility, poly-TPD is applicable as a hole transport layer (HTL) in thin film devices due to its highest occupied molecular orbital (HOMO) energy level providing the hole transport matches well with HOMO levels of hole transport polymers and organic photoconductors. In particular, the polymer was used as an HTL in the development of organic photodiodes [8], organic light-emitting diodes (OLEDs) [4,7], quantum-dot-based light-emitting diodes (QLEDs) [5,9,10], and organic phototransistors [8]. In addition, due to the HOMO level of poly-TPD matching well with the work function of ITO and valence band of perovskite derivatives, the polymer successfully served as an HTL in perovskite solar cells [6,11,12]. Moreover, poly-TPD can exhibit photoconductivity. A blend of an electron donor (D) derivative of poly-TPD and an electron acceptor (A) PCBM (phenyl-C$_{61}$-butyric acid methyl ester) was shown to enhance light absorption and promote the separation of electron–hole pairs in the photoactive film of an inorganic/organic heterojunction-based photodetector [13] and organic photovoltaic cell [14]. The photoconductivity spectrum of the poly-TPD:PCBM blend was additively composed of the absorption spectra of the polymer and PCBM, and to expand the photosensitive spectrum, sensitizing additives were introduced into the D-A composite. Nevertheless, the mobility of charge carriers in D-A composites based on poly-TPD remains poorly studied.

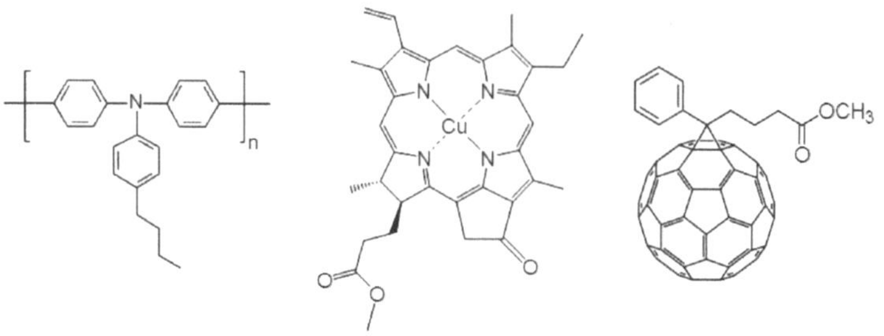

Figure 1. Chemical structures of materials used.

Semisynthetic macrocyclic compounds (such as porphyrins, pheophorbide-a, pheophorbide-b, chlorins, and bacteriochlorins) extracted from animal blood, plant chlorophyll or from bacterial cultures and followed by their chemical modification [15] are of great interest as active components of composite materials for research in photodynamic therapy of cancer diseases [16], water photolysis for hydrogen generation [17], converting solar energy into chemical energy [18], and electroluminescent devices [19,20]. Unlike semisynthetic coproporphyrins, mesoporphyrins, and etioporphyrins [20], pheophorbides and bacteriochlorins contain stereocenter carbon atoms associated with the aromatic cycle. Due to stereocenters, the molecules can absorb photon incidents not only at right angles to the planes of aromatic rings [21,22], but also with a significant deviation from orthogonality increasing the harvest of incident light. In addition, stereocenters in the molecules can exhibit the effect of circularly polarized luminescence, for example, similar to polymers containing chiral fragments of cholesterol [23].

In this work, the copper (II) complex with pyropheophorbide-a (Cu-PP) is used as a component of composites. The copper (II) complex with pyropheophorbide-a exhibits no luminescence in the visible range while the free base of pyropheophorbide-a demonstrates intense fluorescence in the red band of the spectrum. Therefore, we can expect that in a composite containing Cu-PP, the possible luminescent loss of absorbed light energy is not caused by the Cu-PP component. We have experience in forming a polymer composite with chlorin. Chlorin Cu-C-e6, an analog of Cu-PP, incorporated into the polymer composite exhibits electroluminescence and is applicable in light-emitting diodes [20]. In the present work, we develop composite materials based on poly-TPD with the addition of Cu-PP, which is able to enhance absorption in the blue and red ranges of the spectrum. For the first time, we study the mobility of charge carriers in a poly-TPD composite with Cu-PP and in a ternary composite based on poly-TPD, Cu-PP, and PCBM (Figure 1) using the linear increasing voltage (CELIV) technique and space charge limited current (SCLC) mode.

2. Materials and Methods

2.1. Materials and Synthesis

Poly[N,N′-bis(4-butylphenyl)-N,N′-bis(phenyl)-benzidine] (poly-TPD) from American Dye Source, Inc. (ADS254BE) (Baie-D'Urfe, QC, Canada), [6,6]-Phenyl-C61-butyric acid methyl ester (PCBM) from SES Research, C_{60} fullerene from MST ("Modern Synthesis Technology", St. Petersburg, Russia) and bathocuproine (BCP) from Kintec (Hong Kong, China) were used as received. PEDOT:PSS (poly(3,4-ethylene-dioxythiophene):polystyrenesulfonate) from Heraeus (Clevios P VP AI 4083) (Hanau, Germany) was filtered prior to use, The ITO-coated glass (7 Ohm/□) was purchased from Kaivo (Zhuhai, Guangdong, China).

The free base of methyl ester of pyropheophorbide-a (Figure A1) was obtained according to procedure [24]. The structure and purity of the compound were confirmed by NMR, UV–Vis, and luminescent spectroscopy. For the synthesis of Cu(II) methyl ester of pyropheophorbide-a (Cu-PP) shown in Figure 1, 10 mg of the free base of methyl ester of pyropheophorbide-a was dissolved in 25 mL of methylene chloride and a solution of 7 mg of Cu(II) acetate in 15 mL of ethanol was added to the resulting solution. The solutions were mixed, and the resulting solution was stirred at 35 °C for 40 min. The solvents were removed in a vacuum. The product was dissolved in methylene chloride and purified by column chromatography from traces of the original methyl ester of pyropheophorbide-a, and the solvent was removed in vacuum. The yield of Cu-PP was 10.5 mg (95%). The structure of the resulting Cu-PP compound was confirmed by MALDI TOF mass spectrometry and UV–Vis spectroscopy in a chlorobenzene solution.

2.2. Cyclic Voltammetry and Energy Levels

Cyclic voltammetry (CV) measurements were performed for calculation of the lowest unoccupied molecular orbital (LUMO) and HOMO energy levels. The CV experiment was carried out at the scan rate of 20 mV/s in a three-electrode, three-compartment electrochemical cell in the glove box with dry argon atmosphere. Platinum sheets served as working and counter electrodes. A 0.2 M solution of tetrabutylammonium hexafluorophosphate (NBu$_4$PF$_6$, Fluka) (Pittsburgh, PA, USA) in acetonitrile (ACN) was used as an electrolyte. An Ag wire immersed into the electrolyte solution with the addition of 0.1 M AgNO$_3$ was used as a pseudo reference electrode (Ag/Ag$^+$). It was calibrated against ferrocene/ferricenium couple (−0.039 V vs. Ag/Ag$^+$) and its potential was recalculated to the energy scale using −4.988 eV value for Fc/Fc$^+$ in ACN. Thus, the energy level of Ag/Ag$^+$ is as E_{ref} = −5.03 eV. The values of potentials corresponding to the HOMO and LUMO levels were determined by applying a tangent to the onset of anodic and cathodic currents (Figure 2a), with $E_{HOMO} = E_{ref} - E_a$ (typically, $E_a > 0$), $E_{LUMO} = E_{ref} - E_c$ (typically, $E_c < 0$).

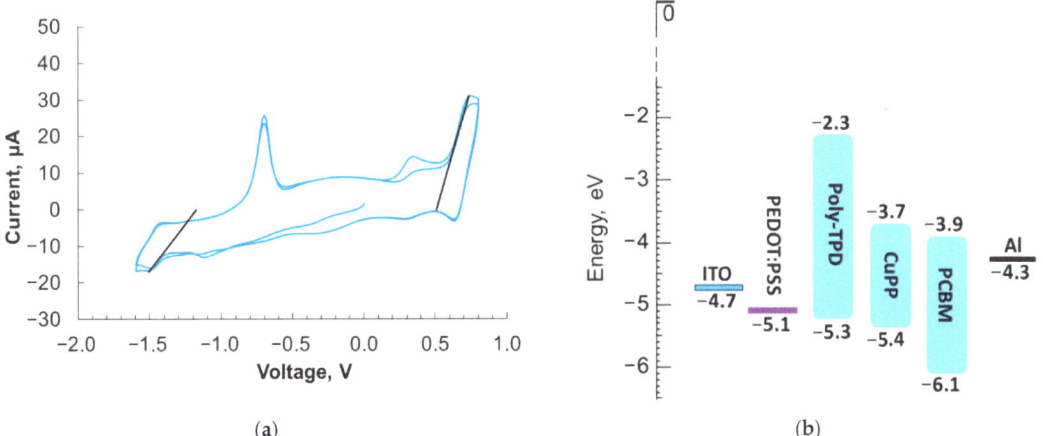

Figure 2. (**a**) Cyclic voltammograms of Cu-PP, the black lines show the tangents to the curves. The accuracy of the CV experiments is ±0.02 V. (**b**) Energy levels of the used materials.

Other details of the CV were the same as described in our previous article [19]. The obtained values of the HOMO and LUMO energy levels for Cu-PP as well as for poly-TPD and PCBM [14] are shown in Figure 2b.

2.3. Thin Films and Device Preparation

The mobility of charge carriers was studied in films of poly-TPD:PCBM, poly-TPD:Cu-PP, and poly-TPD:PCBM:Cu-PP composites in molar ratios of 6.33:1, 6.33:1, and 6.33:1:1, respectively.

To measure the mobility of charge carriers in the transient current mode of the CELIV technique, samples of the ITO/SiO$_2$/composite/Al structure were prepared as follows. A 70 nm thick SiO$_2$ layer was deposited onto a glass substrate with an electrically conductive ITO layer using magnetron sputtering. A solution of the polymer composite in chlorobenzene was deposited on top of the dielectric layer by spin-coating method at a speed of 1500 rpm, then the sample was dried in an argon atmosphere at 80 °C for 2 h. The 80 nm thick Al counter electrode was deposited by the resistive thermal evaporation (RTE) technique using an MB Evap vacuum thermal evaporator (CreaPhys, Dresden, Germany) at a vacuum of 6×10^{-6} mbar as described earlier [19].

To measure electron mobility in the SCLC mode, electron-only and hole-only devices were prepared with a structure of Al/composite/Al and ITO/PEDOT:PSS/composite/MoO$_3$/Al, respectively. A 20 nm thick MoO$_3$ layer was deposited by the RTE technique and 30 nm thick PEDOT:PSS layer was deposited by spin-coating to block electrons. Thicknesses of the poly-TPD composite layers and both Al electrodes were 75 ÷ 80 nm and 80 nm, respectively.

The photoconductivity of the composites was investigated using ITO/PEDOT:PSS/composite/C$_{60}$/BCP/Al devices (Figure 3), with the 30 nm thick PEDOT:PSS layer serves as a hole transport (electron blocking) layer, and the 10 nm/7.5 nm thick C$_{60}$/BCP layers provide electron transport (hole blocking). The C$_{60}$ and BCP layers were deposited by the RTE technique. The thickness of the composite layers ranged from 75 to 80 nm.

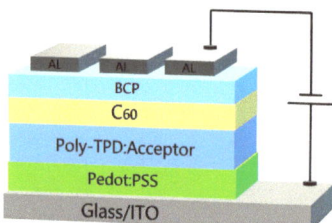

Figure 3. A sketch of device structure with functional layers.

2.4. Electrical Characterization

For mobility measurements, the MIS-CELIV mode was used, which allows recording of the current of monopolar charge carriers. The experimental setup included a USB oscilloscope (DL-Analog Discovery, Digilent Co., Pullman, WA, USA), which generated master pulse and recorded transient current, as described in our works [19,20,25]. The mobility μ was calculated according to the expression:

$$\mu = \frac{2d^2}{3At_{max}^2 \left(1 + 0.36 \frac{\Delta J}{J(0)}\right)},$$

here A is the voltage ramp, d is the composite film thickness, $J(0)$ is the capacitance current, and ΔJ is the conduction current at the time t_{max}.

In the SCLC mode, the charge carrier transport occurs in the nontrapping regime and the current J_{SCLC} obeys the Mott–Gurney equation [26]

$$J_{SCLC} = \frac{9\mu_{SCLC}\varepsilon\varepsilon_0 V^2}{8d^3}$$

where μ_{SCLC} is the charge carrier mobility, ε is the dielectric constant of the composite, ε_0 is the vacuum permittivity, V is the applied voltage. We used $\varepsilon = 3.5$ for the composites studied.

J–V characteristics of the devices were recorded by an SMU Keithley 2400 (Solon, OH, USA, photocurrent was measured under illumination provided by a solar simulator with a 150 W Xe lamp (Newport 67005) (Newport Corp., Irvine, CA, USA). The relative error of J–V measurements was 5%. Processing of J–V curves in SCLC mode for calculation of mobility was shown earlier [19,27].

All the measurements were carried out at room temperature in an inert atmosphere of Ar with an oxygen and water content of <10 ppm.

3. Results and Discussion
3.1. UV–Vis Spectroscopy

UV–Vis absorption spectra were recorded with a Shimadzu UV-3101PC spectrophotometer (Shimadzu Corp., Kyoto, Japan). Absorption spectra of the individual components are shown in Figure 4a. PCBM exhibits typical absorption in the UV range. The UV–visible spectrum of Cu-PP exhibits a Soret band between 340 and 435 nm and a Q band between 590 and 690 nm (Figure 4a), which are characteristic of compounds such as chlorophyll A. In the absorption spectra of the composites in chlorobenzene solution presented in Figure 4b, the absorption bands of the individual components are clearly visible. The absorption of the poly-TPD (Figure 4c) is slightly different from the absorption for TPD small molecules (Figure 4a). TPD molecules are characterized by two absorption bands in the ranges of 300–330 nm and 350–390 nm, corresponding to π-π* transitions for peripheral rings and the biphenyl fragment, respectively [28]. The dominance of the latter is more pronounced for the poly-TPD than for the small molecules. This is typical for polymers with a fairly high degree of polymerization and a small number of terminal groups.

The absorption spectra of the composite films are shown in Figure 4c. The absorption spectra of the poly-TPD:Cu-PP and poly-TPD:PCBM:Cu-PP composite films are identical in the visible range. The absorption spectrum of the poly-TPD:PCBM composite film is simply the addition of the absorption bands of PCBM and poly-TPD, peaking at wavelengths around 340 and 390 nm, respectively. In contrast to the poly-TPD:PCBM composite, in the spectrum of the poly-TPD:CuPP composite film, in addition to the intrinsic absorption bands of poly-TPD and CuPP, a new absorption band appears in the range of 450–500 nm. This band reflects the donor–acceptor interaction between TPD donor and Cu-PP acceptor forming the charge transfer complex. In addition, TPD cation radicals are known to absorb in this range [28]. Thus, Cu-PP can both sensitize poly-TPD in the red spectral range and enhance the absorption of the composite in the blue–cyan range.

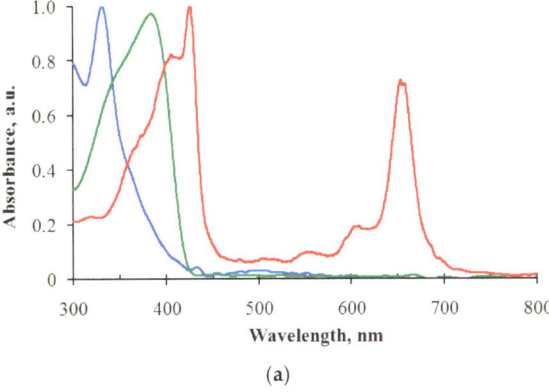

(a)

Figure 4. Cont.

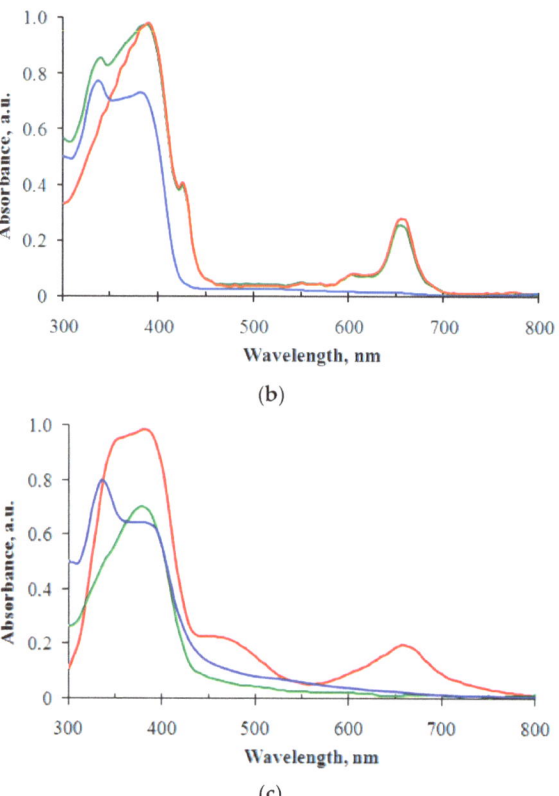

Figure 4. (a) Absorption spectra of CuPP (red), PCBM (blue), and poly-TPD (green) in chlorobenzene (10^{-5} M, optical pass length is 1 cm). (b) Absorption spectra of poly-TPD:CuPP (red), poly-TPD:PCBM (blue), and poly-TPD:PCBM:Cu-PP (green) in chlorobenzene solution. (c) Absorption spectra of poly-TPD:CuPP (red), poly-TPD:PCBM (blue) composite films, and single poly-TPD film (green).

3.2. AFM of Solid Layers

Surfaces of the 80 nm thick composite layers were characterized by AFM (Appendix B). The AFM provides reasonable resolution in z-direction for the surface topography of the studied samples (Figure 5).

The surface image of the ternary composite layer resembles a superposition of the poly-TPD:PCBM and poly-TPD:CuPP images. The surface topography of all layers is quite smooth with the root mean square roughness (RMS) not exceeding 2 nm: 0.8–1.1 nm for the polyTPD:PCBM, 0.8–1.0 nm for the polyTPD:PCBM:CuPP, and 0.6–0.9 nm for the polyTPD:CuPP sample (95% confidence intervals for six measurements). However, in the poly-TPD:PCBM topography, spherical inclusions (up to 10 nm in height) are clearly visible, which are actually invisible in the poly-TPD:CuPP image. The addition of PCBM to a polymer composite with a rough surface layer usually promotes a reduction in the RMS value. Yet, for a polymer composite with a smooth surface layer, the addition of PCBM can lead to the opposite result due to the formation of fullerene clusters in the composite layer [29].

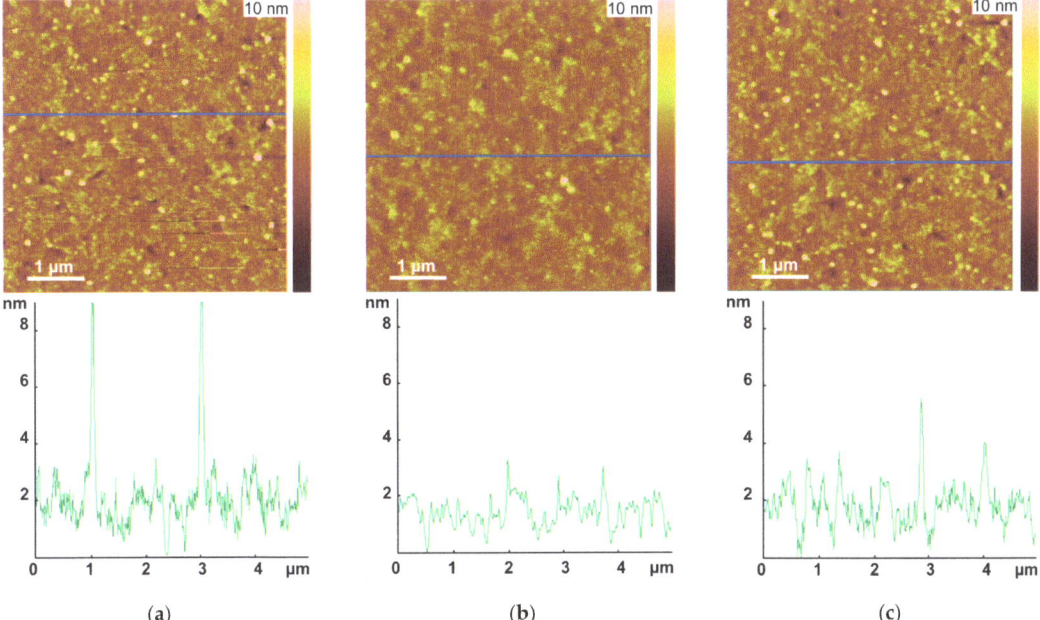

Figure 5. AFM images of the surface topography and sections corresponding to marked lines for (a) poly-TPD:PCBM, (b) poly-TPD:Cu-PP, and (c) poly-TPD:PCBM:Cu-PP composite films.

3.3. Charge Carrier Mobility

In thin films of the studied composites, the mobility of charge carriers measured by the CELIV method is on the orders of 10^{-5} and 10^{-4} cm^{2}V^{-1}s^{-1} (Table 1). The poly-TPD:PCBM:Cu-PP ternary composite possesses the highest electron mobility and improved balance in electron and hole mobility compared to the other two composites.

Table 1. Charge carrier mobility in poly-TPD composites and individual components measured by the CELIV technique [1].

Composite	Mobility, cm^{2}V^{-1}s^{-1}	
	Electrons	Holes
poly-TPD:PCBM	$(8.0 \pm 0.8) \times 10^{-5}$	$(1.4 \pm 0.1) \times 10^{-4}$
poly-TPD:Cu-PP	$(3.5 \pm 0.5) \times 10^{-5}$	$(2.2 \pm 0.1) \times 10^{-4}$
poly-TPD:PCBM:Cu-PP	$(1.0 \pm 0.1) \times 10^{-4}$	$(1.5 \pm 0.1) \times 10^{-4}$
poly-TPD	-	$(2.8 \pm 0.2) \times 10^{-4}$
PCBM	$(8.0 \pm 0.8) \times 10^{-4}$	-
Cu-PP	$(6.0 \pm 0.6) \times 10^{-5}$	$(4.5 \pm 0.5) \times 10^{-5}$

[1] Calculated from 10 replicates, the confidence level is 90%.

Due to the small HOMO energy offsets between the donor poly-TPD and Cu-PP and the small LUMO energy offsets between the acceptor PCBM and Cu-PP (Figure 2b), Cu-PP molecules can provide the transport of both holes and electrons.

The hole mobility in the poly-TPD:Cu-PP composite is higher than in other composites and slightly lower than in neat poly-TPD. The Cu-PP molecule cannot act as a hole trap in the polymer matrix because its HOMO level is only slightly lower than that of poly-TPD (Figure 2b). However, Cu-PP molecules can influence the spatial arrangement of poly-TPD macromolecule fragments, making it difficult to transfer holes between them. Cu-PP molecules can also increase the width of the density of states (DOS) σ in poly-

TPD:Cu-PP compared to that in neat poly-TPD. An increase in σ leads to a decrease in mobility according to expression (1) obtained in the Gaussian disorder model and the correlated disorder model to describe the charge transport properties of organic disordered semiconductors [2,30–32].

$$\mu = \mu_0 exp\left[-(3\sigma/5kT)^2 + C_0\left((\sigma/kT)^{3/2} - \Gamma\right)\sqrt{qRF/\sigma}\right] \quad (1)$$

where $C_0 = 0.78$, $\Gamma = 2$, μ_0 is the pre-exponential factor, R is the distance between transport sites, k is the Boltzmann constant, T is absolute temperature, q is the elementary charge, F is the electric field strength.

The electron mobilities in the ternary composite and in the poly-TPD:PCBM composite are approximately an order of magnitude lower than in a neat PCBM. This is due to the high mobility of electrons in the latter. The electron mobility in the poly-TPD:PCBM composite is approximately two times higher than that in the poly-TPD:Cu-PP composite. A different effect of Cu-PP and PCBM on the electron mobility in poly-TPD composites may be associated with the different dispersion of these molecules in the poly-TPD matrix. It is well known that a mixture of a donor polymer (D) and an acceptor fullerene (A) forms a BHJ, the concept that was proposed for polymer solar cells about three decades ago [33,34]. In a three-dimensional BHJ, molecules D and A form a bicontinuous layer consisting of domains D and A. The interpenetrating network of the domains provides both a greater D–A interface for effective exciton dissociation and two channels for transporting electrons and holes to the corresponding electrodes.

In the AFM images (Figure 5), spherical objects representing fullerene clusters are clearly visible on the surface of the poly-TPD:PCBM and poly-TPD:PCBM:Cu-PP films, respectively. In contrast, on the poly-TPD:Cu-PP film, such objects are invisible. Thus, in contrast to PCBM molecules, Cu-PP molecules do not form a network of domains in the poly-TPD matrix; therefore, they are dispersed in the free volume of the polymer. For the homogeneous dispersion of the Cu-PP molecules in poly-TPD, the average intermolecular distance R estimated according to Equation (2) is equal to 1.48 nm, where $M = 610$ g/mole is the molecular weight of Cu-PP, $\rho \approx 1.3$ g × cm^{-3} is the density of the solid film, $c = 0.242$ is the weight concentration of Cu-PP in poly-TPD and N_A is the Avogadro number. This distance is quite sufficient to implement the transport of charge carriers through the mechanism of intermolecular hopping. Since PCBM molecules in domains are evidently located close to each other, the electron mobility in the poly-TPD:PCBM is higher than in poly-TPD:Cu-PP.

$$R = \sqrt[3]{M/\rho c N_A} \quad (2)$$

In the ternary mixture, the Cu-PP molecules serve as electron transporting centers and thereby facilitate the transfer of electrons from separately located PCBM clusters to a continuous network of PCBM domains.

It is noteworthy that the electron mobility in PCBM, measured by us in the CELIV mode, is lower than that calculated using J–V characteristics of field-effect transistors (FETs): 4.5×10^{-3} cm^{2}V^{-1}s^{-1} [35] and 9×10^{-3} cm^{2}V^{-1}s^{-1} [36]. For hole mobility in neat poly-TPD, the same correlation is observed: the CELIV measurement shows a value of 2.8×10^{-4} (Table 1), which is an order of magnitude less than the FET mobility 2×10^{-3} [37]. Measurements with these two methods exhibit that there always seems to be about one order of magnitude difference in FET and CELIV mobilities for polymer and organic semiconductors [38,39].

In the studied poly-TPD composites, the mobility of electrons and holes, calculated from the J–V curves in the SCLC mode (Table 1), changes with changes in the additives to polymer in the same way as the CELIV mobility does (Table 2). However, to establish a generalized correlation between the absolute values of CELIV mobility and SCLC mobility, a wide array of experimental data is required.

Table 2. Charge carrier mobility in poly-TPD composite layers measured in the SCLC mode [1].

Composite	Mobility, cm^{2}V^{-1}s^{-1}	
	Electrons	Holes
polyTPD:PCBM	$(4.2 \pm 0.5) \times 10^{-6}$	$(6.0 \pm 0.6) \times 10^{-6}$
polyTPD:Cu-PP	$(1.9 \pm 0.2) \times 10^{-6}$	$(7.0 \pm 0.6) \times 10^{-6}$
polyTPD:PCBM:Cu-PP	$(8.3 \pm 0.8) \times 10^{-6}$	$(9.8 \pm 0.8) \times 10^{-6}$

[1] Calculated from 10 replicates, the confidence level is 90%.

3.4. Photoconductivity

The measurements of the J–V characteristics for ITO/PEDOT:PSS/composite/C$_{60}$/BCP/Al diode devices (Figure 3) were carried out in the dark and under illumination with white light from a Xe lamp and at a wavelength of 650 nm at the Cu-PP absorption maximum. Presented in Figures 6 and 7, J–V curves are typical characteristics of a photodiode. Under white light illumination (Figure 6) at a reverse bias of 1 V, the ratio of photo-to-dark current was about 1100, 1300, and 1500 for poly-TPD:Cu-PP (blue curves), poly-TPD:PCBM (black curves), and poly-TPD:PCBM:Cu-PP (green curves) based devices, respectively. Under illumination at a wavelength of 650 nm, (Figure 7), the ratio of photo-to-dark current was between 30 and 50 for Cu-PP-containing devices whereas the photocurrent was negligible in Cu-PP-free devices.

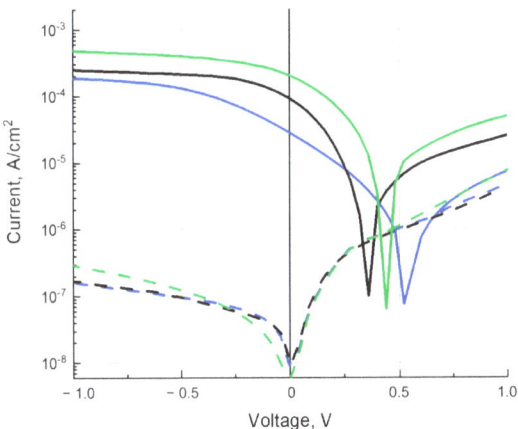

Figure 6. J–V curves of the dark current (dashed line) and photocurrent (solid line) under white light illumination for poly-TPD:Cu-PP- (blue), polyTPD:PCBM- (black), and poly-TPD:PCBM:Cu-PP- (green) based devices.

Photoconductivity σ_{ph} is expressed by the equation $\sigma_{ph} = q(\mu_e \times e + \mu_h \times p)$, where q is the electron charge, μ_e and μ_h are the mobility of electrons and holes, respectively, e and p are the concentration of photogenerated electrons and holes, respectively. Additional absorption bands of Cu-PP in the red spectral band 590–690 nm and the charge transfer complex between Cu-PP and poly-TPD in the 450–500 nm (Figure 4c) band provide an increase in light harvesting and, as a consequence, an increase in e and p-values.

Due to the small HOMO energy offsets between the donor poly-TPD and Cu-PP and the small LUMO energy offsets between the acceptor PCBM and Cu-PP as seen in the energy diagram (Figure 2b), it is favorable for electrons and holes photogenerated on Cu-PP molecules to transfer to PCBM and poly-TPD, respectively. In turn, as shown in Section 3.3, the mobilities μ_e and μ_h are more balanced in the ternary composite than in the binary ones (Tables 1 and 2). The strong mobility imbalance in the poly-TPD:Cu-PP film is evident in that the J–V curve increases slowly with increasing reverse bias compared to the

J–V curves of the TPD:Cu-PP and TPD composites, which have a more balanced mobility (Figures 6 and 7). If the mobilities of electrons and holes are approximately equal, then the probability of bimolecular recombination of charge carriers decreases, and, consequently, the loss of charge carrier concentration decreases.

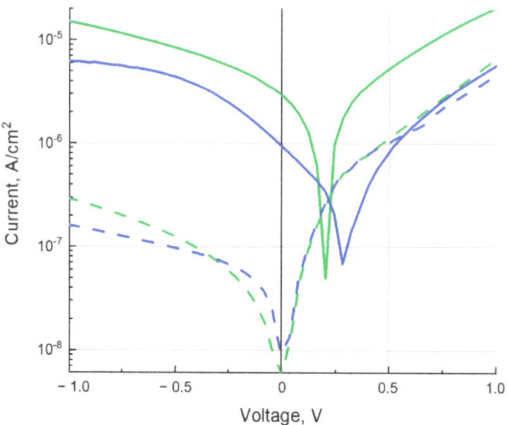

Figure 7. J–V curves of the dark current (dashed line) and photocurrent (solid line) under illumination at a wavelength of 650 nm (a power of 3 mW × cm^{-2}) for poly-TPD:Cu-PP- (blue) and poly-TPD:PCBM:Cu-PP- (green) based devices.

The J–V curves show that the studied composites exhibit the photovoltaic effect. The open circuit voltage V_{oc} decreases in the sequence poly-TPD:Cu-PP-, poly-TPD:PCBM:Cu-PP-, and poly-TPD: PCBM-based devices. The result agrees with the empirical relation [40]:

$$V_{OC} = (1/q)\left(\left|E^D_{HOMO}\right| - \left|E^A_{LUMO}\right|\right) - 0.3V \qquad (3)$$

here E^D_{HOMO} is the HOMO level energy of the donor and E^A_{LUMO} is the LUMO level energy of the acceptor. In (3), the V_{OC} loss of 0.3 eV is empirical, and the loss could be greater or lesser. The reasons for voltage losses are still discussed [41,42].

The photoconductivity of the ternary composite demonstrates the promise of using the material in photodetector devices, in particular, those with sensitivity in the red spectral band. At the absorption maximum of Cu-PP, the external quantum efficiency (EQE) is equal to 0.5% for an incident flux of 1.87×10^{16} photons × s^{-1} × cm^{-2} at a wavelength of 650 nm and a photocurrent of 1.5×10^{-5} A × cm^{-2}. In turn, the calculation of the internal quantum efficiency (IQE) gives a value of 5.7%, taking into account the decadic absorption coefficient (0.003 nm^{-1}) of Cu-PP.

4. Conclusions

Based on the polymer poly-TPD having excellent thin film formation and hole transport abilities, we have developed a ternary composite with fullerene derivative PCBM and Cu-PP, a copper(II) complex with pyropheophorbide-a. We used the Cu-PP complex obtained from natural raw materials. The Cu-PP component blended with poly-TPD can (1) expand the photoconductivity of the composite in the blue–cyan and red bands of the spectrum, (2) serve as an electron acceptor with respect to the polymer in the same way as fullerene, and (3) provide the transport of both electrons and holes due to the small offset between HOMO levels of Cu-PP and poly-TPD and between LUMO levels of Cu-PP and PCBM. We focused on studying the charge transport. For the first time, the mobility of charge carriers in a poly-TPD composite with Cu-PP and in a ternary composite based on poly-TPD, Cu-PP, and PCBM was investigated, with a reasonably balanced mobility found in the ternary composite compared to the binary composites.

Based on charge mobility data and AFM topography images, we concluded that Cu-PP molecules are dispersed homogenously in the free volume of poly-TPD, while PCBM molecules form a network of domains in the polymer matrix (bulk heterojunction) serving as electron transporting pathways. Indeed, the electron mobility in poly-TPD:PCBM:Cu-PP is higher than in poly-TPD:PCBM, so the Cu-PP molecules, providing the hopping transport of electrons, can facilitate the transfer of electrons from separately dispersed PCBM clusters to the continuous network of the PCBM domains.

By blending poly-TPD with Cu-PP, the polymer composites are sensitized in the red spectral range and absorb more photons in the blue range. The results obtained on the photoconductivity of the composites demonstrate that the ternary composite is promising for use in photodetectors, in particular, those sensitive in the red range of the spectrum.

Author Contributions: Conceptualization, A.R.T. and D.A.L.; methodology, A.R.T., A.E.A., D.A.L., A.Y.C. and S.I.P.; validation, D.A.L., A.Y.C. and A.E.A.; formal analysis, A.Y.T.; investigation, A.E.A., I.R.S., D.A.L., S.I.P., A.Y.C., A.V.D. and N.V.N.; data curation, A.R.T.; writing—original draft preparation, A.R.T. and D.A.L.; writing—review and editing, A.R.T.; supervision, A.Y.T.; project administration, A.Y.T.; funding acquisition, A.Y.T. All authors have read and agreed to the published version of the manuscript.

Funding: This work was supported by the Ministry of Science and Higher Education of the Russian Federation within the framework of the topic "Physical and chemical problems of creating effective nano- and supramolecular systems" (registration number 122011300052-1).

Institutional Review Board Statement: Not applicable.

Data Availability Statement: Data are contained within the article.

Conflicts of Interest: The authors declare no conflicts of interest. The funders had no role in the design of the study; in the collection, analyses, or interpretation of the data; in the writing of the manuscript; or in the decision to publish the results.

Appendix A

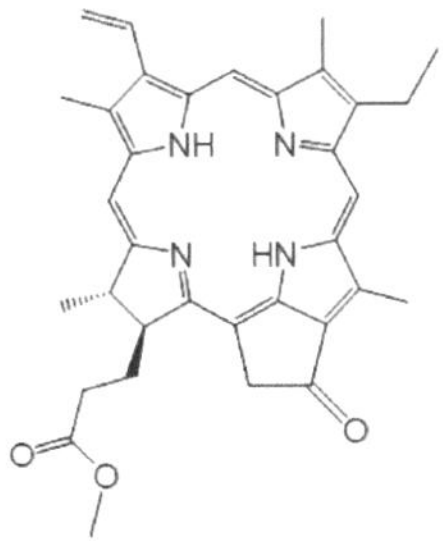

Figure A1. Chemical structure of free base of methyl ester of pyropheophorbide-a.

Appendix B

AFM observations were carried out in the tapping mode (repulsive mode) in the air. EnviroScope (NanoScope5 controller) was used. Silicon cantilevers with spring constants of 5–40 N × m^{-1} and resonant frequencies of 150–350 kHz were used. During the measurement process, we made sure that the static charge did not blur the image. The measurements were carried out directly on working samples between the electrodes.

References

1. Murad, A.R.; Iraqi, A.; Aziz, S.B.; Abdullah, S.N.; Brza, M.A. Conducting Polymers for Optoelectronic Devices and Organic Solar Cells: A Review. *Polymers* **2020**, *12*, 2627. [CrossRef]
2. Bässler, H.; Köhler, A. *Electronic Processes in Organic Semiconductors: An Introduction*; Wiley-VCH: Weinheim, Germany, 2015.

3. Yang, K.; Wang, J.; Zhao, Z.; Zhou, Z.; Liu, M.; Zhang, J.; He, Z.; Zhang, F. Smart Strategy: Transparent Hole-Transporting Polymer as a Regulator to Optimize Photomultiplication-type Polymer Photodetectors. *ACS Appl. Mater. Interfaces* **2021**, *13*, 21565–21572. [CrossRef]
4. Ha, H.; Shim, Y.J.; Lee, D.H.; Park, E.Y.; Lee, I.-H.; Yoon, S.-K.; Suh, M.C. Highly Efficient Solution-Processed Organic Light-Emitting Diodes Containing a New Cross-linkable Hole Transport Material Blended with Commercial Hole Transport Materials. *ACS Appl. Mater. Interfaces* **2021**, *13*, 21954–21963. [CrossRef]
5. Chen, H.; Ding, K.; Fan, L.; Liu, W.; Zhang, R.; Xiang, S.; Zhang, Q.; Wang, L. All-Solution-Processed Quantum Dot Light Emitting Diodes Based on Double Hole Transport Layers by Hot Spin-Coating with Highly Efficient and Low Turn-On Voltage. *ACS Appl. Mater. Interfaces* **2018**, *10*, 29076–29082. [CrossRef]
6. Hu, X.; Tao, C.; Liang, J.; Chen, C.; Zheng, X.; Li, J.; Li, J.; Liu, Y.; Fang, G. Molecular weight effect of poly-TPD hole-transporting layer on the performance of inverted perovskite solar cells. *Sol. Energy* **2021**, *218*, 368–374. [CrossRef]
7. Thesen, M.W.; Höfer, B.; Debeaux, M.; Janietz, S.; Wedel, A.; Köhler, A.; Johannes, H.-H.; Krueger, H. Hole-Transporting Host-Polymer Series Consisting of Triphenylamine Basic Structures for Phosphorescent Polymer Light-Emitting Diodes. *J. Polym. Sci. Part A Polym. Chem.* **2010**, *48*, 3417–3430. [CrossRef]
8. Wang, C.; Zhang, Z.; Hu, W. Organic photodiodes and phototransistors toward infrared detection: Materials, devices, and applications. *Chem. Soc. Rev.* **2020**, *49*, 653–670. [CrossRef]
9. Dai, X.; Zhang, Z.; Jin, Y.; Niu, Y.; Cao, H.; Liang, X.; Chen, L.; Wang, J.; Peng, X. Solution-processed, high-performance light-emitting diodes based on quantum dots. *Nature* **2014**, *515*, 96–99. [CrossRef] [PubMed]
10. Lan, K.; Liu, B.; Tao, H.; Zou, J.; Jiang, C.; Xu, M.; Wang, L.; Penga, J.; Cao, Y. Preparation of efficient quantum dot light-emitting diodes by balancing charge injection and sensitizing emitting layer with phosphorescent dye. *J. Mater. Chem. C* **2019**, *7*, 5755–5763. [CrossRef]
11. Zhao, D.; Sexton, M.; Park, H.-Y.; Baure, G.; Nino, J.C.; So, F. High-Efficiency Solution-Processed Planar Perovskite Solar Cells with a Polymer Hole Transport Layer. *Adv. Energy Mater.* **2015**, *5*, 1401855. [CrossRef]
12. Malinkiewicz, O.; Yella, A.; Lee, Y.H.; Espallargas, G.M.; Graetzel, M.; Nazeeruddin, M.K.; Bolink, H.J. Perovskite solar cells employing organic charge-transport layers. *Nat. Photonics* **2014**, *8*, 128–132. [CrossRef]
13. Huang, F.; Li, J.Z.; Xu, Z.H.; Liu, Y.; Luo, R.P.; Zhang, S.W.; Nie, P.B.; Lv, Y.F.; Zhao, S.X.; Su, W.T.; et al. A Bilayer 2D-WS2/Organic-Based Heterojunction for High-Performance Photodetectors. *Nanomaterials* **2019**, *9*, 1312. [CrossRef]
14. Blankenburg, L.; Sensfuss, S.; Schache, H.; Marten, J.; Milker, R.; Schrödner, M. TPD wide-bandgap polymers for solar cell application and their sensitization with small molecule dyes. *Synth. Met.* **2015**, *199*, 93–104. [CrossRef]
15. Kadish, K.M.; Smith, K.M.; Guilard, R. *The Porphyrin Handbook*; Academic Press: New York, NY, USA, 2000.
16. Josefsen, L.B.; Boyle, R.W. Photodynamic Therapy and the Development of Metal-Based Photosensitisers. *Met. Based Drugs* **2008**, *2008*, 276109. [CrossRef]
17. Beyene, B.B.; Hung, C.-H. Photocatalytic hydrogen evolution from neutral aqueous solution by a water-soluble cobalt(II) porphyrin. *Sustain. Energy Fuels* **2018**, *2*, 2036–2043. [CrossRef]
18. Saide, A.; Lauritano, C.; Ianora, A. Pheophorbide a: State of the Art. *Mar. Drugs* **2020**, *18*, 257. [CrossRef]
19. Chernyadyev, A.Y.; Aleksandrov, A.E.; Lypenko, D.A.; Tyurin, V.S.; Tameev, A.R.; Tsivadze, A.Y. Copper (II) Me-so-Tetraphenyl and Meso-Tetafluorenyl Porphyrinates as Charge Carrier Transporting and Electroluminescent Compounds. *ACS Omega* **2022**, *7*, 8613–8622. [CrossRef]
20. Lypenko, D.A.; Aleksandrov, A.E.; Chernyadyev, A.Y.; Pozin, S.I.; Tsivadze, A.Y.; Tameev, A.R. Photoconduction and Electroluminescence of Copper (II) Protoporphyrin and Chlorin Cu-C-e6. *Int. J. Mol. Sci.* **2023**, *24*, 3178. [CrossRef] [PubMed]
21. Parker, C.A. *Photoluminescence of Solutions*; Elsevier: Amsterdam, The Netherlands; London, UK; New York, NY, USA, 1968.
22. Lakowicz, J.R. *Principles of Fluorescence Spectroscopy*, 3rd ed.; Springer: New York, NY, USA, 2006.
23. Bobrovsky, A.; Piryazev, A.; Ivanov, D.; Kozlov, M.; Utochnikova, V. Temperature-Dependent Circularly Polarized Luminescence of a Cholesteric Copolymer Doped with a Europium Complex. *Polymers* **2023**, *15*, 1344. [CrossRef]
24. Scheer, H.; Gross, E.; Nitsche, B.; Cmiel, E.; Schneider, S.; Schäfer, W.; Schiebel, H.-M.; Schulten, H.-R. Structure of methylpheophorbide-RCI. *Photochem. Photobiol.* **1986**, *43*, 559–571. [CrossRef]
25. Malov, V.V.; Tanwistha, G.; Nair, V.C.; Maslov, M.M.; Katin, K.P.; Unni, K.N.N.; Tameev, A.R. Hole mobility in thieno [3,2-b]thiophene oligomers]. *Mendeleev Commun.* **2019**, *29*, 218–219. [CrossRef]
26. Lampert, M.A.; Mark, P. *Current Injection in Solids*; Academic: New York, NY, USA, 1970.
27. Tameev, A.R.; Licea, J.L.; Pereshivko, L.Y.; Rychwalski, R.W.; Vannikov, A.V. Charge carrier mobility in films of carbon-nanotube-polymer composites. *J. Phys. Conf. Ser.* **2007**, *61*, 1152–1156. [CrossRef]
28. Low, P.J.; Paterson, M.A.J.; Puschmann, H.; Goeta, A.E.; Howard, J.A.K.; Lambert, C.; Cherryman, J.C.; Tackley, D.R.; Leeming, S.; Brown, B. Crystal, molecular and electronic structure of N,N′-diphenyl-N,N′-bis(2,4-dimethylphenyl)-(1,1′-biphenyl)-4,4′-diamine and the corresponding radical cation. *Chem. Eur. J.* **2004**, *10*, 83–91. [CrossRef]
29. Liu, F.; Li, C.; Li, J.; Wang, C.; Xiao, C.; Wu, Y.; Li, W. Ternary organic solar cells based on polymer donor, polymer acceptor and PCBM components. *Chinese Chem. Lett.* **2020**, *31*, 865–868. [CrossRef]
30. Bässler, H. Charge Transport in Disordered Organic Photoconductors: A Monte Carlo Simulation Study. *Phys. Status Solidi B* **1993**, *175*, 15–56. [CrossRef]

31. Novikov, S.V.; Dunlap, D.H.; Kenkre, V.M.; Parris, P.E.; Vannikov, A.V. Essential Role of Correlations in Governing Charge Transport in Disordered Organic Materials. *Phys. Rev. Lett.* **1998**, *81*, 4472–4475. [CrossRef]
32. Baranovskii, S.D. Theoretical description of charge transport in disordered organic semiconductors. *Phys. Status Solidi B* **2014**, *251*, 487–525. [CrossRef]
33. Yu, G.; Gao, J.; Hummelen, J.C.; Wudl, F.; Heeger, A.J. Polymer Photovoltaic Cells: Enhanced Efficiencies via a Network of Internal Donor-Acceptor Heterojunctions. *Science* **1995**, *270*, 1789–1791. [CrossRef]
34. Halls, J.J.M.; Walsh, C.A.; Greenham, N.C.; Marseglia, E.A.; Friend, R.H.; Moratti, S.C.; Holmes, A.B. Efficient photodiodes from interpenetrating polymer networks. *Nature* **1995**, *376*, 498–500. [CrossRef]
35. Waldauf, C.; Schilinsky, P.; Perisutti, M.; Hauch, J.; Brabec, C.J. Solution-Processed Organic n-Type Thin-Film Transistors. *Adv. Mater.* **2003**, *15*, 2084–2088. [CrossRef]
36. Anthopoulos, T.D.; Tanase, C.; Setayesh, S.; Meijer, E.J.; Hummelen, J.C.; Blom, P.W.M.; de Leeuw, D.M. Ambipolar Organic Field-Effect Transistors Based on a Solution-Processed Methanofullerene. *Adv. Mater.* **2004**, *16*, 2174–2179. [CrossRef]
37. Zhu, W.W.; Xiao, S.; Shih, I. Field-effect mobilities of polyhedral oligomeric silsesquioxanes anchored semiconducting polymers. *Appl. Surf. Sci.* **2004**, *221*, 358–363. [CrossRef]
38. Andersson, L.M.; Zhang, F.; Inganäs, O. Bipolar transport observed through extraction currents on organic photovoltaic blend materials. *Appl. Phys. Lett.* **2006**, *89*, 142111. [CrossRef]
39. Andersson, L.M.; Zhang, F.; Inganäs, O. Stoichiometry, mobility, and performance in bulk heterojunction solar cells. *Appl. Phys. Lett.* **2007**, *91*, 071108. [CrossRef]
40. Scharber, M.C.; Mühlbacher, D.; Koppe, M.; Denk, P.; Waldauf, C.; Heeger, A.J.; Brabec, C.J. Design Rules for Donors in Bulk-Heterojunction Solar Cells—Towards 10% Energy-Conversion Efficiency. *Adv. Mater.* **2006**, *18*, 789–794. [CrossRef]
41. Wang, J.; Yao, H.; Xu, Y.; Ma, L.; Hou, J. Recent Progress in Reducing Voltage Loss in Organic Photovoltaic Cells. *Mater. Chem. Front.* **2021**, *5*, 709–722. [CrossRef]
42. Chen, X.K.; Qian, D.; Wang, Y.; Kirchartz, T.; Tress, W.; Yao, H.; Yuan, J.; Hülsbeck, M.; Zhang, M.; Zou, Y.; et al. A unified description of non-radiative voltage losses in organic solar cells. *Nat. Energy* **2021**, *6*, 799–806. [CrossRef]

Disclaimer/Publisher's Note: The statements, opinions and data contained in all publications are solely those of the individual author(s) and contributor(s) and not of MDPI and/or the editor(s). MDPI and/or the editor(s) disclaim responsibility for any injury to people or property resulting from any ideas, methods, instructions or products referred to in the content.

Article

A Novel and Green Method for Preparing Highly Conductive PEDOT:PSS Films for Thermoelectric Energy Harvesting

Fuwei Liu [1,2,3,*,†], **Luyao Gao** [1,2,3,†], **Jiajia Duan** [1], **Fuqun Li** [1], **Jingxian Li** [1], **Hongbing Ge** [1], **Zhiwei Cai** [1], **Huiying Li** [1], **Mengke Wang** [1], **Ruotong Lv** [1] and **Minrui Li** [1]

1. College of Physics and Electronic Engineering, Xinyang Normal University, Xinyang 464000, China
2. Key Laboratory of Advanced Micro/Nano Functional Materials of Henan Province, Xinyang Normal University, Xinyang 464000, China
3. Energy-Saving Building Materials Innovative Collaboration Center of Henan Province, Xinyang Normal University, Xinyang 464000, China
* Correspondence: liufw@xynu.edu.cn
† These authors contributed equally to this work.

Citation: Liu, F.; Gao, L.; Duan, J.; Li, F.; Li, J.; Ge, H.; Cai, Z.; Li, H.; Wang, M.; Lv, R.; et al. A Novel and Green Method for Preparing Highly Conductive PEDOT:PSS Films for Thermoelectric Energy Harvesting. *Polymers* 2024, *16*, 266. https://doi.org/10.3390/polym16020266

Academic Editors: Vineet Kumar and Md Najib Alam

Received: 29 November 2023
Revised: 14 January 2024
Accepted: 16 January 2024
Published: 18 January 2024

Copyright: © 2024 by the authors. Licensee MDPI, Basel, Switzerland. This article is an open access article distributed under the terms and conditions of the Creative Commons Attribution (CC BY) license (https://creativecommons.org/licenses/by/4.0/).

Abstract: As a π-conjugated conductive polymer, poly(3,4-ethylenedioxythiophene):poly(styrene sulfonate) (PEDOT:PSS) is recognized as a promising environmentally friendly thermoelectric material. However, its low conductivity has limited applications in the thermoelectric field. Although thermoelectric efficiency can be significantly enhanced through post-treatment doping, these processes often involve environmentally harmful organic solvents or reagents. In this study, a novel and environmentally benign method using purified water (including room temperature water and subsequent warm water) to treat PEDOT:PSS film has been developed, resulting in improved thermoelectric performance. The morphology data, chemical composition, molecular structure, and thermoelectric performance of the films before and after treatment were characterized and analyzed using a scanning electron microscope (SEM), Raman spectrum, XRD pattern, X-ray photoelectron spectroscopy (XPS), and a thin film thermoelectric measurement system. The results demonstrate that the water treatment effectively removes nonconductive PSS from PEDOT:PSS composites, significantly enhancing their conductivity. Treated films exhibit improved thermoelectric properties, particularly those treated only 15 times with room temperature water, achieving a high electrical conductivity of 62.91 S/cm, a Seebeck coefficient of 14.53 µV K^{-1}, and an optimal power factor of 1.3282 µW·m^{-1}·K^{-2}. In addition, the subsequent warm water treatment can further enhance the thermoelectric properties of the film sample. The underlying mechanism of these improvements is also discussed.

Keywords: PEDOT:PSS; water treatment; thermoelectrics

1. Introduction

With growing concerns about the environment and energy, there is increasing demand for renewable material-based energy technologies [1–3]. In particular, energy conversion materials that harness sustainable energy from the surrounding environment are garnering significant interest [4]. Thermoelectric materials, which facilitate the direct conversion of heat energy to electric power through the movement of charge carriers in solids, are seen as having wide-ranging applications in sustainable and clean energy [5,6]. The thermoelectric conversion efficiency of a material is primarily associated with its Seebeck coefficient (S), electrical conductivity (σ), thermal conductivity (κ), and absolute temperature (T), evaluated by the dimensionless thermoelectric figure of merit, ZT = S^{2}σT/κ. Excluding thermal conductivity, a parameter named power factor (represented as PF = S^{2}σ) is another metric for assessing thermoelectric performance.

Inorganic thermoelectric materials, known for their high output power density, have been extensively studied. However, their high cost and fragility limit their application in flexible, portable, and wearable devices. Conversely, organic thermoelectric materials, marked by their low density, excellent flexibility, abundant resources, and inherent low thermal conductivity, have recently been the focus of numerous studies. Among various conducting polymers, PEDOT:PSS stands out for its high energy conversion efficiency, making it an ideal organic TE material. For instance, Xu et al. prepared a doped PEDOT:PSS aqueous solution using acids and alkalis, investigating the effect of pH variation on TE performance. Their results indicated that decreasing the pH value enhances TE properties, achieving an optimal power factor of 1.35 $\mu W \cdot m^{-1} \cdot K^{-2}$ [7]. Kumar et al. enhanced the wettability of PEDOT:PSS on a glass substrate using methanol, followed by treatment with 1 M sulfuric acid, simultaneously increasing its electrical conductivity and Seebeck coefficient [8]. Similarly, sulfuric acid vapor was also employed to treat PEDOT:PSS thin films, and the obtained samples presented significantly enhanced electrical conductivity with an improved Seebeck coefficient, yielding a maximum power factor of 17.0 $\mu W \cdot m^{-1} \cdot K^{-2}$ [9]. Moreover, the TE properties of PEDOT:PSS films can be significantly enhanced by adding, during and/or post treatment, organic acids [10,11], organic solvents [12–15], and other chemical reagents (such as surfactants [16,17], salts solutions [18], and ionic liquids [19,20]). Additionally, incorporating inorganic fillers with a high conductivity or Seebeck coefficient has proven effective in enhancing TE properties. For example, Du et al. fabricated highly conductive CNTs/PEDOT:PSS composite films through a DMSO treatment, achieving a power factor of 108.7 $\mu W \cdot m^{-1} \cdot K^{-2}$ [21]. Chen et al. developed layer-like PEDOT:PSS/SWCNT composite films with honeycomb-like structures, which achieved a high power factor value of around 45.72 $\mu W \cdot m^{-1} \cdot K^{-2}$ after H_2SO_4 treatment [22]. See et al. created a novel composite film composed of Te nanorods and PEDOT:PSS, demonstrating an exceptional power factor of approximately 70 $\mu W \cdot m^{-1} \cdot K^{-2}$ [23]. These findings are heartening and suggest further research is warranted.

Despite these advancements, the use of toxic solvents and fillers, along with complex preparation processes, poses environmental concerns. Therefore, simple, green approaches are urgently needed. This study employs purified water to treat PEDOT:PSS films through a straightforward washing process. As is well known, water is rich in resources, and the preparation of DI water is relatively simple, convenient, and low-cost. Notably, this treatment can also selectively remove nonconductive PSS and effectively modify the crystallinity of PEDOT molecules. The electrical conductivity of the samples significantly increases with the water washing cycle number, reaching 76.35 S cm^{-1} after 20 cycles. Simultaneously, the Seebeck coefficient decreases marginally. Continuous optimization with warm water yields a high room temperature power factor of approximately 4.4344 $\mu W \cdot m^{-1} \cdot K^{-2}$, which is 672 times larger than that of the pristine sample. The mechanism of this significant enhancement of TE properties due to the water washing process is also examined.

2. Materials and Methods

2.1. Materials

A commercial PEDOT:PSS aqueous solution (PH1000, Mw = 326.388) was procured from Heraeus Company (Hanau, Germany). Deionized (DI) water with a relatively high resistance of approximately 18.2 MΩ cm was used to dilute the PEDOT:PSS solution during film preparation and to treat film samples during the water washing process.

2.2. Pre-Treatment and Film Preparation

Initially, the acquired PH1000 was diluted with DI water to a concentration of 6 mg/mL PEDOT:PSS solution. Subsequently, 150 µL of this solution was drop-coated onto a pre-cleaned glass substrate (1.2 × 1.2 cm^2) and allowed to dry under ambient conditions. The as-prepared PEDOT:PSS films were then dipped into DI water (around 17 °C) for 10 s, removed, and dried on a heating platform at 60 °C. This immersion and drying procedure was repeated 20 times. PEDOT:PSS film samples treated 5, 10, 15, and 20 times were selected for further characterizations. These as-prepared films were labelled as PH-5, PH-10, PH-15, and PH-20, respectively. To further improve its performance, warm water (30 °C) was used in the subsequent water washing process. The immersion and drying procedure was repeated until the electrical conductivity reached a relatively stable value (around 60 times).

2.3. Characterizations and Measurements

Film morphologies were obtained by employing a field emission scanning electron microscope (FE-SEM, S-4800, Hitachi Limited, Tokyo, Japan). Film thickness was determined using a step profiler (ET-4000M, Kosaka Laboratory Ltd., Tokyo, Japan). Molecular structure changes in the films were analyzed using a Nicolelis5 Fourier transform infrared spectroscopy (FTIR) instrument (ThermoFisher, Waltham, MA, USA) scanning in the range of 400~4000 cm^{-1}. Raman spectroscopy was conducted with a LabRAM HR Evolution instrument using an 514 nm excitation laser (HORIBA, Kyoto, Japan). XRD spectra were obtained with a Smartlab 9 (Rigaku Corporation, Tokyo, Japan) X-ray Diffractometer, using a scanning range $2\theta = 5°$–$60°$. Electron-binding energies of the films were measured using a Thermo KAlpha X-ray Photoelectron Spectroscope (XPS, Thermo Fisher Scientific). The UV-Vis absorption spectra were collected using a LAMBDA950 UV/Vis/NIR Spectrophotometer (PerkinElmer, Waltham, MA, USA). The conductivity and Seebeck coefficient of PEDOT:PSS films were tested using a MRS-3 thin-film thermoelectric tested system (Wuhan Joule Yacht Science & Technology Co., Ltd., Wuhan, China).

3. Results and Discussion

3.1. Fabrication of PEDOT:PSS Film and the Water Treatment

Figure 1 displays the schematic illustration of the water treatment process for the PEDOT:PSS film. A typical drop coating method was used to prepare the pristine PEDOT:PSS film. After drying at room temperature, the prepared films were soaked in DI water to selectively remove PSS and improve the molecular arrangement of PEDOT. Due to the solubilizing effect of the water on the film, the soaking duration was limited to 10 s, followed by a drying process. To thoroughly understand the impact of the water on the film, the immersion and drying steps were repeated multiple times. Notably, the water treatment led to the formation of an interconnected porous network. The effective removal of PSS suggests that the resultant films would exhibit enhanced electrical conductivities and thermoelectric properties, which are discussed subsequently.

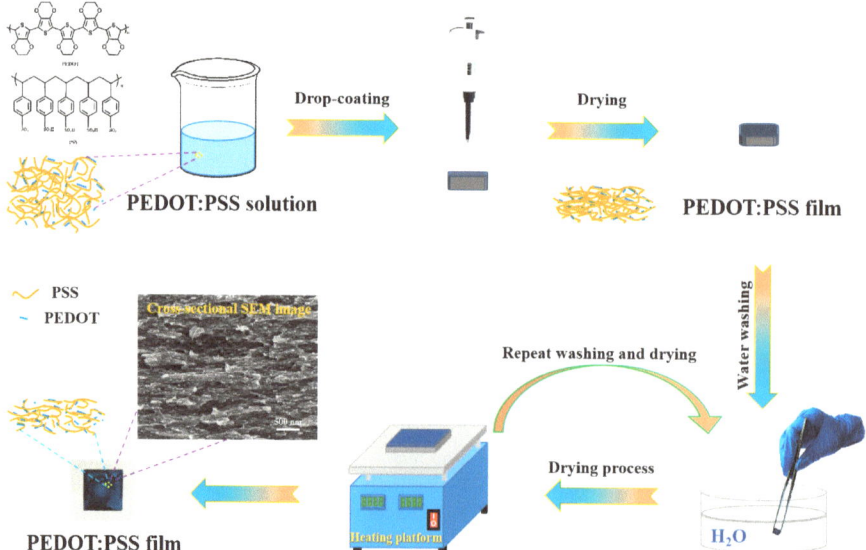

Figure 1. Schematic representation of the formation mechanism of water-treated PEDOT:PSS film.

3.2. Structure Characterization and Analysis

The films were examined using SEM, with the results presented in Figure 2. The pristine film displayed a relatively compact structure. In contrast, water treatment significantly altered the film's cross-sectional morphology, revealing sheet-like structures with microporous morphology. Notably, the film thickness increased following the water treatment, aligning with the measurements obtained via the step profiler. The emergence of porous structures and increased film thickness can be attributed to DI water penetration and the resultant microstructural changes. The film surfaces became non-uniform due to PEDOT:PSS dissolution in water (Figure 2f). Despite these changes, all water-treated films exhibited highly interconnected microporous structures within the PEDOT:PSS, offering efficient channels for carrier transportation.

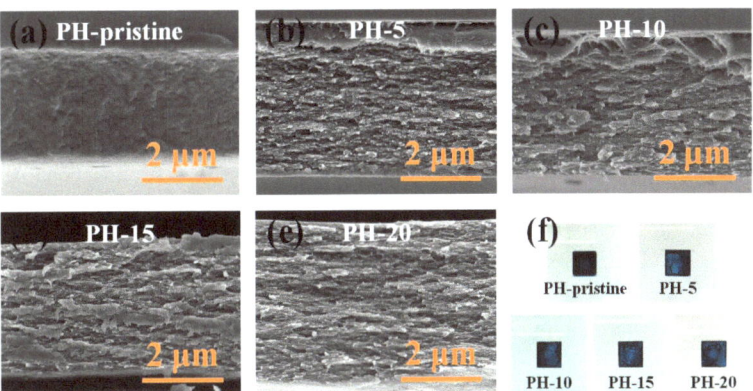

Figure 2. Cross-section SEM images of (**a**) pristine PEDOT:PSS film; (**b**) film water treated 5 times; (**c**) film water treated 10 times; (**d**) film water treated 15 times; (**e**) film water treated 20 times. (**f**) Optical photograph of all samples.

Figure S1 presents the FTIR spectra of the samples before and after different cycles of the water washing treatment. In the original PEDOT:PSS film, the transmission peaks centered at 1161 cm^{-1} and 1130 cm^{-1} are assigned to the S–O of PSS; the peak observed at 1011 cm^{-1} is related to the S-phenyl stretching of PSS. Peaks centered at 856 cm^{-1} and 943 cm^{-1} correspond to the C–S stretching of PEDOT, while peaks located at 1261 cm^{-1}, 1371 cm^{-1}, and 1060 cm^{-1} are ascribed to C–O–C, C–C, and EDOT ring structures, respectively [24]. It can be seen that the peaks associated with PSS at 1161 cm^{-1} and 1130 cm^{-1} decreased slowly in intensity with water treatments. To quantify this variation, a ratio of A_{1130}/A_{1060} was defined as the ratio of the area of the PSS peak centered at 1130 cm^{-1} to the area of the PEDOT peak centered at 1060 cm^{-1}. The calculated values are presented in Figure S2 and a declining trend was observed, indicating the removal of PSS from PEDOT:PSS. It is also noteworthy that a characteristic peak of the C=C quinoidal structure in PEDOT appeared at 1450 cm^{-1} after 15 cycles of water washing, and became more apparent after 20 cycles of the water treatment. The observed changes in PEDOT:PSS contribute to the improvement of the crystallinity of the polymer, thus increasing its electrical conductivity.

As a supplement to the infrared spectra, Raman spectroscopy, a nondestructive testing method, was continuously employed to identify structural variations in the organic polymers and carbon-based materials [25–27]. Figure 3a displays the Raman spectra of the PEDOT:PSS films including the pristine as-prepared one and those treated with different cycles of the water washing treatment. Characteristic peaks associated with PEDOT:PSS structures were observed, with the peak attributions summarized in Table 1 [28,29]: peaks centered at 1566 cm^{-1} and 1493 cm^{-1}, corresponding to asymmetric $C_\alpha=C_\beta$ stretching, became increasingly prominent after the water treatment; the peak at 1428 cm^{-1}, attributed to symmetric $C_\alpha=C_\beta(-O)$ stretching vibrations of the five-membered thiophene ring, became sharper and more intense. Similar trends were observed at 1367 cm^{-1} (C_β–C_β stretching), 1254 cm^{-1} (inter-ring C_α–C_α stretching), 1117 cm^{-1} (C–C ring bending vibration), 1094 cm^{-1} (C–O–C deformation), 701 cm^{-1} (symmetric C–S–C deformation), 437 cm^{-1} (SO$_2$ bending), 988 cm^{-1}, and 576 cm^{-1} (deformation of oxyethylene ring). The peak centered at 522 cm^{-1}, indicating the presence of the PSS component, exhibited a significant reduction in intensity after the water treatment, suggesting the effective removal of the PSS component. To quantify these changes, a ratio of I_{522}/I_{988} was defined as the ratio of the intensity of the PSS peak centered at 522 cm^{-1} to the intensity of the PEDOT peak located at 988 cm^{-1} [27]. The calculated values are displayed in Figure S3 and a declining trend was observed, further confirming the above assertion. The peak related to the stretching vibration of symmetric $C_\alpha=C_\beta(-O)$ at 1425 cm^{-1} exhibited a red shift, similar to treatments using dielectric solvents. Generally, shifts in symmetric $C_\alpha=C_\beta(-O)$ stretching are influenced by the relative ratio of benzene and quinoid rings, correlating with the oxidation degree of the molecular structures and/or the configuration of the PEDOT molecule. In a quinoid-dominated structure, conjugated π-electrons are more easily delocalized across the entire PEDOT chain, with adjacent thiophene rings nearly parallel to one another, thereby enhancing charge carrier mobility compared to a coiled benzoid-dominated structure. Additionally, after the water treatment a significant increase in corresponding band intensities was observed, possibly due to PSS removal and/or the reorienting effect of water on PEDOT chains. Further details regarding the variation of PEDOT:PSS films during treatment are discussed in subsequent sections.

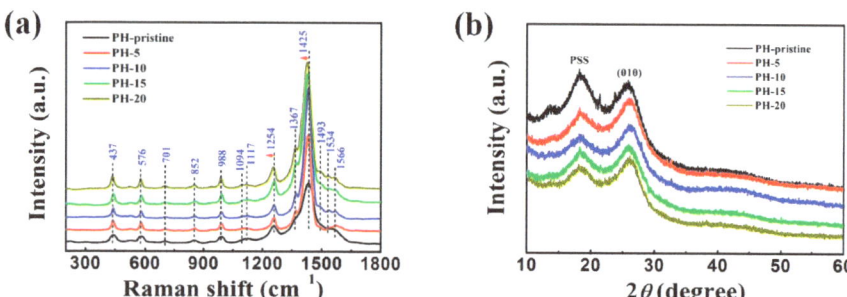

Figure 3. Variations in PEDOT:PSS structures before and after different cycles of water washing treatment: (a) Raman spectra; (b) XRD spectra.

Table 1. The characteristic Raman bands and their assignments for pristine PEDOT:PSS film.

Raman Bands (cm^{-1})	Assignments
1566, 1493	asymmetrical $C\alpha=C\beta$ vibrations
1428	symmetric $C\alpha=C\beta(-O)$ stretching vibrations of the five-membered thiophene ring
1367	$C\beta$–$C\beta$ stretching
1254	$C\alpha$–$C\alpha$ inter ring stretching
1117	bending vibration of C–C ring
1094	C–O–C stretching
988	deformation of the oxyethylene ring
701	deformation of the symmetric C–S–C
576	deformation of the oxyethylene ring
437	SO$_2$ bending

X-ray diffraction (XRD) patterns were analyzed to further comprehend the impact of the water treatment on PEDOT:PSS structures, with results displayed in Figure 3b. The PEDOT:PSS film without water washing revealed two distinct diffraction peaks at 2θ = 26.4°

and 18.5°, corresponding to the interchain planar π-π stacking distance ($d_{(010)}$) of aromatic rings in PEDOT and the amorphous halo of PSS, respectively [30,31]. Post water treatment, the intensity of the PSS peaks noticeably decreased, indicating either the effective removal of PSS or increased conformational disorder. Conversely, the peak related to the planar ring-stacking of PEDOT showed enhanced intensity, signifying increased lamellar stacking and improved crystallinity. This ordered and oriented pattern with relatively high reflective intensities should facilitate carrier transport in the continuous porous framework, consequently boosting electrical conductivity. These alterations are attributed to the selective removal of nonconductive PSS and subsequent molecular rearrangement in PEDOT. It is worth mentioning that these structural changes are analogous to the results reported in those studies [13].

UV-Vis-NIR spectra were also applied to investigate the compositional and conformational variation in PEDOT:PSS structures. No noticeable changes can be observed during the water washing process, until the cycle of treatments reached 20. For analytic simplicity, the data from the pristine PEDOT:PSS were compared with those from the film treated 20 times, and the results are presented in Figure S4. As depicted, the absorption peak at 256 nm can be assigned to the aromatic ring of PSS [24,29]. After a water treatment for 20 cycles, this peak decreased in density, clear evidence of PSS removal. However, the two curves almost coincide in the visible and NIR regions, demonstrating no obvious variation of the oxidation degree of the molecular structures [32].

To confirm the compositional changes in the treated films further, an X-ray electron spectroscopy (XPS) analysis was also performed, and its results are presented in Figure 4. The elements C, O, and S were identified in the full-range XPS spectra. The reduction in the insulating PSS content from PEDOT:PSS films during the H_2O treatment was confirmed by analyzing the S2p spectrum in the XPS. As illustrated in Figure 4b–f, the binding energies between 166 and 172 eV corresponded to sulfur atoms S2p bands in PSS, while the ones between 162 and 166 eV were related to the S2p bands of sulfur in PEDOT [33]. Moreover, PEDOT bands featured two peaks at approximately 164 eV and 165 eV for $S2p_{3/2}$ and $S2p_{1/2}$ thiophene sulfur, while two peaks at around 168 eV and 169 eV corresponded to the $S2p_{3/2}$ and $S2p_{1/2}$ sulfur of PSS. Based on the changes in peak areas corresponding to PSS and PEDOT, the structural modifications of the films were inferred. Calculations revealed that the PSS/PEDOT ratio decreased from 3.02 to 2.84, 2.42, 2.13, and 1.55 after 5, 10, 15, and 20 water washing cycles, respectively. This reduction can be ascribed to the conformational changes resulting from the removal of PSS during the H_2O washing process, indicating that PSS was progressively eliminated with an increasing number of H_2O treatments. The selective removal of the nonconductive-phase PSS effectively enhanced the conductivity of the PEDOT:PSS films. Additionally, the XPS results were consistent with the observed increase in electrical conductivity.

3.3. Thermoelectric Performance of PEDOT:PSS Film Prepared via Water Treatment

In the experiments, the pristine PEDOT:PSS films were prepared using a drop coating method, and the obtained samples exhibited a relatively low conductivity of 0.25 S cm^{-1} (Figure 5a), aligning with other research reports and attributed to the high content of PSS in the original films. After water washing, film conductivity increased substantially. For instance, conductivity rose to 26.88 S cm^{-1} after 5 wash cycles, and to 76.35 S cm^{-1} after 20 cycles, representing a nearly threefold increase. The ratio of PSS to PEDOT decreased from 3.02 to 1.55 over 20 wash cycles, indicating the effective removal of PSS and the resultant increase in carrier concentration in the films. Enhanced peak intensity in the XRD patterns of PEDOT's inter-chain planar ring stacking suggests improved crystallization, which is generally known to enhance the charge transfer within and between chains, thereby improving polymer conductivity. As shown in Figure 2, a porous structure was generated during the water washing process, which may partly reduce the number of the conductive pathways. Notably, as discussed above, networks for charge transport have become more efficient following the water washing treatment. As a result, the obtained films showed much higher electric conductivities when compared to untreated ones.

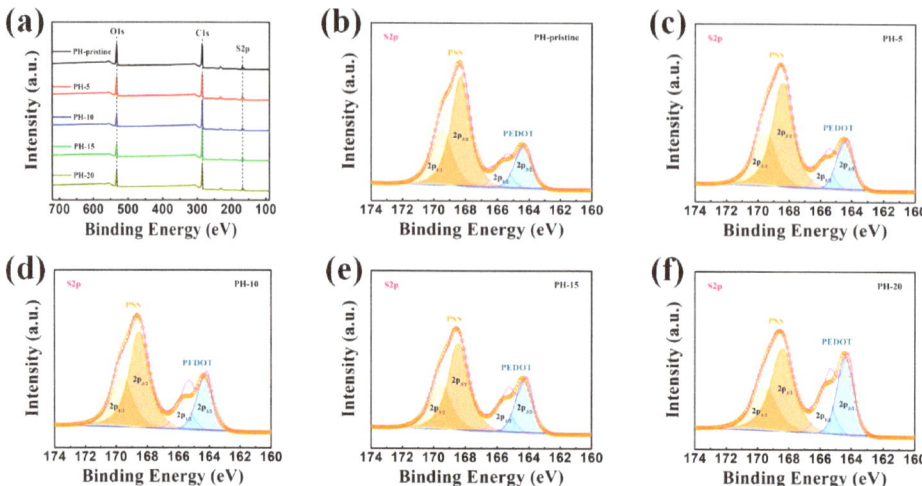

Figure 4. XPS spectra of PEDOT:PSS films with and without water treatment: (**a**) full-scale spectra of PEDOT:PSS; (**b**) S2p spectra of pristine PEDOT:PSS film; (**c**) S2p spectra of PEDOT:PSS film washed 5 times; (**d**) S2p spectra of PEDOT:PSS film washed 10 times; (**e**) S2p spectra of PEDOT:PSS film washed 15 times; (**f**) S2p spectra of PEDOT:PSS film washed 20 times.

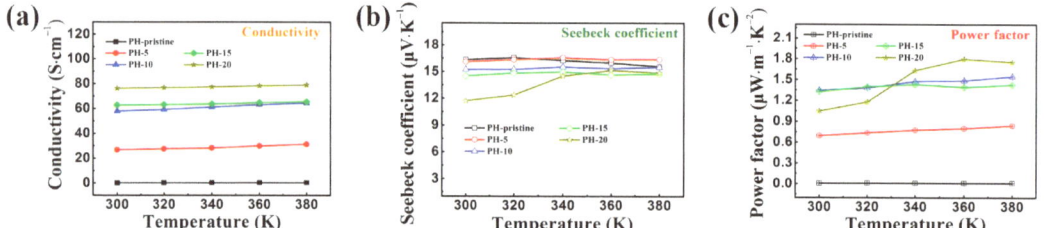

Figure 5. Variation of the TE performance of PEDOT:PSS film before and after treatment using room temperature water: (**a**) conductivity, (**b**) Seebeck coefficients, and (**c**) power factor.

Figure 5b illustrates the Seebeck coefficient variation in PEDOT:PSS films following different water treatment cycles. The untreated film maintained a Seebeck coefficient of ~16.34 µV K^{-1}, which gradually decreased to 16.10, 15.24, 14.53, and 11.73 µV K^{-1} after 5, 10, 15, and 20 wash cycles, respectively. This decrease likely correlates with the marked increase in carrier concentration due to PSS reduction. Figure 5c displays the power factor variation, showing that the pristine PEDOT:PSS film had a power factor of 0.0066 µW·m^{-1}·K^{-2} due to low conductivity. Post water treatment for 5, 10, 15, and 20 cycles, the power factors reached 0.6971, 1.3473, 1.3282, and 1.0514 µW·m^{-1}·K^{-2}, respectively. During heating, the power factor of the thin film exhibited an upward trend, reaching 1.7986 µW·m^{-1}·K^{-2} at 360 K after 20 wash cycles, which is at least 200 times higher than that of the pristine film. This upward trend of the power factor with temperature may be associated with the changes in the PEDOT:PSS structure, such as a transformation from a benzenoid type to a quinoidal one and the subsequent increase in the crystallinity [26,34].

Certainly, this method still has the potential for further developments. We also explored the relevant strategies in the experimental part, including repeating the water treatment at room temperature (water temperature remains at around 17 °C) for the first

20 times, with a subsequent warm water (30 °C) washing process for further modification. The results are shown in Figure S5. During the warm water washing process, the conductivities of the samples continued their previous incremental growth state until they reached a maximum of 221.52 S cm^{-1} after 60 cycles of further treatment (these conductivity values in Figure S5 were also confirmed by the film resistance shown in Figure S6). This conductivity was comparable to or even higher than most films modified using polar solvents, such as EG, DMSO, THF, etc. For example, PEDOT:PSS films doped with THF (25 vol%), DMF (25 vol%), and DMSO (25 vol%) possessed conductivities of ~4 S cm^{-1}, 30 S cm^{-1}, and 80 S cm^{-1}, respectively [12]. Additionally, the electrical conductivity could be raised to 3.5 S cm^{-1} when 20% of the film was EG [35]. This great enhancement in conductivity may be mainly induced by a significant structural change in PEDOT:PSS (e.g., a better crystallinity of PEDOT and/or more ordered molecular arrangement), rather than the removal of a large amount of PSS component (Figures S7 and S8). It is also worth noting that the Seebeck coefficient has been kept at around 14 µV K^{-1}, which is very similar to that of the sample treated using DMSO (~15 µV K^{-1}) [36]. Therefore, an optimal power factor of 4.4344 µW·m^{-1}·K^{-2} can be achieved, which is three times higher than that of the sample treated only using room temperature water (a PF value of 1.3282 µW·m^{-1}·K^{-2}, as reported in this work before warm water washing), 672 times larger than that of the pristine sample (0.0066 µW·m^{-1}·K^{-2}). More than that, the value of the power factor achieved is comparable to or even higher than those of some composites, such as PEDOT:PSS/Ca$_3$Co$_4$O$_9$ composite films (3.77 µW·m^{-1}·K^{-2}) [37], some multilayered film systems like PEDOT:PSS/PANI (3.0 µW·m^{-1}·K^{-2}) [38], and PEDOT:PSS/P3HT (5.79 µW·m^{-1}·K^{-2}) [39]. Although the power factor is not as good as that of the H$_2$SO$_4$-treated samples (as reported in our previous work, a high power factor of around 47 µW·m^{-1}·K^{-2} can be obtained because of the strong effect of concentrated sulfuric acid on the polymer structure, including the removal of PSS, better crystallinity of the conductive PEDOT:PSS, and variation of the microstructures) [29], the process of water treatment was very environmentally friendly, and no harmful reagent was used.

Another crucial factor that affects the thermoelectric properties of TE materials is thermal conductivity. In general, most conducting polymers, including PEDOT:PSS, possess inherent low thermal conductivities, which remain in a narrow range of 0.1–0.4 W·m^{-1}·K^{-1}, substantially lower (around 2–4 orders of magnitude) than those of inorganic TE materials [40]. For example, a DMSO-doped PEDOT:PSS film showed a thermal conductivity of 0.15 W·m^{-1}·K^{-1}, and the original sample showed a thermal conductivity of 0.17 W·m^{-1}·K^{-1} [36,41]. It is worth noting that the introduction of an inorganic material would lead to an increase in thermal conductivity to a certain extent. In our experiment, no inorganic component was incorporated, and during the water washing process only a portion of the PSS component was removed. Considering that the PSS component is essentially a polymer, it may have little effect on the variation of the thermal conductivity. In addition, after multiple cycles of water treating, the PEDOT:PSS films present porous structures (as shown in Figure 2), which can effectively promote the scattering of photons, thus reducing their thermal conductivity without significant influence on their electrical performance. Therefore, it is reasonable to speculate that the thermal conductivities of all samples prepared in this work are kept at a relatively low level.

4. Conclusions

In conclusion, a green and innovative method was developed for treating PEDOT:PSS films with water to enhance their thermoelectric (TE) properties. Structural characterization confirmed the selective removal effect of water on PSS, promoting the formation of ordered PEDOT molecular chains and a continuous conductive network, thereby enhancing electrical conductivities. By varying the number of water treatments, PEDOT:PSS films achieved a high electrical conductivity of up to 76.35 S cm^{-1}, with only a slight decline in the Seebeck coefficient. With 20 water wash cycles, the film's power factor reached a high value of 1.0514 µW·m^{-1}·K^{-2}, escalating to 1.7986 µW·m^{-1}·K^{-2} at 360 K, indicating its potential as a TE material. When further treating the film with warm water (60 cycles), a higher conductivity of 221.52 S cm^{-1} and a Seebeck coefficient of 14.15 µV K^{-1} can be achieved. As a result, the power factor reached up to 4.4344 µW·m^{-1}·K^{-2} at room temperature. This study presents a straightforward, green, and efficient method for improving the conductivity and TE performance of PEDOT:PSS films, paving the way for expanding their applications in various fields.

Supplementary Materials: The following supporting information can be downloaded at: https://www.mdpi.com/article/10.3390/polym16020266/s1. Ref. [42] is cited in the supplementary materials.

Author Contributions: Conceptualization, F.L. (Fuwei Liu) and L.G.; methodology, investigation, and validation, L.G., J.D., F.L. (Fuqun Li), J.L. and Z.C.; writing—original draft preparation, F.L. (Fuwei Liu) and L.G.; writing—review and editing, F.L. (Fuwei Liu); discussion of experiments, H.G., H.L., M.W., R.L. and M.L.; supervision, F.L. (Fuwei Liu) All authors have read and agreed to the published version of the manuscript.

Funding: This research was funded by the National Natural Science Foundation of China (Grant No. U2004174) and Nanhu Scholars Program for Young Scholars of XYNU.

Institutional Review Board Statement: Not applicable.

Data Availability Statement: The data that support the findings of this study are available from the corresponding author.

Acknowledgments: We appreciate the support and help from the Analysis Testing Center of Xinyang Normal University for their assistance during sample characterization. The authors would like to thank Home for Researchers (www.home-for-researchers.com) for their assistance. The authors also appreciate Gaofeng Tian and Jiaxiu Zhou from Shiyanjia Lab (www.shiyanjia.com) for Raman measurements as well. The first and the second authors made equal contributions to this work.

Conflicts of Interest: The authors declare no conflicts of interest.

References

1. Østergaard, P.A.; Duic, N.; Noorollahi, Y.; Mikulcic, H.; Kalogirou, S. Sustainable development using renewable energy technology. *Renew. Energy* **2020**, *146*, 2430–2437. [CrossRef]
2. You, S.; Bi, S.; Huang, J.; Jia, Q.; Yuan, Y.; Xia, Y.; Xiao, Z.; Sun, Z.; Liu, J.; Sun, S.; et al. Additive-Enhanced Crystallization of Solution Process for Planar Perovskite Solar Cells with Efficiency Exceeding 19%. *Chem. A Eur. J.* **2017**, *23*, 18140–18145. [CrossRef]
3. Cheng, N.; Liu, Z.; Li, W.; Yu, Z.; Lei, B.; Zi, W.; Xiao, Z.; Sun, S.; Zhao, Z.; Zong, P.-A. Cu$_2$ZnGeS$_4$ as a novel hole transport material for carbon-based perovskite solar cells with power conversion efficiency above 18%. *Chem. Eng. J.* **2023**, *454*, 140146. [CrossRef]
4. Gielen, D.; Boshell, F.; Saygin, D.; Bazilian, M.D.; Wagner, N.; Gorini, R. The role of renewable energy in the global energy transformation. *Energy Strategy Rev.* **2019**, *24*, 38–50. [CrossRef]
5. Zhang, Q.; Sun, Y.; Xu, W.; Zhu, D. Organic Thermoelectric Materials: Emerging Green Energy Materials Converting Heat to Electricity Directly and Efficiently. *Adv. Mater.* **2014**, *26*, 6829–6851. [CrossRef]
6. Zeng, Y.J.; Wu, D.; Cao, X.H.; Zhou, W.X.; Tang, L.M.; Chen, K.Q. Nanoscale Organic Thermoelectric Materials: Measurement, Theoretical Models, and Optimization Strategies. *Adv. Funct. Mater.* **2019**, *30*, 1903873. [CrossRef]
7. Kong, F.; Liu, C.; Song, H.; Xu, J.; Huang, Y.; Zhu, H.; Wang, J. Effect of solution pH value on thermoelectric performance of free-standing PEDOT:PSS films. *Synth. Met.* **2013**, *185–186*, 31–37. [CrossRef]
8. Kumar, S.R.S.; Kurra, N.; Alshareef, H.N. Enhanced high temperature thermoelectric response of sulphuric acid treated conducting polymer thin films. *J. Mater. Chem. C* **2016**, *4*, 215–221. [CrossRef]
9. Kim, J.; Jang, J.G.; Hong, J.-I.; Kim, S.H.; Kwak, J. Sulfuric acid vapor treatment for enhancing the thermoelectric properties of PEDOT:PSS thin-films. *J. Mater. Sci. Mater. Electron.* **2016**, *27*, 6122–6127. [CrossRef]
10. Song, H.; Kong, F.; Liu, C.; Xu, J.; Jiang, Q.; Shi, H. Improved thermoelectric performance of PEDOT:PSS film treated with camphorsulfonic acid. *J. Polym. Res.* **2013**, *20*, 316. [CrossRef]
11. Ouyang, J. Solution-Processed PEDOT:PSS Films with Conductivities as Indium Tin Oxide through a Treatment with Mild and Weak Organic Acids. *ACS Appl. Mater. Interfaces* **2013**, *5*, 13082–13088. [CrossRef] [PubMed]
12. Kim, J.Y.; Jung, J.H.; Lee, D.E.; Joo, J. Enhancement of electrical conductivity of poly(3,4-ethylenedioxythiophene)/poly(4-styrenesulfonate) by a change of solvents. *Synth. Met.* **2002**, *126*, 311–316. [CrossRef]
13. Kim, G.H.; Shao, L.; Zhang, K.; Pipe, K.P. Engineered doping of organic semiconductors for enhanced thermoelectric efficiency. *Nat. Mater.* **2013**, *12*, 719–723. [CrossRef] [PubMed]

14. Takano, T.; Masunaga, H.; Fujiwara, A.; Okuzaki, H.; Sasaki, T. PEDOT Nanocrystal in Highly Conductive PEDOT:PSS Polymer Films. *Macromolecules* **2012**, *45*, 3859–3865. [CrossRef]
15. Lee, S.H.; Park, H.; Kim, S.; Son, W.; Cheong, I.W.; Kim, J.H. Transparent and flexible organic semiconductor nanofilms with enhanced thermoelectric efficiency. *J. Mater. Chem. A* **2014**, *2*, 7288–7294. [CrossRef]
16. Kishi, N.; Kondo, Y.; Kunieda, H.; Hibi, S.; Sawada, Y. Enhancement of thermoelectric properties of PEDOT:PSS thin films by addition of anionic surfactants. *J. Mater. Sci. Mater. Electron.* **2017**, *29*, 4030–4034. [CrossRef]
17. Zhu, Z.; Liu, C.; Jiang, F.; Xu, J.; Liu, E. Effective treatment methods on PEDOT:PSS to enhance its thermoelectric performance. *Synth. Met.* **2017**, *225*, 31–40. [CrossRef]
18. Xia, Y.; Ouyang, J. Salt-Induced Charge Screening and Significant Conductivity Enhancement of Conducting Poly(3,4-ethylenedioxythiophene):Poly(styrenesulfonate). *Macromolecules* **2009**, *42*, 4141–4147. [CrossRef]
19. Badre, C.; Marquant, L.; Alsayed, A.M.; Hough, L.A. Highly Conductive Poly(3,4-ethylenedioxythiophene):Poly (styrenesulfonate) Films Using 1-Ethyl-3-methylimidazolium Tetracyanoborate Ionic Liquid. *Adv. Funct. Mater.* **2012**, *22*, 2723–2727. [CrossRef]
20. Liu, C.; Xu, J.; Lu, B.; Yue, R.; Kong, F. Simultaneous Increases in Electrical Conductivity and Seebeck Coefficient of PEDOT:PSS Films by Adding Ionic Liquids into a Polymer Solution. *J. Electron. Mater.* **2012**, *41*, 639–645. [CrossRef]
21. Du, Y.; Shi, Y.; Meng, Q.; Shen, S.Z. Preparation and thermoelectric properties of flexible SWCNT/PEDOT:PSS composite film. *Synth. Met.* **2020**, *261*, 116318. [CrossRef]
22. Wei, S.; Huang, X.; Deng, L.; Yan, Z.-C.; Chen, G. Facile preparations of layer-like and honeycomb-like films of poly(3,4-ethylenedioxythiophene)/carbon nanotube composites for thermoelectric application. *Compos. Sci. Technol.* **2021**, *208*, 108759. [CrossRef]
23. See, K.C.; Feser, J.P.; Chen, C.E.; Majumdar, A.; Urban, J.J.; Segalman, R.A. Water-Processable Polymer−Nanocrystal Hybrids for Thermoelectrics. *Nano Lett.* **2010**, *10*, 4664–4667. [CrossRef] [PubMed]
24. Liu, F.; Xie, L.; Wang, L.; Chen, W.; Wei, W.; Chen, X.; Luo, S.; Dong, L.; Dai, Q.; Huang, Y.; et al. Hierarchical Porous RGO/PEDOT/PANI Hybrid for Planar/Linear Supercapacitor with Outstanding Flexibility and Stability. *Nanomicro. Lett.* **2020**, *12*, 17. [CrossRef] [PubMed]
25. Liu, H.; He, Y.; Cao, K. Flexible Surface-Enhanced Raman Scattering Substrates: A Review on Constructions, Applications, and Challenges. *Adv. Mater. Interfaces* **2021**, *8*, 2100982. [CrossRef]
26. Zhou, X.; Pan, C.; Liang, A.; Wang, L.; Wong, W.-Y. Thermoelectric properties of composite films prepared with benzodithiophene derivatives and carbon nanotubes. *Compos. Sci. Technol.* **2017**, *145*, 40–45. [CrossRef]
27. Jang, D.; Yoon, H.N.; Seo, J.; Cho, H.J.; Kim, G.M.; Kim, Y.-K.; Yang, B. Improved electromagnetic interference shielding performances of carbon nanotube and carbonyl iron powder (CNT@CIP)-embedded polymeric composites. *J. Mater. Res. Technol.* **2022**, *18*, 1256–1266. [CrossRef]
28. Gao, L.; Liu, F.; Wei, Q.; Cai, Z.; Duan, J.; Li, F.; Li, H.; Lv, R.; Wang, M.; Li, J.; et al. Fabrication of Highly Conductive Porous Fe_3O_4@RGO/PEDOT:PSS Composite Films via Acid Post-Treatment and Their Applications as Electrochemical Supercapacitor and Thermoelectric Material. *Polymers* **2023**, *15*, 3453. [CrossRef]
29. Liu, F.-W.; Zhong, F.; Wang, S.-C.; Xie, W.-H.; Chen, X.; Hu, Y.-G.; Ge, Y.-Y.; Gao, Y.; Wang, L.; Liang, Z.-Q. Facile fabrication of highly flexible, porous PEDOT:PSS/SWCNTs films for thermoelectric applications. *Chin. Phys. B* **2022**, *31*, 027303. [CrossRef]
30. Kim, N.; Kee, S.; Lee, S.H.; Lee, B.H.; Kahng, Y.H.; Jo, Y.R.; Kim, B.J.; Lee, K. Highly Conductive PEDOT:PSS Nanofibrils Induced by Solution-Processed Crystallization. *Adv. Mater.* **2013**, *26*, 2268–2272. [CrossRef]
31. Kim, N.; Lee, B.H.; Choi, D.; Kim, G.; Kim, H.; Kim, J.-R.; Lee, J.; Kahng, Y.H.; Lee, K. Role of Interchain Coupling in the Metallic State of Conducting Polymers. *Phys. Rev. Lett.* **2012**, *109*, 106405. [CrossRef] [PubMed]
32. Wang, J.; Cai, K.; Shen, S. Enhanced thermoelectric properties of poly(3,4-ethylenedioxythiophene) thin films treated with H_2SO_4. *Org. Electron.* **2014**, *15*, 3087–3095. [CrossRef]
33. Zhao, Z.; Liu, Q.; Zhang, W.; Yang, S. Conductivity enhancement of PEDOT:PSS film via sulfonic acid modification: Application as transparent electrode for ITO-free polymer solar cells. *Sci. China Chem.* **2018**, *61*, 1179–1186. [CrossRef]
34. Zhou, X.; Pan, C.; Gao, C.; Shinohara, A.; Yin, X.; Wang, L.; Li, Y.; Jiang, Q.; Yang, C.; Wang, L. Thermoelectrics of two-dimensional conjugated benzodithiophene-based polymers: Density-of-states enhancement and semi-metallic behavior. *J. Mater. Chem. A* **2019**, *7*, 10422–10430. [CrossRef]
35. Ashizawa, S.; Horikawa, R.; Okuzaki, H. Effects of solvent on carrier transport in poly(3,4-ethylenedioxythiophene)/poly(4-styrenesulfonate). *Synth. Met.* **2005**, *153*, 5–8. [CrossRef]
36. Jiang, Q.; Lan, X.; Liu, C.; Shi, H.; Zhu, Z.; Zhao, F.; Xu, J.; Jiang, F. High-performance hybrid organic thermoelectric SWNTs/PEDOT:PSS thin-films for energy harvesting. *Mater. Chem. Front.* **2018**, *2*, 679–685. [CrossRef]
37. Liu, C.; Jiang, F.; Huang, M.; Lu, B.; Yue, R.; Xu, J. Free-Standing PEDOT-PSS/$Ca_3Co_4O_9$ Composite Films as Novel Thermoelectric Materials. *J. Electron. Mater.* **2010**, *40*, 948–952. [CrossRef]
38. Andrei, V.; Bethke, K.; Madzharova, F.; Bronneberg, A.C.; Kneipp, J.; Rademann, K. In Situ Complementary Doping, Thermoelectric Improvements, and Strain-Induced Structure within Alternating PEDOT:PSS/PANI Layers. *ACS Appl. Mater. Interfaces* **2017**, *9*, 33308–33316. [CrossRef]
39. Shi, H.; Liu, C.; Xu, J.; Song, H.; Lu, B.; Jiang, F.; Zhou, W.; Zhang, G.; Jiang, Q. Facile Fabrication of PEDOT:PSS/Polythiophenes Bilayered Nanofilms on Pure Organic Electrodes and Their Thermoelectric Performance. *ACS Appl. Mater. Interfaces* **2013**, *5*, 12811–12819. [CrossRef]

40. Song, H.; Liu, C.; Xu, J.; Jiang, Q.; Shi, H. Fabrication of a layered nanostructure PEDOT:PSS/SWCNTs composite and its thermoelectric performance. *RSC Adv.* **2013**, *3*, 22065–22071. [CrossRef]
41. Jiang, F.-X.; Xu, J.-K.; Lu, B.-Y.; Xie, Y.; Huang, R.-J.; Li, L.F. Thermoelectric Performance of Poly(3,4-ethylenedioxythiophene):Poly-(styrenesulfonate). *Chin. Phys. Lett.* **2008**, *25*, 2202–2205.
42. Ely, F.; Matsumoto, A.; Zoetebier, B.; Peressinotto, V.S.; Hirata, M.K.; de Sousa, D.A.; Maciel, R. Handheld and automated ultrasonic spray deposition of conductive PEDOT:PSS films and their application in AC EL devices. *Org. Electron.* **2014**, *15*, 1062–1070. [CrossRef]

Disclaimer/Publisher's Note: The statements, opinions and data contained in all publications are solely those of the individual author(s) and contributor(s) and not of MDPI and/or the editor(s). MDPI and/or the editor(s) disclaim responsibility for any injury to people or property resulting from any ideas, methods, instructions or products referred to in the content.

Article

Self-Rotation of Electrothermally Responsive Liquid Crystal Elastomer-Based Turntable in Steady-State Circuits

Zongsong Yuan [1], Junxiu Liu [1,2,*], Guqian Qian [1], Yuntong Dai [1] and Kai Li [1,2,*]

1. College of Civil Engineering, Anhui Jianzhu University, Hefei 230601, China; ys2523578175@163.com (Z.Y.); qianguqian1215@163.com (G.Q.); daiytmechanics@ahjzu.edu.cn (Y.D.)
2. Anhui Province Key Laboratory of Building Structure and Underground Engineering, Anhui Jianzhu University, Hefei 230601, China
* Correspondence: tjuliu@ahjzu.edu.cn (J.L.); kli@ahjzu.edu.cn (K.L.)

Abstract: Self-excited motions, characterized by their ability to harness energy from a consistent environment and self-regulate, exhibit significant potential in micro-devices, autonomous robotics, sensor technology, and energy generation. This study introduces an innovative turntable system based on an electrothermally responsive liquid crystal elastomer (LCE). This system facilitates self-rotation within a steady-state circuit. Employing an electrothermal LCE model, we have modeled and numerically analyzed the nonlinear dynamics of an LCE-rope within steady-state circuits, utilizing the four-order Runge–Kutta method for calculations. The numerical results reveal the emergence of two distinct motion patterns in the turntable system under steady-state conditions: a self-rotation pattern and a static pattern. The self-rotation is initiated when the system's absorbed energy surpasses the energy lost due to damping effects. Furthermore, this paper delves into the critical conditions necessary for initiating self-rotation and examines the influence of various key dimensionless parameters on the system's rotation amplitude and frequency. These parameters include gravitational acceleration, the initial position of the mass ball, elastic stiffness of the LCE and spring, limiting temperature, heating zone angle, thermal shrinkage coefficient, and damping factor. Our computational findings establish that these parameters exert a modulatory impact on the rotation amplitude and period. This research enhances the understanding of self-excited motions and offers promising avenues for applications in energy harvesting, monitoring, soft robotics, medical devices, and micro- and nano-devices.

Keywords: liquid crystal elastomers; self-excited motion; rotation; electrothermally responsive; rope

1. Introduction

Self-excited motion is characterized by a system's generation of periodic motion in response to a uniform external stimulus, as documented in various studies [1–5]. This phenomenon significantly diminishes the need for complex system control, as it relies on a constant, rather than a periodic, external stimulus for its periodic motion [6,7]. A key attribute of self-excited motion is its capacity to actively assimilate energy from a stable external environment to sustain its periodic activity. The system's intrinsic properties, including amplitude and period of the self-excited motion, contribute to its enhanced robustness [8]. Notably, self-excited motion is inherently passive and operates without external control. This attribute facilitates more streamlined system design, fostering intelligence, automation, and resource efficiency, thereby enhancing overall system efficiency [9,10]. The versatile nature of self-excited motion extends its applicability across diverse fields such as energy control [11], autonomous robotics [12–14], micro-nano devices [15], and medical technology [16–18].

During self-excited motion, energy is dissipated due to damping effects in the system. In order to maintain this self-excited motion, it is usually necessary to provide energy compensation by means of a nonlinear feedback mechanism, which is used to counteract the energy loss of system damping [19–21]. For example, energy replenishment can be realized through a self-shading mechanism [22], a mechanism of chemical reaction coupled with large deformation [23], and the coupled motion mechanisms of air expansion and liquid column formation [24,25]. These nonlinear feedback mechanisms are crucial in maintaining the stability of self-excited motion. There has been a notable increase in the reports on self-excited motion systems constructed from active materials, including hydrogels [26,27], dielectric elastomers, ionogels [23], liquid crystal elastomers (LCEs) [28–32], and temperature-sensitive polymers [33–36]. Meanwhile, numerous efforts have been brought into proposing and constructing a variety of self-excited motion patterns using these active materials, including bending [33–35], twisting [36,37], stretching and contracting [38], rolling [22], oscillating [39,40], jumping [41], rotating [42], and even achieving synchronized motions of multiple coupled self-resonators [43].

LCE is a type of electrothermally responsive material that consists of rod-like mesocrystalline monomers with main or side chains of flexible cross-linked polymers to combine rubbery elasticity with liquid crystal anisotropy. When encountering external stimuli such as electricity, heat, light, and magnetism, liquid crystal monomer molecules will rotate or go through phase transitions, modifying their structures and thereby resulting in macroscopic deformation [44–49]. Compared to other types of active materials like temperature-sensitive gels, moisture-sensitive gels, pneumatic artificial muscles, and polyelectrolyte gels, LCE is able to achieve self-excited motion and exhibits superior responsiveness and controllability, bringing innovative solutions to relevant applications. The advantageous properties of the LCE material make these LCE-based self-excited motion systems highly stable and reliable, and they can be driven and controlled in a wireless and non-contact manner. As a consequence, the LCE-based self-excited motion systems are of great interest, with a wide range of potential applications in energy control [50,51], autonomous robotics [52,53], micro- and nano-devices [15], and medical devices.

Recent advancements have led to a notable increase in the prevalence of self-excited motion systems utilizing LCE materials [54]. The majority of these systems operate primarily through mechanisms like direct ambient heating or photothermal and photochemical effects [11,24,28,31,55–64]. Despite this progress, the variety of self-excited motion patterns remains limited, constraining their application potential in active motor systems. Electronic actuation, particularly in practical applications, offers significant advantages in terms of system control and integration. Recent research has shown success in integrating scalable resistive heaters within LCEs, enabling system operation through controlled electrical potentials [49]. This innovation broadens the scope for both control and application of self-excited motion systems. The present study introduces a novel LCE-based turntable system, designed for self-rotation within a steady-state circuit. It explores both the dynamic rotation behavior and the impact of key system parameters. This investigation lays the groundwork for designing more efficient and responsive control systems.

This paper is organized in the following manner. Firstly, in Section 2, we model the nonlinear dynamics of the turntable system based on an electrothermally responsive LCE in a steady-state circuit, and derive the corresponding control equations. In Section 3, we numerically calculate the rotation of the system using the four-order Runge–Kutta method. We investigate the two motion patterns of the system and the corresponding mechanism. Then, in Section 4, we explore the triggering conditions of the system rotation and the influence of the system parameters on the rotation frequency, and we analyze the rotation behavior under different parameter settings as well. Finally, some conclusions and outlooks are given.

2. Model and Formulation

This section delineates the construction of an LCE-based turntable system within a steady-state circuit, accompanied by the development of a theoretical model predicated on the dynamic behavior of LCEs. Key areas of focus encompass the dynamics of the

LCE-based turntable system, the progression of molecules within the LCE material, the approach to resolving differential control equations with variable coefficients, and the dimensionless quantification of system parameters.

2.1. Dynamics of an LCE-Based Turntable System

As shown in Figure 1, this model is an electrothermally responsive LCE-based turntable system rotating around a point O'. The turntable as a whole comprises two layers of discs, with the upper disc being the driving layer, and the discs are equipped with n motion tubes, each of which contains a small mass ball of mass m, a spring of initial length L_s, and an LCE-rope of initial length L_i. The mass ball is placed in the middle of the tube, the inner side connects the mass ball to the turntable with a spring, and the outer side connects the mass ball to the turntable with an LCE-rope. The initial position of the mass ball is at a distance of L_0 from the center O' of the turntable. The lower disc is the electric heating layer, the blue region is the heat insulation zone, whose temperature is always stable and unchanged, and the red region denotes the heating zone, the angle of which is θ_0. The resistance wires are evenly laid in the heating zone and connected to the steady-state circuit. The heat generated due to the thermal effect of the electric current raises the temperature of the entire heating zone. The upper boundary of the heating zone is the starting position of the turntable, and the angle between the first mass ball and the starting position is θ_i. The current length of LCE-rope is denoted as $L(t)$. A right-angled coordinate system Oxy is established with the initial position of the mass ball as the coordinate origin, and the turntable is given an initial angular velocity w_0 of clockwise rotation.

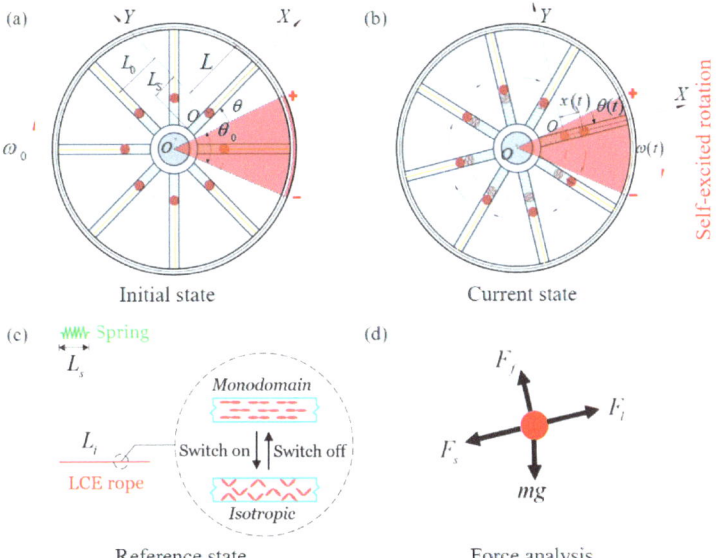

Figure 1. Schematic of an LCE-based turntable system in steady-state circuits, including two layers of discs and n motion tubes. Each motion tube contains an electrothermally responsive LCE-rope, a spring of conventional material, and a mass ball. The red region represents the heating zone and the blue region is the heat insulation zone. (**a**) Initial state; (**b**) current state; (**c**) reference state; (**d**) force analysis of the mass ball, which is subjected to gravity mg, damping force F_f, spring force F_s of the spring, and tensile force F_l of the LCE-rope. The mass ball keeps entering the heating zone, and the mass ball entering the heating zone will move outward under the tension of the LCE-rope, thus generating a torque difference and sustaining the system in periodic motion.

Figure 1c illustrates the modeling of electrothermally responsive LCE-rope in a steady-state circuit. This LCE material is composed of anisotropic rod-shaped liquid crystal molecules and stretchable long-chain polymers that are capable of achieving large and reversible actuation strains, which are mainly caused by the phase transition from the nematic to isotropic phase and the reorientation of the liquid crystal mesogens [28,64]. The monodomain LCE-rope is produced by initially loosely cross-linking the mixture within a circular mold, followed by fully cross-linking through uniaxial stretching and exposure to ultraviolet light. The LCE-rope is in a monodomain state in which the liquid crystal mesogens are arranged axially. When the circuit is connected, the temperature rises, and the LCE-rope will shrink along its axial direction. When the circuit is disconnected, the temperature decreases, and the rope is able to fully recover to its initial length. In the disconnected state, the length of the LCE-rope is L_i, which serves as a reference state.

During the rotation of the turntable, the following control equation can be derived from the momentum moment theorem:

$$\sum_{i=1}^{n} \left(\dot{J}_i \dot{\theta} + J_i \ddot{\theta} \right) = \sum_{i=1}^{n} \left\{ mg \cos\left[\frac{\theta_0}{2} - \theta - \frac{2\pi}{n}(i-1) \right] x_i - \beta \dot{\theta}(t)(L_0 + x_i)^2 \right\} \quad (1)$$

where $J_i = m(L_0 + x_i)^2$, β is the damping factor, g is the gravitational acceleration, $x_i(t)$ is the displacement of the ith mass ball, $\dot{\theta} = \frac{d\theta(t)}{dt}$ and $\ddot{\theta} = \frac{d^2\theta(t)}{dt^2}$ represent the rotation angular velocity and acceleration of the system, respectively.

As described in Figure 1b, when the motion tube enters the heating zone, the LCE-rope experiences an electrothermally driven shrinkage and pulls the mass ball towards the outside of the turntable, and the displacement moved by the mass ball is recorded as $x(t)$, which generates a gravitational moment difference to promote the rotation of the system. The mass ball enters the heating zone and is subjected to four forces, i.e., the spring force F_s of the spring, the tension F_l of the LCE-rope, the gravity mg of the ball itself, and the damping force F_f. It is assumed that the gravity is negligible relative to the elastic force. Since the ball is in equilibrium at any moment in the x-axis direction, we can obtain its equilibrium equation as

$$F_s = F_l \quad (2)$$

where $\theta(t)$ is the rotation angle. In accordance with Hooke's law, we can further simplify its equilibrium equation as

$$F_s = k_s x(t), \quad F_l = k_l [x(t) + L_i \varepsilon_T(t)] \quad (3)$$

where k_s refers to the elastic stiffness of the spring, k_l refers to the elastic stiffness of the LCE-rope, and $\varepsilon_T(t)$ is the thermally driven contraction strain of the LCE-rope.

Inserting Equation (3) into Equation (2) leads to

$$x_i(t) = -\frac{k_l \varepsilon_T(t) L_i L_s}{k_s L_i + k_l L_s} (i = 1, 2, 3, \ldots, n) \quad (4)$$

where $x_i(t)$ is the displacement of the ith mass ball.

The thermally driven contraction strain of the LCE-rope can be calculated as

$$\varepsilon_T(t) = \alpha T \quad (5)$$

where T is the temperature difference between the LCE-rope and the environment, and α is the coefficient of thermal contraction of the LCE-rope.

Based on Equations (2)–(4), Equation (1) can be simplified as

$$\ddot{\theta} = \frac{\sum_{i=1}^{n}\left\{mg\cos\left[\frac{\theta_0}{2}-\theta-\frac{2\pi}{n}(i-1)\right]x_i - \beta\dot{\theta}(t)(L_0+x_i)^2 - J_i\ddot{\theta}\right\}}{\sum_{i=1}^{n} J_i} \quad (6)$$

2.2. Temperature Field in LCE

In this section, the thermal dynamics of an electrothermally responsive system, encompassing both heated and unheated zones, are detailed. The assumption is made that heat exchange within the heating zone occurs rapidly, resulting in a uniform temperature across the LCE-rope. The presence of electric current induces a thermal effect, leading to heat generation in the resistance wire of the turntable's heating zone upon activation. The quantity of heat produced per second due to the electrothermal effect is represented by the parameter q. The LCE-rope is capable of thermal interaction with the turntable's heating zone, and it is presumed that the density of heat flow is linearly correlated to the temperature differential T between the LCE-rope and its surrounding environment. Within the heating zone, this temperature difference T is defined as per equation [42].

$$\dot{T} = \frac{q - KT}{\rho c} \quad (7)$$

$$T = T_L - T_e \quad (8)$$

where ρc refers to the specific heat capacity, K indicates the heat transfer coefficient, T_L is the LCE-rope temperature, and T_e is the ambient temperature.

By solving Equation (7), it can be seen that, in the heating zone, the temperature difference T is expressed as

$$T_{n+1} = T_n + \frac{t}{\tau}(T_0 - T_n) \quad (9)$$

where $T_0 = \frac{q}{K}$ denotes the limiting temperature difference of the electrothermally responsive LCE-rope in the case of long-time energization, $\tau = \frac{\rho c}{K}$ is the characteristic time of heat exchange between the LCE-rope and the environment, and the larger τ is, the longer the time required for the LCE to reach the limiting temperature difference T_0.

In the unheated zone, when $T_0 = 0$, the temperature difference T is

$$T_{n+1} = T_n - \frac{t}{\tau}T_n \quad (10)$$

2.3. Nondimensionalization and Solution Method

For convenience, the dimensionless quantities are introduced as follows: $\bar{x}_1(t) = \frac{x_1(t)}{L_i}$, $\bar{x}_2(t) = \frac{x_2(t)}{L_i}$, $\bar{g} = \frac{g\tau^2}{L_i}$, $\bar{\beta} = \frac{\beta\tau}{m}$, $\bar{t} = \frac{t}{\tau}$, $\bar{L}_0 = \frac{L_0}{L_i}$, $\bar{L}_s = \frac{L_s}{L_i}$, $\bar{L}(t) = \frac{L(t)}{L_i}$, $\bar{T} = \frac{T}{T_e}$, $\bar{T}_0 = \frac{T_0}{T_e}$, $\bar{k}_s = \frac{k_s\tau^2}{m}$, and $\bar{k}_l = \frac{k_l\tau^2}{m}$.

Re-substituting the above dimensionless parameters into the aforementioned Equations (3) and (6) yields the dimensionless control equations

$$\bar{x}_i(\bar{t}) = \frac{\bar{k}_l\varepsilon_T(t)}{\bar{k}_s + \bar{k}_l} \quad (11)$$

$$\ddot{\bar{\theta}} = \frac{\sum_{i=1}^{n}\left\{\bar{g}\cos\left[\frac{\theta_0}{2}-\theta-\frac{2\pi}{n}(i-1)\right]\bar{x}_i - \bar{\beta}\dot{\bar{\theta}}(\bar{x}_i+\bar{L}_0)^2 - 2\dot{\bar{\theta}}\sum_{i=1}^{n}(\bar{x}_i+\bar{L}_0)\frac{\bar{k}_l\alpha(\bar{T}_0-\bar{T})}{\bar{k}_s+\bar{k}_l}\right\}}{\sum_{i=1}^{n}(\bar{x}_i+\bar{L}_0)^2} \quad (12)$$

where $\ddot{\bar{\theta}} = \frac{d^2\theta(\bar{t})}{d\bar{t}^2}$ and $\dot{\bar{\theta}} = \frac{d\theta(\bar{t})}{d\bar{t}}$ represent the dimensionless rotation angular acceleration and velocity of the system, respectively.

Since Equation (12) is a differential equation with variable coefficients, it has no analytic solution. In order to solve this differential equation, we utilize the classical fourth-order Runge–Kutta method with the help of Matlab software (R2016b). The fourth-order Runge–Kutta formula involved is as follows

$$\theta(\bar{t}+h) = \theta(\bar{t}) + \frac{1}{6}(K_1 + 2K_2 + 2K_3 + K_4) \quad (13)$$

where $K_i (i = 1, 2, 3, 4)$ is expressed as

$$\begin{cases} K_1 = f(\bar{t}, \theta) \\ K_2 = f\left(\bar{t} + \frac{h}{2}, \theta + \frac{h}{2}K_1\right) \\ K_3 = f\left(\bar{t} + \frac{h}{2}, \theta + \frac{h}{2}K_2\right) \\ K_4 = f(\bar{t} + h, \theta + hK_3) \end{cases} \quad (14)$$

where h is the time step. Ultimately, the dynamic response of the angle and angular velocity for the system rotation over time can be obtained by iterative analysis.

The initial condition of the turntable system can be given as

$$\theta = 1.75\pi \text{ and } \bar{w} = 1 \text{ at } \bar{t} = 0. \quad (15)$$

Taking into account the dimensionless parameters including $\bar{T}_0$, θ_0, $\bar{\alpha}$, $\bar{g}$, $\bar{\beta}$, $\bar{L}_0$, $\bar{k}_l$, $\bar{k}_s$, θ, and $\bar{w}$, Equations (3), (4), (6) and (9) can be solved programmatically in Matlab by the four-order Runge–Kutta method. In the calculation, with the position x_{i-1} of the mass ball at the previous moment and the temperature change $\bar{T}_i$ of the LCE, the current shrinkage strain $\bar{\varepsilon}_i$ of the LCE can be estimated by Equation (4), which is then combined with Equation (10) to calculate the current position x_i of the mass ball. On the basis of $\bar{\varepsilon}_i$, x_i, and Equation (12), the current rotation angular velocity of the system can be calculated. When $0 \leq \mod(\theta_i, 2\pi) \leq \theta_0$, the mass ball is within the heating zone; otherwise, it is in the heat insulation zone. Repeating the above process, the time course of the rotation angle of the LCE-based turntable system is obtained via iterative calculation.

3. Two Motion Patterns and Mechanism of Self-Excited Motion

Utilizing the control equations established earlier and employing numerical analysis, this section is dedicated to examining the dynamic behavior of the system within a steady-state circuit. Initially, the two predominant motion patterns, specifically the static and self-rotation patterns, are introduced. Subsequently, a detailed exploration of the mechanisms driving self-rotation is conducted through a comprehensive parametric analysis.

3.1. Two Motion Patterns

To investigate the self-rotation of LCE-based turntable systems, the dimensionless parameters in the theoretical model must first be determined. Based on the available experimental data [11,65,66], the material properties and geometrical parameters of the system, as well as the relevant dimensionless parameters, are listed in detail in Tables 1 and 2, respectively. Table 1 illustrates the basic material properties and geometrical parameters of the LCE-based turntable system, which are fundamental and indispensable for analyzing the system, while Table 2 shows the associated dimensionless parameters, which are derived based on the fundamental data and the associated dimensionless formulas in Figure 1. These parameters are essential for the study of the effect of self-rotation on the turntable system. In this study, these parameter values will be utilized to investigate the self-rotation characteristics of the LCE-based turntable system in a steady-state circuit.

Table 1. Material properties and geometric parameters.

Parameter	Definition	Value	Unit
α	Thermal shrinkage coefficient of the LCE material	0–0.5	/
g	Gravitational acceleration	10	m/s^2
β	Damping factor	0.001~0.01	kg/s
θ_0	Heating zone angle	0.4π~0.8π	/
θ	Initial angle of the mass ball	0~2π	/
w_0	Initial angular velocity	0.4~2	rad/s
k_s	Elastic stiffness of the spring	0.005~50	N/m
k_l	Elastic stiffness of the LCE-rope	0.005~50	N/m
L_0	Distance from mass ball to turntable center	0.04~0.16	m
L_s	Initial length of spring	1.6~20	mm
L_i	Initial length of LCE-rope	0.02	m
τ	Characteristic time of heat exchange between LCE-rope and the environment	0.001~0.1	s
m	Mass of small mass ball	0.01	kg
T_0	Limit temperature difference of LCE-rope	0–20	°C
ρ_c	Specific heat capacity of LCE material	1000~4500	J/(kg°C)
q	Heat generated by the thermal effect of electric current	0~10	J/s

Table 2. Dimensionless parameters.

Parameter	$\overline{T}_0$	$\overline{g}$	$\overline{\beta}$	$\overline{L}_0$	$\overline{\alpha}$	$\overline{k}_l$	$\overline{k}_s$	θ_0
Value	0.1~0.5	6~20	0.01~0.2	0.5~3	0.38~0.45	10~100	10~100	0.5π~1.5π

From Equations (11) and (12), the phase trajectories and time histories of the system during the rotation in a steady-state circuit can be obtained. In the calculation, we first set $\overline{T}_0 = 0$, $\theta_0 = 0.5\pi$, $\overline{\alpha} = 0.35$, $\overline{g} = 10$, $\overline{\beta} = 0.01$, $\overline{L}_0 = 1$, $\overline{k}_l = 10$, $\overline{k}_s = 10$, $\theta = 0.25\pi$, and $\overline{w}_0 = 1$. With this group of parameters, the turntable begins to rotate with an initial speed of $\overline{w}_0 = 1$. And since $\overline{T}_0 = 0$ indicates that the circuit is in a power-off state, the LCE-rope does not change after entering the heating zone and the turntable continues to rotate. Due to the air damping, the rotation velocity of the turntable decreases and finally remains stable, which is termed the static pattern, as plotted in Figure 2a,b. When the parameters are set to $\overline{T}_0 = 0.35$, $\theta_0 = 0.5\pi$, $\overline{\alpha} = 0.35$, $\overline{g} = 10$, $\overline{\beta} = 0.01$, $\overline{L}_0 = 1$, $\overline{k}_l = 10$, $\overline{k}_s = 10$, $\theta = 0.25\pi$, and $\overline{w}_0 = 1$, as depicted in Figure 2c,d, the turntable can rotate continuously and eventually develop into a self-rotation, named the self-rotation. Similar to other self-excited motion systems, the LCE-based turntable system demonstrates the capability to execute rotation motion within a steady-state circuit. This phenomenon is primarily due to the external energy input offsetting damping losses, thereby sustaining the self-rotation. Section 3.2 will delve into the intricate mechanisms underlying this self-rotation.

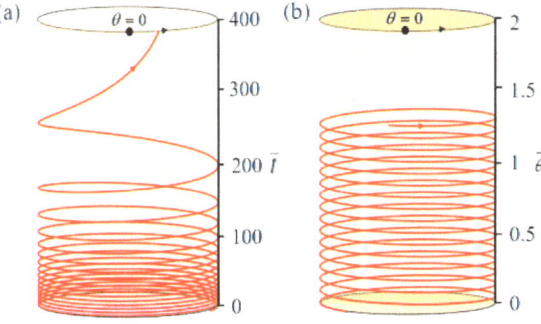

Figure 2. Cont.

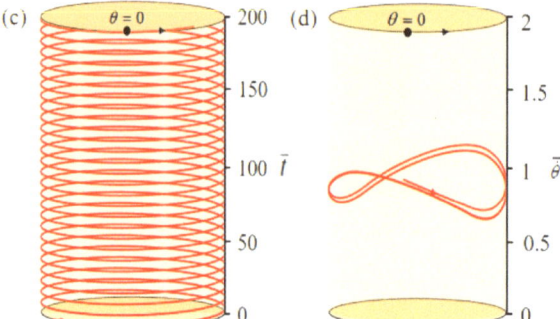

Figure 2. Time histories and phase trajectories of the LCE-based turntable. (**a,b**) Static pattern with parameters of $\bar{T}_0 = 0$, $\theta_0 = 0.5\pi$, $\bar{\alpha} = 0.35$, $\bar{g} = 10$, $\bar{\beta} = 0.01$, $\bar{L}_0 = 1$, $\bar{k}_l = 10$, $\bar{k}_s = 10$, $\theta = 0.25\pi$, and $\bar{w}_0 = 1$. (**c,d**) Self-rotation with parameters of $\bar{T}_0 = 0.35$, $\theta_0 = 0.5\pi$, $\bar{\alpha} = 0.35$, $\bar{g} = 10$, $\bar{\beta} = 0.01$, $\bar{L}_0 = 1$, $\bar{k}_l = 10$, $\bar{k}_s = 10$, $\theta = 0.25\pi$, and $\bar{w}_0 = 1$. The LCE-based turntable system involves two motion patterns in a steady-state circuit, i.e., static pattern and self-rotation.

3.2. Mechanism of the Self-Excited Motion

Figure 3 displays the evolution of several key parameters related to the self-rotation depicted in Figure 2c,d in order to facilitate the exploration of the self-rotation mechanism. Figure 4 illustrates the self-rotation of the LCE-based turntable system during one cycle. The time dependence of the electrothermally driven shrinkage strain in the LCE material during the self-rotation is plotted in Figure 3a, showing a periodic variation with time. The time dependences of the elastic force $\bar{F}_l$ and the damping force $\bar{F}_f$ on the mass ball are plotted in Figure 3b,c, respectively, both of which vary periodically with time. In Figure 3d, the driving torque is plotted versus time, which gradually increases in the heating zone and decreases in the heat insulation zone. Figure 3e,f reflect the relationship between driving torque, damping torque, and rotation angle, respectively. The areas enclosed by the two curves represent the net work done by the driving torque and the work consumed by the system damping during one cycle, both of which are 0.51. It is the positive net work done by the driving torque that compensates for the energy dissipated by the damping, allowing the system to maintain a continuous and stable motion pattern of self-rotation.

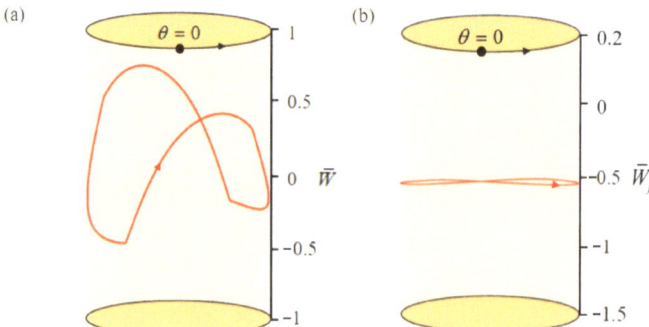

Figure 3. Cont.

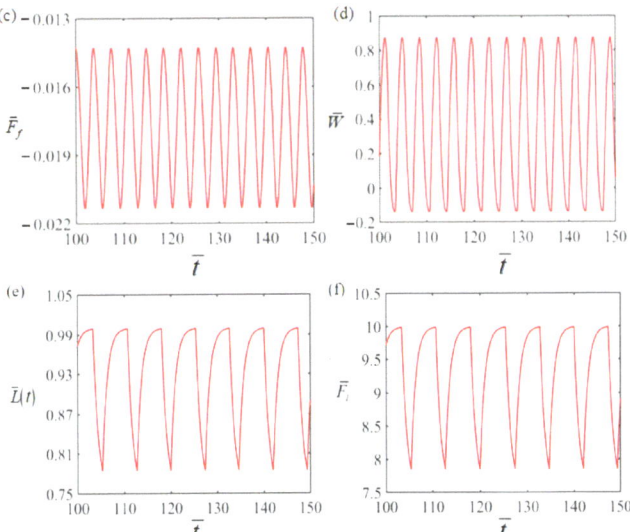

Figure 3. (a) Time dependence of the electrothermally driven shrinkage strain in the LCE-rope; (b) time dependence of the elastic force; (c) time dependence of the damping force; (d) time dependence of the driving torque; (e) driving torque vs. rotation angle; (f) damping torque vs. rotation angle.

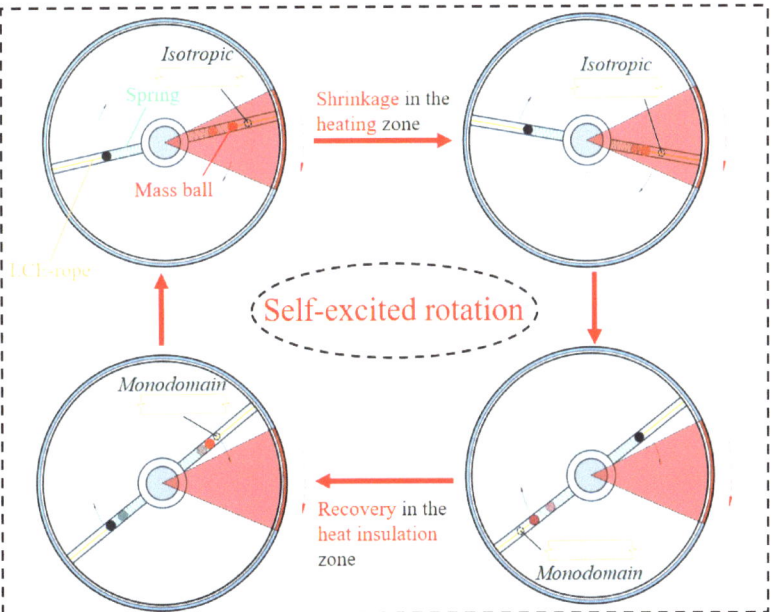

Figure 4. Self-rotation of the turntable system during one cycle. The LCE-rope contracts axially in the heating zone, pulling the mass ball outward, and the electrothermally driven contraction recovers axially in the heat insulation zone. In the steady-state circuit, the LCE-based turntable system exhibits continuous periodic rotation due to the periodic variation of the electrothermally driven contraction of the LCE-rope. The red shaded region indicates the heating zone and the blue shaded region indicates the heat insulation zone.

4. Parametric Study

The equations under consideration encompass eight dimensionless parameters: $\bar{g}$, $\bar{T}_0$, $\bar{\beta}$, $\bar{L}_0$, $\bar{\alpha}$, $\bar{k}_l$, $\bar{k}_s$, and θ_0. These parameters play a pivotal role in modulating the self-rotation dynamics of the LCE-based turntable system, as demonstrated in Figure 1. This segment of the study is dedicated to examining the influence of these crucial parameters on the critical conditions, periodicity, and magnitude of self-excited motion in a system equipped with just two mass balls. The objective is to provide insights applicable to various domains such as energy harvesting, power generation, monitoring, soft robotics, medical devices, and micro- and nano-devices. In this context, A and f are employed to denote the dimensionless amplitude and frequency of the system, respectively.

4.1. Effect of the Gravitational Acceleration

Figure 5 illustrates the effect of the dimensionless gravitational acceleration $\bar{g}$ on the rotation of the system. In the numerical calculation, we set $\bar{T}_0 = 0.35$, $\theta_0 = 0.5\pi$, $\bar{\alpha} = 0.35$, $\bar{\beta} = 0.01$, $\bar{L}_0 = 1$, $\bar{k}_l = 10$, $\bar{k}_s = 10$, $\theta = 0.25\pi$, and $\bar{w}_0 = 1$. Figure 5a illustrates the boundary conditions for the rotation behavior of the system under varying gravitational accelerations $\bar{g}$. The analysis reveals a critical gravitational acceleration threshold, specifically $\bar{g} = 5$, which initiates rotation motion in the system. At this juncture, $\bar{g} \geq 5$, the system transitions into a rotation mode. This phenomenon is interpretable through the lens of energy balance, considering the interplay between the net energy input and damping losses. Under conditions of low gravitational acceleration, the system can only generate a minimal torque in the heated zone, rendering the net electrothermal energy imparted to the LCE-rope insufficient to offset the energy loss due to damping. In scenarios where the energy input fails to counterbalance the damping losses, the LCE-based turntable system will ultimately cease motion, settling into a static equilibrium in a static state. Figure 5b demonstrates the impact of varying dimensionless gravitational accelerations on both the amplitude of change in the LCE-rope and the system's rotation frequency. As can be seen in Figure 5, both the amplitude and the rotation frequency increase gradually and more significantly as the gravitational acceleration $\bar{g}$ increases, which is consistent with physical intuition.

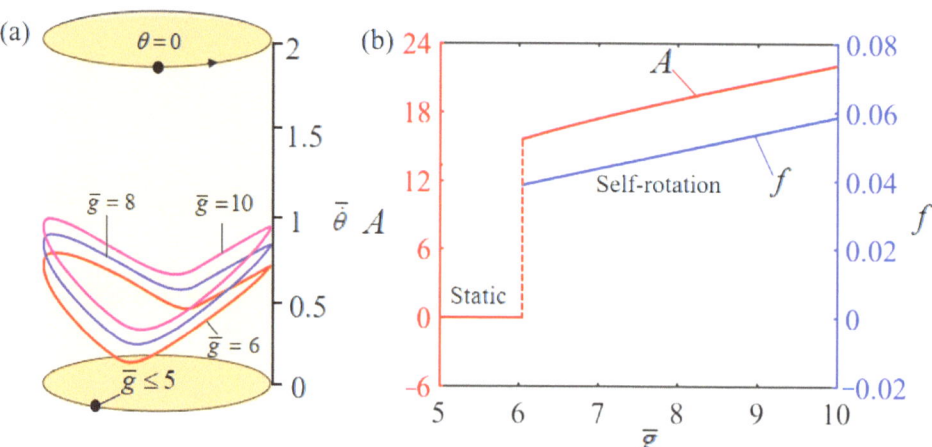

Figure 5. Effect of dimensionless gravitational acceleration on the self-rotation of the LCE-based turntable system, with $\bar{T}_0 = 0.35$, $\theta_0 = 0.5\pi$, $\bar{\alpha} = 0.35$, $\bar{\beta} = 0.01$, $\bar{L}_0 = 1$, $\bar{k}_l = 10$, $\bar{k}_s = 10$, $\theta = 0.25\pi$, and $\bar{w}_0 = 1$. (**a**) Limit cycle; (**b**) amplitude and frequency.

4.2. Effect of Initial Position of the Mass Ball

Figure 6 describes how the initial position $\bar{L}_0$ of the mass ball affects the rotation of the system. In the numerical calculation, we set $\bar{T}_0 = 0.35$, $\theta_0 = 0.5\pi$, $\bar{\alpha} = 0.35$, $\bar{\beta} = 0.01$, $\bar{g} = 10$, $\bar{k}_l = 10$, $\bar{k}_s = 10$, $\theta = 0.25\pi$, and $\bar{w}_0 = 1$. Figure 6a plots the limit cycles of rotation for different initial positions, where a critical initial position of about $\bar{L}_0 = 3$ exists for the phase transition between static and self-rotation. When the initial position exceeds or equals the critical value, the energy input to the system from external sources is unable to compensate for the damping loss, which results in the system rotating slower and slower and eventually stopping at the static equilibrium position. When $\bar{L}_0 = 1$, $\bar{L}_0 = 2$, and $\bar{L}_0 = 3$, the self-rotation is triggered with the limit cycles being plotted in Figure 6a. Figure 6b shows how the initial position of the mass ball influences the rotation amplitude and frequency. As the initial distance from the mass ball to the turntable center increases, there are significant decreases in the amplitude and frequency. This result indicates that the increase in initial position is not conducive to the system rotation, but instead makes the system rotate slower and slower, and finally stop at the static equilibrium position.

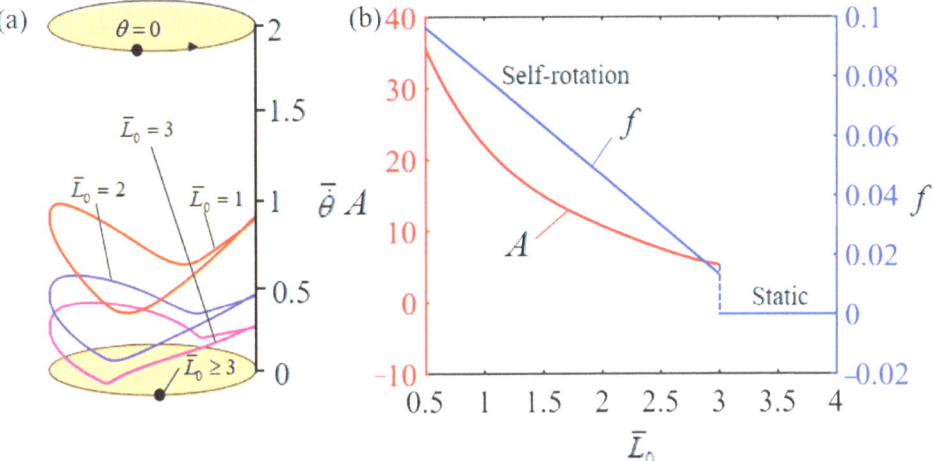

Figure 6. Effect of dimensionless initial position $\bar{L}_0$ of the mass ball on the self-rotation of the LCE-based turntable system, with $\bar{T}_0 = 0.35$, $\theta_0 = 0.5\pi$, $\bar{\alpha} = 0.35$, $\bar{\beta} = 0.01$, $\bar{g} = 10$, $\bar{k}_l = 10$, $\bar{k}_s = 10$, $\theta = 0.25\pi$, and $\bar{w}_0 = 1$. (**a**) Limit cycles; (**b**) amplitude and frequency.

4.3. Effect of Damping Factor

Figure 7 illustrates the effect of dimensionless damping factor $\bar{\beta}$ on the rotation, with other parameters being set as $\bar{T}_0 = 0.35$, $\theta_0 = 0.5\pi$, $\bar{\alpha} = 0.35$, $\bar{g} = 10$, $\bar{L}_0 = 1$, $\bar{k}_l = 10$, $\bar{k}_s = 10$, $\theta = 0.25\pi$, and $\bar{w}_0 = 1$. Figure 7a presents the rotation limit cycles corresponding to varying damping factors, identifying a critical damping factor, denoted as $\bar{\beta} = 0.2$, which signifies the transition threshold between static and self-rotation states. Beyond this critical value of $\bar{\beta}$, the system's damping dissipation becomes excessively pronounced to be compensated by the mechanical energy derived from the input thermal energy, consequently stabilizing the turntable system in a static equilibrium state. In contrast, for $\bar{\beta} = 0.05$, $\bar{\beta} = 0.1$, and $\bar{\beta} = 0.15$, the self-rotation can be triggered. Figure 7b displays the effect of the dimensionless damping factor $\bar{\beta}$ on the rotation amplitude and frequency. The larger the damping factor, the more energy loss is generated and, therefore, relatively less energy is available for the system to rotate. Hence, the time required for the system to rotate one revolution increases, indicating that the frequency is decreasing. However, the damping factor has almost no effect on the rotation amplitude.

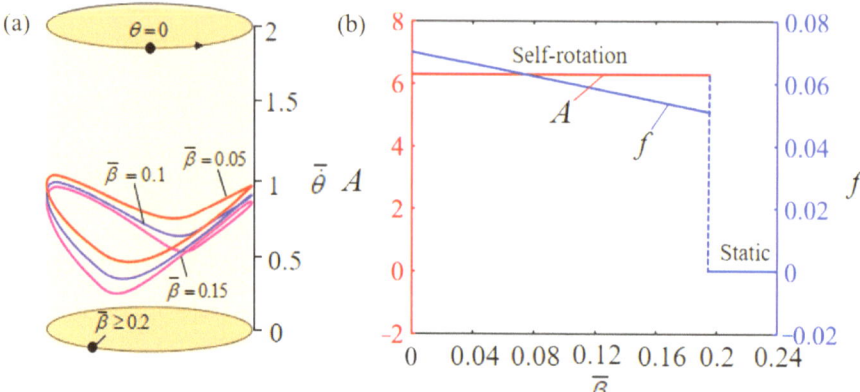

Figure 7. Effect of dimensionless damping factor on self-rotation of the LCE-based turntable system, with $\bar{T}_0 = 0.35$, $\theta_0 = 0.5\pi$, $\bar{\alpha} = 0.35$, $\bar{g} = 10$, $\bar{L}_0 = 1$, $\bar{k}_l = 10$, $\bar{k}_s = 10$, $\theta = 0.25\pi$, and $\bar{w}_0 = 1$. (a) Limit cycles; (b) amplitude and frequency.

4.4. Effect of Limit Temperature

Figure 8 illustrates the effect of dimensionless limit temperature $\bar{T}_0$ on the rotation, in which the other parameters are set to $\bar{g} = 10$, $\theta_0 = 0.5\pi$, $\bar{\alpha} = 0.35$, $\bar{\beta} = 0.01$, $\bar{L}_0 = 1$, $\bar{k}_l = 10$, $\bar{k}_s = 10$, $\theta = 0.25\pi$, and $\bar{w}_0 = 1$. Figure 8a illustrates the limit cycles of rotation at varying threshold temperatures, revealing a critical threshold temperature of approximately $\bar{T}_0 = 0.1$, essential for initiating self-rotation. Below this critical temperature, the energy derived from heat is inadequate to offset the damping losses, leading to the system stabilizing in a static equilibrium state. In contrast, at $\bar{T}_0 = 0.2$, $\bar{T}_0 = 0.5$, and $\bar{T}_0 = 1$, the system demonstrates the capability of self-rotation. Figure 8b presents the influence of the dimensionless threshold temperature $\bar{T}_0$ on the rotation amplitude and frequency. It is noted that the amplitude escalates in tandem with an increase in the dimensionless threshold temperature. This trend is attributable to the fact that higher threshold temperatures facilitate greater thermal energy generation, resulting in increased energy input to the system and, consequently, an elevated rotation amplitude. And the rotation frequency increases gradually with the increase in limit temperature as well.

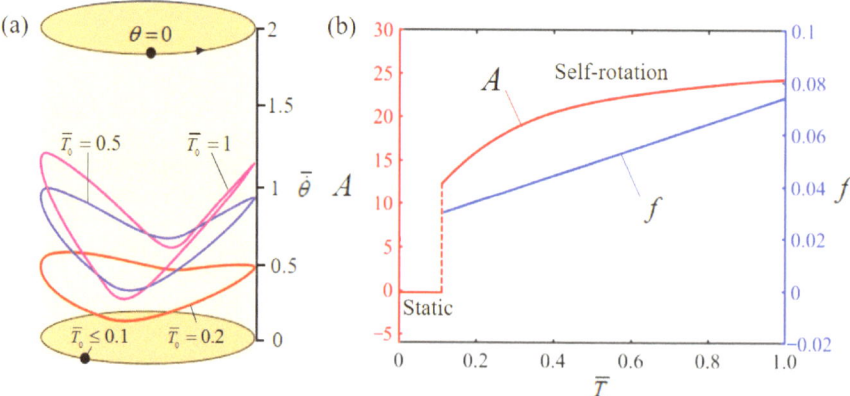

Figure 8. Effect of dimensionless limit temperature $\bar{T}_0$ on the self-rotation of the LCE-based turntable system, with $\bar{g} = 10$, $\theta_0 = 0.5\pi$, $\bar{\alpha} = 0.35$, $\bar{\beta} = 0.01$, $\bar{L}_0 = 1$, $\bar{k}_l = 10$, $\bar{k}_s = 10$, $\theta = 0.25\pi$, and $\bar{w}_0 = 1$. (a) Limit cycles; (b) amplitude and frequency.

4.5. Effect of Thermal Shrinkage Coefficient

The impact of the dimensionless thermal shrinkage coefficient $\bar{\alpha}$ on rotation behavior is delineated in Figure 9. During the numerical analysis, several parameters were held constant, specified as $\bar{T}_0 = 0.35$, $\theta_0 = 0.5\pi$, $\bar{g} = 10$, $\bar{\beta} = 0.01$, $\bar{L}_0 = 1$, $\bar{k}_l = 10$, $\bar{k}_s = 10$, $\theta = 0.25\pi$, and $\bar{w}_0 = 1$. Figure 9a illustrates the rotation limit cycles corresponding to various values of $\bar{\alpha}$, revealing a pivotal thermal shrinkage coefficient, approximately $\bar{\alpha} = 0.1$, which demarcates the transition from static to self-rotation patterns. For $\bar{\alpha} = 0.1$ values below this critical threshold, the system's thermal energy input is insufficient to counterbalance the energy losses due to damping, leading to a steady-state equilibrium. Conversely, for $\bar{\alpha} = 0.2$, $\bar{\alpha} = 0.3$, and $\bar{\alpha} = 0.4$, the system exhibits self-rotation. Figure 9b demonstrates how the rotation amplitude and frequency are influenced by $\bar{\alpha}$. Notably, the amplitude exhibits a progressive increase with higher values of $\bar{\alpha}$. In contrast, the frequency remains relatively unaffected by changes in the thermal shrinkage coefficient.

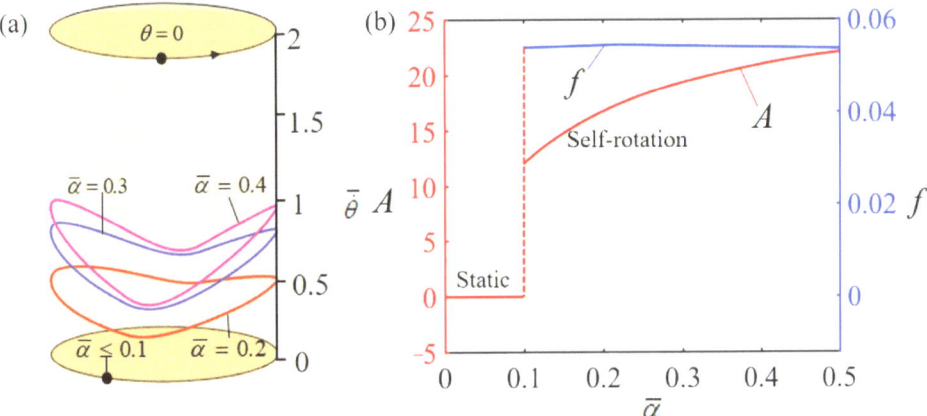

Figure 9. Effect of dimensionless thermal shrinkage coefficient $\bar{\alpha}$ on the self-rotation of the LCE-based turntable system, with $\bar{T}_0 = 0.35$, $\theta_0 = 0.5\pi$, $\bar{g} = 10$, $\bar{\beta} = 0.01$, $\bar{L}_0 = 1$, $\bar{k}_l = 10$, $\bar{k}_s = 10$, $\theta = 0.25\pi$, and $\bar{w}_0 = 1$. (**a**) Limit cycles; (**b**) amplitude and frequency.

4.6. Effect of Elastic Stiffness of LCE-Rope

Figure 10a illustrates how the dimensionless elastic stiffness $\bar{k}_l$ of the LCE-rope influences the rotation. In the numerical calculation, the other parameters are chosen as $\bar{T}_0 = 0.35$, $\theta_0 = 0.5\pi$, $\bar{\alpha} = 0.35$, $\bar{\beta} = 0.01$, $\bar{L}_0 = 1$, $\bar{g} = 10$, $\bar{k}_s = 10$, $\theta = 0.25\pi$, and $\bar{w}_0 = 1$. Figure 10 delineates the limit cycles of rotation for the LCE-rope across a range of dimensionless elastic stiffness values, denoted as $\bar{k}_l$. A critical threshold for elastic stiffness, identified at $\bar{k}_l = 10$, marks the phase transition from static to self-rotation patterns. When the dimensionless elastic stiffness of the LCE-rope falls below this critical value, the energy converted from heat into the system proves inadequate to counterbalance the energy dissipated through damping effects. This insufficiency results in the turntable ultimately achieving a state of static equilibrium. Instead, for $\bar{k}_l = 10$, $\bar{k}_l = 30$, and $\bar{k}_l = 50$, the system can exhibit self-rotation. Figure 10b depicts how the elastic stiffness of the LCE-rope affects the rotation amplitude and frequency. Only a slight increase in both amplitude and frequency occurs as the elastic stiffness of the LCE-rope is gradually improved. This is attributed to the fact that the elastic stiffness has almost no effect on the deformation of the LCE-rope, and, thus, has a negligible effect on the rotation frequency.

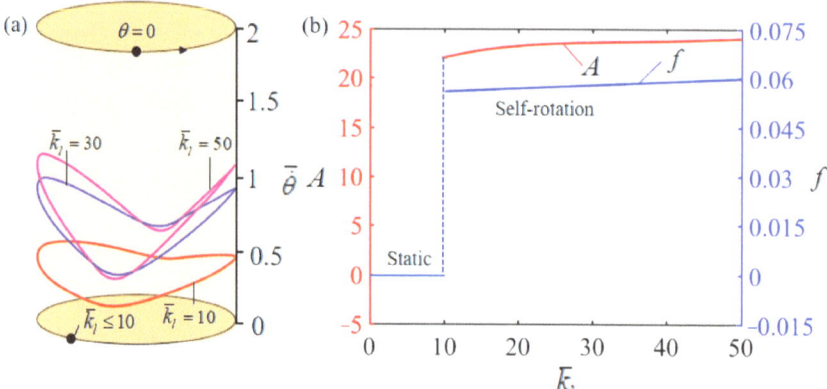

Figure 10. Effect of dimensionless LCE-rope elastic stiffness $\bar{k}_l$ on the self-rotation of the LCE-based turntable system, with $\bar{T}_0 = 0.35$, $\theta_0 = 0.5\pi$, $\bar{\alpha} = 0.35$, $\bar{\beta} = 0.01$, $\bar{L}_0 = 1$, $\bar{g} = 10$, $\bar{k}_s = 10$, $\theta = 0.25\pi$, and $\bar{w}_0 = 1$. (a) Limit cycles; (b) amplitude and frequency.

4.7. Effect of Elastic Stiffness of Spring

Figure 11 shows the effect of the dimensionless elastic stiffness of the spring, $\bar{k}_s$, on the rotation. In the numerical calculation, we set $\bar{T}_0 = 0.35$, $\theta_0 = 0.5\pi$, $\bar{\alpha} = 0.35$, $\bar{\beta} = 0.01$, $\bar{L}_0 = 1$, $\bar{k}_l = 10$, $\bar{g} = 10$, $\theta = 0.25\pi$, and $\bar{w}_0 = 1$. Figure 11a plots the limit cycles of rotation for different dimensionless elastic stiffnesses of the spring $\bar{k}_s$. And there presents a critical elastic stiffness of about $\bar{k}_s = 10$ for the phase transition between static and self-rotation patterns. When the elastic stiffness of the spring is below the critical value, the increase in time for the mass ball to return to its initial position within the heat insulation zone results in a decrease in the net work produced by the driving torque, which is insufficient to offset the energy consumed by the damping effect, and the turntable will eventually remain in a static equilibrium position. Conversely, for $\bar{k}_s = 10$, $\bar{k}_s = 50$, and $\bar{k}_s = 100$, the system can undergo self-rotation. Figure 11b illustrates how the elastic stiffness of the spring affects the rotation amplitude and frequency. As the elastic stiffness of the spring is improved, both the amplitude and frequency experience declines. This is due to the fact that an increase in the elastic stiffness of the spring reduces the displacement of the mass ball in the heating zone, thereby diminishing the rotation frequency of the system.

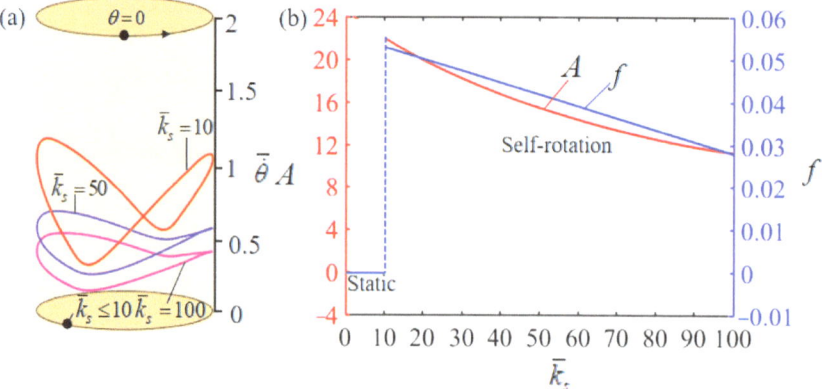

Figure 11. Effect of dimensionless spring elastic stiffness $\bar{k}_s$ on the self-rotation of the LCE-based turntable system, with $\bar{T}_0 = 0.35$, $\theta_0 = 0.5\pi$, $\bar{\alpha} = 0.35$, $\bar{\beta} = 0.01$, $\bar{L}_0 = 1$, $\bar{k}_l = 10$, $\bar{g} = 10$, $\theta = 0.25\pi$, and $\bar{w}_0 = 1$. (a) Limit cycles; (b) amplitude and frequency.

4.8. Effect of Heating Zone Angle

Figure 12a illustrates the limit cycles of rotation behavior across varying ranges of the heating zone. It identifies two pivotal heating zone angles, approximately $\theta_0 = 0.5\pi$ and $\theta_0 = 1.75\pi$, that demarcate the phase transition from static to self-rotation patterns. For heating zone angles less than the critical threshold of $\theta_0 = 0.5\pi$, the duration of LCE exposure to the heating zone is insufficient. This brevity in exposure results in inadequate thermal energy input to offset the energy losses attributed to damping effects, leading to the turntable maintaining a static equilibrium. Conversely, when the heating zone angle exceeds 1.75π, the LCE-rope's time in the heat insulation zone is too brief to revert it to its initial state. Consequently, the thermal energy supplied to the system fails to counterbalance the damping dissipation, culminating in the turntable's persistence in a static equilibrium state. Whereas, for $\theta_0 = 0.5\pi$, $\theta_0 = 0.75\pi$, and $\theta_0 = \pi$, the system can initiate the self-rotation. The effects of the heating zone angle on the rotation amplitude and frequency are displayed in Figure 12b. As the heating zone angle increases, both the amplitude and frequency present a tendency to increase and then decrease, which is similar to the reason mentioned above that both smaller and larger heating zone ranges negatively affect the rotation of the system.

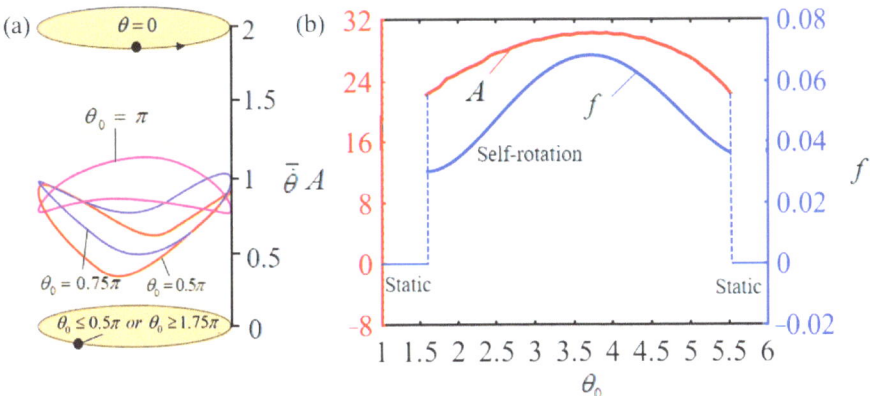

Figure 12. Effect of heating zone angle θ_0 on the self-rotation of the LCE-based turntable system, with $\overline{T}_0 = 0.35$, $\overline{g} = 10$, $\overline{\alpha} = 0.35$, $\overline{\beta} = 0.01$, $\overline{L}_0 = 1$, $\overline{k}_l = 10$, $\overline{k}_s = 10$, $\theta = 0.25\pi$, and $\overline{w}_0 = 1$. (**a**) Limit cycles; (**b**) amplitude and frequency.

In summary, the influence of various critical dimensionless parameters on the rotation amplitude and frequency, as explored in this section, is systematically compiled in Table 3. This data offers essential guidance for the engineering of self-excited motion systems, facilitating the precise control of self-excited motion attributes in real-world applications.

Table 3. Effects of several key dimensionless parameters.

Dimensionless Parameter	Amplitude	Frequency
$\overline{g}$	increases with increasing $\overline{g}$	increases with increasing $\overline{g}$
$\overline{L}_0$	decreases with increasing $\overline{L}_0$	decreases with increasing $\overline{L}_0$
$\overline{\beta}$	not affected by $\overline{\beta}$	decreases with increasing $\overline{\beta}$
$\overline{T}_0$	increases with increasing $\overline{T}_0$	increases with increasing $\overline{T}_0$
$\overline{\alpha}$	increases with increasing $\overline{\alpha}$	not affected by $\overline{\alpha}$
$\overline{k}_l$	increases slightly with increasing $\overline{k}_l$	increases slightly with increasing $\overline{k}_l$
$\overline{k}_s$	decreases with increasing $\overline{k}_s$	decreases with increasing $\overline{k}_s$
θ_0	increases and then decreases with increasing θ_0	increases and then decreases with increasing θ_0

5. Conclusions

Self-rotating systems exhibit the ability to autonomously harness energy and sustain periodic self-rotation under consistent external stimuli. These systems are increasingly relevant for applications in micro-devices, autonomous robotics, sensors, and energy generation. This study introduces a turntable system based on an electrothermally responsive LCE, capable of self-rotation in a steady-state electrical circuit. By employing an electrothermal LCE model, a nonlinear dynamical model of the system is formulated, and numerical simulations are conducted using the four-order Runge–Kutta method. The simulations reveal two distinct motion regimes for the LCE-based turntable in a steady-state circuit: a static state and a self-rotation state. Critical conditions for initiating self-rotation in a two-mass-ball LCE-based turntable system are quantitatively evaluated, considering various system parameters and their impact on rotation amplitude and frequency. Key parameters influencing self-rotation include gravitational acceleration, initial position of the mass ball, elastic properties of the LCE-rope and spring, limiting temperature, heating zone angle range, thermal shrinkage coefficient, and damping factor. These parameters also govern the system's rotation frequency and amplitude. These results of this paper are expected to be validated in future experimental works, and the findings of this research offer innovative design perspectives for self-rotating systems, contributing to the understanding of self-rotation principles and broadening potential applications in areas such as energy harvesting, monitoring, soft robotics, medical devices, and micro- and nano-devices. Additionally, these insights provide valuable references for research and development in related disciplines, fostering technological innovation and advancing practical applications. The deployment of self-rotating systems paves the way for more efficient, sustainable, and autonomous energy conversion and utilization, enhancing the scope and efficacy of their use across various domains.

Author Contributions: The contributions of the authors are as follows: Software, Methodology, and Writing—Original draft, Z.Y.; Conceptualization, Writing—Review and editing, and Supervision, J.L.; Validation, and Writing—Original draft, G.Q.; Validation, Writing—Review and editing, Y.D.; Conceptualization, Writing—Review and editing, K.L. All authors have read and agreed to the published version of the manuscript.

Funding: This study is supported by National Natural Science Foundation of China (Nos. 51608005 and 12172001), University Natural Science Research Project of Anhui Province (Nos. 2022AH030035, 2022AH020029 and KJ2021ZD0066), Outstanding Talents Cultivation Project of Universities in Anhui (No. gxyq2022029), Anhui Provincial Natural Science Foundation (Nos. 2208085Y01 and 2008085QA23) and Housing and Urban-Rural Development Science and Technology Project of Anhui Province (No. 2023-YF129).

Institutional Review Board Statement: Not applicable.

Data Availability Statement: The data that support the findings of this study are available upon reasonable request from the authors.

Conflicts of Interest: The authors declare no conflict of interest.

References

1. Ding, W.J. *Self-Excited Vibration*; Tsing-Hua University Press: Beijing, China, 2009.
2. Sun, Z.; Shuang, F.; Ma, W. Investigations of vibration cutting mechanisms of Ti_6Al_4V alloy. *Int. J. Mech. Sci.* **2018**, *148*, 510–530. [CrossRef]
3. Liu, Z.; Qi, M.; Zhu, Y.; Huang, D.; Yan, X. Mechanical response of the isolated cantilever with a floating potential in steady electrostatic field. *Int J. Mech. Sci.* **2019**, *161*, 105066. [CrossRef]
4. Zhang, Z.; Duan, N.; Lin, C.; Hua, H. Coupled dynamic analysis of a heavily-loaded propulsion shafting system with continu-ous bearing-shaft friction. *Int. J. Mech. Sci.* **2020**, *172*, 105431. [CrossRef]
5. Charroyer, L.; Chiello, O.; Sinou, J.J. Self-excited vibrations of a nonsmooth contact dynamical system with planar friction based on the shooting method. *Int. J. Mech. Sci.* **2018**, *144*, 90–101. [CrossRef]

6. Hu, W.; Lum, G.Z.; Mastrangeli, M.; Sitti, M. Small-scale soft-bodied robot with multimodal locomotion. *Nature* **2018**, *554*, 81–85. [CrossRef]
7. Iliuk, I.; Balthazar, J.M.; Tusset, A.M.; Piqueira, J.R.; Pontes, D.B.R.; Felix, J.L.; Bueno, A.M. Application of passive control to energy harvester efficiency using a nonideal portal frame structural support system. *J. Intell. Mater. Syst. Struct.* **2014**, *25*, 417–429. [CrossRef]
8. Korner, K.; Kuenstler, A.S.; Hayward, R.C.; Audoly, B.; Bhattacharya, K. A nonlinear beam model of photomotile structures. *Proc. Natl. Acad. Sci. USA* **2020**, *117*, 9762–9770. [CrossRef]
9. Martella, D.; Nocentini, S.C.; Parmeggiani, C.; Wiersma, D.S. Self-regulating capabilities in photonic robotics. *Adv. Mater. Technol.* **2019**, *4*, 1800571. [CrossRef]
10. Sangwan, V.; Taneja, A.; Mukherjee, S. Design of a robust self-excited biped walking mechanism. *Mech. Theory* **2004**, *39*, 1385–1397. [CrossRef]
11. Serak, S.; Tabiryan, N.; Vergara, R.; White, T.J.; Vaia, R.A.; Bunning, T.J. Liquid crystalline polymer cantilever oscillators fueled by light. *Soft Matter* **2010**, *6*, 779–783. [CrossRef]
12. Chen, Y.; Zhao, H.; Mao, J.; Chirarattananon, P.; Helbling, E.F.; Hyun, N.P.; Clarke, D.R.; Wood, R.J. Controlled flight of a microrobot powered by soft artificial muscles. *Nature* **2019**, *575*, 324–329. [CrossRef] [PubMed]
13. Wie, J.J.; Shankar, M.R.; White, T.J. Photomotility of polymers. *Nat. Commun.* **2016**, *7*, 13260. [CrossRef]
14. Maeda, S.; Hara, Y.; Sakai, T.; Yoshida, R.; Hashimoto, S. Self-walking gel. *Adv. Mater.* **2007**, *19*, 3480–3484. [CrossRef]
15. Jenkins, A. Self-oscillation. *Phys. Rep.* **2013**, *525*, 167–222. [CrossRef]
16. Baumann, A.; Sánchez-Ferrer, A.; Jacomine, L.; Martinoty, P.; le Houerou, V.; Ziebert, F.; Kulić, I.M. Motorizing fibres with geometric zero-energy modes. *Nat. Mater.* **2018**, *17*, 523–527. [CrossRef] [PubMed]
17. Dawson, N.J.; Kuzyk, M.G.; Neal, J.; Luchette, P.; Palffy-Muhoray, P. Cascading of liquid crystal elastomer photomechanical optical devices. *Opt. Commun.* **2011**, *284*, 991–993. [CrossRef]
18. Kim, Y.; Tamate, R.; Akimoto, A.; Yoshida, R. Recent developments in self-oscillating polymeric systems as smart materials: From polymers to bulk hydrogels. *Mater. Horiz.* **2017**, *4*, 38–54. [CrossRef]
19. Chakrabarti, A.; Choi, G.P.T.; Mahadevan, L. Self-Excited Motions of Volatile Drops on Swellable Sheets. *Phys. Rev. Lett.* **2020**, *124*, 258002. [CrossRef]
20. Cunha, M.; Peeketi, A.R.; Ramgopal, A.; Annabattula, R.K.; Schenning, A. Light-driven continual oscillatory rocking of a polymer film. *Chem. Open* **2020**, *9*, 1149–1152.
21. Cheng, Y.; Lu, H.; Lee, X.; Zeng, H.; Priimagi, A. Kirigami-based light-induced shape-morphing and locomotion. *Adv. Mater.* **2019**, *32*, 1906233. [CrossRef]
22. Lv, X.; Yu, M.; Wang, W.; Yu, H. Photothermal pneumatic wheel with high loadbearing capacity. *Compos. Commun.* **2021**, *24*, 100651. [CrossRef]
23. Boissonade, J.; Kepper, P.D. Multiple types of spatio-temporal oscillations induced by differential diffusion in the Landolt reaction. *Phys. Chem. Chem. Phys.* **2011**, *13*, 4132–4137. [CrossRef] [PubMed]
24. Gelebart, A.H.; Mulder, D.J.; Varga, M.; Konya, A.; Vantomme, G.; Meijer, E.W.; Selinger, R.S.; Broer, D.J. Making waves in a photoactive polymer film. *Nature* **2017**, *546*, 632–636. [CrossRef] [PubMed]
25. Deng, H.; Zhang, C.; Su, J.; Xie, Y.; Zhang, C.; Lin, J. Bioinspired multi-responsive soft actuators controlled by laser tailored graphene structures. *J. Mater. Chem. B* **2018**, *6*, 5415–5423. [CrossRef]
26. Yoshida, R. Self-oscillating gels driven by the Belousov-Zhabotinsky reaction as novel smart materials. *Adv. Mater.* **2010**, *22*, 3463–3483. [CrossRef]
27. Yashin, V.V.; Balazs, A.C. Pattern formation and shape changes in self-oscillating polymer gels. *Science* **2006**, *314*, 798–801. [CrossRef] [PubMed]
28. He, Q.G.; Wang, Z.J.; Wang, Y.; Wang, Z.J.; Li, C.H.; Annapooranan, R.; Zeng, J.; Chen, R.K.; Cai, S. Electrospun liquid crystal elastomer microfiber actuator. *Sci. Robot.* **2021**, *6*, eabi9704. [CrossRef] [PubMed]
29. Wang, Y.; Liu, J.; Yang, S. Multi-functional liquid crystal elastomer composites. *Appl. Phys. Rev.* **2022**, *9*, 011301. [CrossRef]
30. Zhou, L.; Dai, Y.; Fang, J.; Li, K. Light-powered self-oscillation in liquid crystal elastomer auxetic metamaterials with large volume change. *Int. J. Mech. Sci.* **2023**, *254*, 108423. [CrossRef]
31. Rešetič, A.; Milavec, J.; Domenici, V.; Zupančič, B.; Bubnov, A.; Zalar, B. Deuteron NMR investigation on orientational order parameter in polymer dispersed liquid crystal elastomers. *Phys. Chem. Chem. Phys.* **2020**, *22*, 23064–23072. [CrossRef]
32. Park, S.; Oh, Y.; Moon, J.; Chung, H. Recent Trends in Continuum Modeling of Liquid Crystal Networks: A Mini-Review. *Polymers* **2023**, *15*, 1904. [CrossRef]
33. Xu, P.; Wu, H.; Dai, Y.; Li, K. Self-sustained chaotic floating of a liquid crystal elastomer balloon under steady illumination. *Heliyon* **2023**, *9*, e14447. [CrossRef] [PubMed]
34. Ge, D.; Dai, Y.; Li, K. Self-Sustained Euler Buckling of an Optically Responsive Rod with Different Boundary Constraints. *Polymers* **2023**, *15*, 316. [CrossRef]
35. Hu, Z.; Li, Y.; Lv, J. Phototunable self-oscillating system driven by a self-winding fiber actuator. *Nat. Commun.* **2021**, *12*, 3211. [CrossRef] [PubMed]

36. Zhao, Y.; Chi, Y.; Hong, Y.; Li, Y.; Yang, S.; Yin, J. Twisting for soft intelligent autonomous robot in unstructured environments. *Proc. Natl. Acad. Sci. USA* **2022**, *119*, e2200265119. [CrossRef]
37. Graeber, G.; Regulagadda, K.; Hodel, P.; Küttel, C.; Landolf, D.; Schutzius, T.; Poulikakos, D. Leidenfrost droplet trampolining. *Nat. Commun.* **2021**, *12*, 1727. [CrossRef]
38. Tang, R.; Liu, Z.; Xu, D.; Liu, J.; Yu, L.; Yu, H. Optical pendulum generator based on photomechanical liquid-crystalline actuators. *ACS Appl. Mater. Interfaces* **2015**, *7*, 8393–8397. [CrossRef] [PubMed]
39. Xu, T.; Pei, D.; Yu, S.; Zhang, X.; Yi, M.; Li, C. Design of MXene compo-sites with biomimetic rapid and Self-oscillating actuation under ambient circumstances. *ACS Appl. Mater. Interfaces* **2021**, *13*, 31978–31985. [CrossRef]
40. Lahikainen, M.; Zeng, H.; Priimagi, A. Reconfigurable photoactuator through synergistic use of photochemical and photo thermal effects. *Nat. Commun.* **2018**, *9*, 4148. [CrossRef]
41. Kim, Y.; Berg, J.; Crosby, A.J. Autonomous snapping and jumping polymer gels. *Nat. Mater.* **2021**, *20*, 1695–1701. [CrossRef]
42. Ge, D.; Dai, Y.; Li, K. Light-powered Self-spinning of a button spinner. *Int. J. Mech. Sci.* **2023**, *238*, 107824. [CrossRef]
43. Li, K.; Zhang, B.; Cheng, Q.; Dai, Y.; Yu, Y. Light-Fueled Synchronization of Two Coupled Liquid Crystal Elastomer Self-Oscillators. *Polymers* **2023**, *15*, 2886. [CrossRef] [PubMed]
44. Yu, Y.; Li, L.; Liu, E.; Han, X.; Wang, J.; Xie, Y.; Lu, C. Light-driven core-shell fiber actuator based on carbon nanotubes/liquid crystal elastomer for artificial muscle and phototropic locomotion. *Carbon* **2022**, *187*, 97–107. [CrossRef]
45. Li, M.E.; Lv, S.; Zhou, J.X. Photo-thermo-mechanically actuated bending and snapping kinetics of liquid crystal elastomer cantilever. *Smart Mater. Struct.* **2014**, *23*, 125012. [CrossRef]
46. Zeng, H.; Lahikainen, M.; Liu, L.; Ahmed, Z.; Wani, O.M.; Wang, M.; Yang, H.; Priimagi, A. Light-fuelled freestyle Self-oscillators. *Nat. Commun.* **2019**, *10*, 5057. [CrossRef]
47. Chun, S.; Pang, C.; Cho, S.B. A micropillar-assisted versatile strategy for highly sensitive and efficient triboelectric energy generation under in-plane stimuli. *Adv. Mater.* **2020**, *32*, 1905539. [CrossRef]
48. Marshall, J.E.; Terentjev, E.M. Photo-sensitivity of dye-doped liquid crystal elastomers. *Soft Matter* **2013**, *9*, 8547–8551. [CrossRef]
49. He, Q.; Wang, Z.; Wang, Y.; Minori, A.; Tolley, M.T.; Cai, S. Electrically controlled liquid crystal elastomer–based soft tubular actuator with multimodal actuation. *Sci. Adv.* **2019**, *5*, eaax5746. [CrossRef]
50. Liao, B.; Zang, H.; Chen, M.; Wang, Y.; Lang, X.; Zhu, N.; Yang, Z.; Yi, Y. Soft rod-climbing robot inspired by winding locomotion of snake. *Soft Robot.* **2020**, *7*, 500–511. [CrossRef]
51. Haberl, J.M.; Sanchez-Ferrer, A.; Mihut, A.M.; Dietsch, H.; Hirt, A.M.; Mezzenga, R. Liquid-crystalline elastomer-nanoparticle hybrids with reversible switch of magnetic memory. *Adv. Mater.* **2013**, *25*, 1787–1791. [CrossRef]
52. Li, M.H.; Keller, P.; Li, B.; Wang, X.; Brunet, M. Light-driven side-on nematic elastomer actuators. *Adv. Mater.* **2003**, *15*, 569–572. [CrossRef]
53. Qian, X.; Chen, Q.; Yang, Y.; Xu, Y.; Li, Z.; Wang, Z.; Wu, Y.; Wei, Y.; Ji, Y. Untethered recyclable tubular actuators with versatile locomotion for soft continuum robots. *Adv. Mater.* **2018**, *30*, 1801103. [CrossRef] [PubMed]
54. Corbett, D.; Warner, M. Linear and nonlinear photoinduced deformations of can-tilevers. *Phys. Rev. Lett.* **2007**, *99*, 174302. [CrossRef]
55. Palagi, S.; Mark, A.G.; Reigh, S.Y.; Melde, K.; Qiu, T.; Zeng, H.; Parmeggiani, C.; Martella, D.; Sanchez-Castillo, A.; Kapernaum, N.; et al. Structured light enables biomimetic swimming and versatile locomotion of photoresponsive soft microrobots. *Nat. Mater.* **2016**, *15*, 647–653. [CrossRef] [PubMed]
56. Tian, H.; Wang, Z.; Chen, Y.; Shao, J.; Gao, T.; Cai, S. Polydopamine-coated main-chain liquid crystal elastomer as optically driven artificial muscle. *ACS Appl. Mater. Interfaces* **2018**, *10*, 8307–8316. [CrossRef]
57. Liu, L.; Liu, M.-H.; Deng, L.-L.; Lin, B.-P.; Yang, H. Near-infrared chromophore functionalized soft actuator with ultrafast photo responsive speed and superior mechanical property. *J. Am. Chem. Soc.* **2017**, *139*, 11333–11336. [CrossRef]
58. Pei, Z.; Yang, Y.; Chen, Q.; Wei, Y.; Ji, Y. Regional shape control of strategically assembled multishape memory vitrimers. *Adv. Mater.* **2016**, *28*, 156–160. [CrossRef] [PubMed]
59. Bisoyi, H.K.; Urbas, A.M.; Li, Q. Soft materials driven by photo thermal effect and their applications. *Adv. Opt. Mater.* **2018**, *6*, 1800458. [CrossRef]
60. Schuhladen, S.; Preller, F.; Rix, R.; Petsch, S.; Zentel, R.; Zappe, H. Iris-like tunable aperture employing liquid-crystal elastomers. *Adv. Mater.* **2014**, *26*, 7247–7251. [CrossRef]
61. Petsch, S.; Rix, R.; Reith, P.; Khatri, B.; Schuhladen, S.; Ruh, D.; Zappe, H. A thermotropic liquid crystal elastomer micro-actuator with integrated deformable micro-heater. In Proceedings of the 2014 IEEE 27th International Conference on Micro Electro Mechanical Systems (MEMS), San Francisco, CA, USA, 26–30 January 2014; pp. 905–908.
62. Wang, C.; Sim, K.; Chen, J.; Kim, H.; Rao, Z.; Li, Y.; Hen, W.; Song, J.; Verduzco, R.; Yu, C. Soft ultrathin electronics innervated adaptive fully soft robots. *Adv. Mater.* **2018**, *30*, 1706695. [CrossRef]
63. Liu, Y.; Zhao, J.; Wu, H.; Dai, Y.; Li, K. Self-Oscillating Curling of a Liquid Crystal Elastomer Beam under Steady Light. *Polymers* **2023**, *15*, 344. [CrossRef] [PubMed]
64. Li, K.; Du, C.; He, Q.; Cai, S.Q. Thermally driven self-oscillation of an elastomer fiber with a hanging weight. *Extrem. Mech. Lett.* **2021**, *50*, 101547. [CrossRef]

65. Braun, L.B.; Hessberger, T.; Pütz, E.; Müller, C.; Giesselmann, F.; Serra, C.A.; Zentel, R. Actuating thermo- and photo-responsive tubes from liquid crystalline elastomers. *J. Mater. Chem. C* **2018**, *6*, 9093–9101. [CrossRef]
66. Camacho-Lopez, M.; Finkelmann, H.; Palffy-Muhoray, P.; Shelley, M. Fast liquid-crystal elastomer swims into the dark. *Nat. Mater.* **2004**, *3*, 307–310. [CrossRef]

Disclaimer/Publisher's Note: The statements, opinions and data contained in all publications are solely those of the individual author(s) and contributor(s) and not of MDPI and/or the editor(s). MDPI and/or the editor(s) disclaim responsibility for any injury to people or property resulting from any ideas, methods, instructions or products referred to in the content.

Article

Flexible Actuators Based on Conductive Polymer Ionogels and Their Electromechanical Modeling

Jiawei Xu [1], Hongwei Hu [1,*], Shengtao Zhang [1], Guanggui Cheng [1] and Jianning Ding [1,2,*]

1. School of Mechanical Engineering, Jiangsu University, Zhenjiang 212000, China
2. Technological Institute of Carbon Neutralization, School of Mechanical Engineering, Yangzhou University, Yangzhou 225000, China
* Correspondence: hwhu@ujs.edu.cn (H.H.); dingjn@ujs.edu.cn (J.D.)

Abstract: High-performance flexible actuators, integral components of soft robotics, hold promise for advancing applications in safe human–robot interactions, healthcare, and various other fields. Notable among these actuators are flexible electrochemical systems, recognized for their merits in low-voltage manipulation, rapid response speed, and cost-effectiveness. However, the optimization of output strain, response speed, and stability presents a significant challenge in this domain. Despite the application of diverse electrochemically active materials to enhance actuation performance, a critical need persists for corresponding electrical-mechanical models to comprehensively grasp actuation mechanisms. In this study, we introduce a novel electrochemical actuator that utilizes conductive polymer ionogel as active electrodes. This ionogel exhibits exceptional properties, including high conductivity, flexibility, and electrochemical activity. Our electrochemical actuators exhibit noteworthy bending strain capabilities and rapid response rates, achieving frequencies up to 10 Hz at a modest voltage of 1 V. An analytical model integrating ion migration and dynamic processes has been established to elucidate actuator behavior. Simulation results highlight that electrodes characterized by low resistance and high capacitance are optimal for simultaneous enhancement of bending strain and blocking force. However, the augmentation of Young's modulus, while increasing blocking force, compromises bending strain. Furthermore, a larger aspect ratio proves beneficial for unidirectional stress output, leading to increased bending strain, while actuator blocking force diminishes with greater length. These findings underscore the intricate interplay between material properties and dimensions in optimizing the performance of flexible electrochemical actuators. This work provides important practical and theoretical guidance for the manufacture of high-performance flexible actuators and the search for new smart materials.

Keywords: flexible actuator; electromechanical model; conductive polymer; drive travel; load capacity

Citation: Xu, J.; Hu, H.; Zhang, S.; Cheng, G.; Ding, J. Flexible Actuators Based on Conductive Polymer Ionogels and Their Electromechanical Modeling. *Polymers* **2023**, *15*, 4482. https://doi.org/10.3390/polym15234482

Academic Editors: Md Najib Alam and Vineet Kumar

Received: 24 October 2023
Revised: 16 November 2023
Accepted: 16 November 2023
Published: 22 November 2023

Copyright: © 2023 by the authors. Licensee MDPI, Basel, Switzerland. This article is an open access article distributed under the terms and conditions of the Creative Commons Attribution (CC BY) license (https://creativecommons.org/licenses/by/4.0/).

1. Introduction

Biology has long been an inspiration for engineers to build more capable machines. By emulating the adaptability and agility of soft-bodied animals and muscles, soft actuators have emerged with diverse applications encompassing soft grippers, artificial muscles, and smart medical rehabilitation devices [1–3]. In this domain, electrochemical actuators assert themselves as formidable contenders due to their inherent conformable and elastic nature, cost-effectiveness, low operational voltage, and air-working capabilities [4]. Notably, extensive research has focused on ionic polymer-metal composites (IPMCs) over the past few decades; however, their practicality was hindered by the reliance on aqueous electrolytes [5,6]. Fukushima et al. introduced a dry-type bucky-gel actuator that utilizes carbon nanotubes (CNTs) as electrodes and embeds an ionic liquid within a polymer matrix as the electrolyte [7,8]. Ionic liquid-based actuators offer the advantage of operating in air with prolonged stability, attributed to the non-volatile nature of ionic liquids. Subsequent research has delved into the utilization of various materials, such as graphene, graphdiyne, and two-dimensional nanomaterials, as active electrodes for ionic liquid-based airworking electrochemical actuators [9–11]. Notably, these materials demonstrate elevated electrochemical activities, leading to a substantial enhancement in actuation performance. However, the utilization of these electrochemically active electrode materials typically involves the compounding of these materials with polymers to achieve stable and flexible electrodes, which adds complexity to the manufacturing process, with potential implications for factors like production efficiency and cost. The pursuit of exceptional figures of merit, encompassing actuation strain, response speed, and output force, remains an ongoing challenge.

Conductive polymers (CPs) have earned significant utilization as electrodes for electrochemical energy storage, owing to their heightened electrochemical activity and electronic conductivity—two pivotal attributes for enabling robust electrochemical actuation [12,13]. Ionogels have garnered significant interest in the realm of flexible electronics by combining solid-state networks with ionic liquids. This innovative approach not only offers outstanding mechanical properties and conductivity to address the challenges in flexible electronics but also brings another distinctive feature: high electrochemical activity for electrochemical actuators. Previously, bucky-gel type ionogels, comprising carbon nanotubes (CNT) and ionic liquids, have demonstrated remarkable electrochemical actuation capabilities. The incorporation of conductive fillers in ionogels has proven effective in elevating electrochemical activities and augmenting actuation performance. Consequently, the exploration of conductive polymer-based ionogels for electrochemical actuation presents a promising avenue for advancing the current capabilities of conductive polymer-based electrochemical actuators.

In this study, we introduce CP ionogels characterized by hierarchical layered structures that integrate ionic liquids into polymer matrixes. These ionogels exhibit superior electrical conductivity (500 S/cm) and electrochemical activity. Electrochemical actuators built upon CP ionogels demonstrate remarkable strain capabilities (1.2%) and rapid response rates (up to 10 Hz) while operating at a modest voltage of 1 V. Furthermore, we have established an electromechanical model for the ionogel actuator, closely aligning the electrochemical response with mechanical properties flexible actuators. Leveraging this model, we systematically scrutinized crucial parameters influencing actuator performance, encompassing electrode conductivity, Young's modulus, capacitive characteristics, and actuator dimensions. These results indicate that the optimal characteristics of electrodes featuring low resistance and high capacitance facilitate a concurrent improvement in bending strain and blocking force. However, the elevation of Young's modulus, while bolstering blocking force, comes at the expense of bending strain. Additionally, a larger aspect ratio proves advantageous for unidirectional stress output, leading to amplified bending strain, whereas actuator blocking force undergoes a reduction with increased length. This comprehensive investigation offers insight into the intricate interplay among these physical factors and their influence on actuator displacement and blocking force. This research forms an essential foundation, offering significant guidance for the ensuing design and optimization of novel ionogel-based actuators.

2. Experimental and Simulation Method
2.1. Design and Operating Principle

The fabricated electrochemical actuators consist of three layers: two electroactive layers (conductive polymer ionogel films), which serve as both electrodes and actuation layers, with an electrolyte gel layer in between. When the two electrodes are connected to drive voltages, cations, and anions migrate within the electrolyte and ion gel electrodes due to internal electric fields, leading to the swelling of one conductive polymer ionogel and the contraction of the other. Different deformations of the two electroactive layers produce strain mismatch, which has been studied in our previous work [14,15]. This strain mismatch mechanism causes the actuator to convert strain energy into deformation, resulting in the trilayer actuator bending in one direction.

2.2. Materials

Ionic liquids including EMIMTFSI (1-Ethyl-3-Methylimidazolium Bis(Trifluoromethylsulfonyl)Imide and BMImOTs 1-butyl-3-methylimidazolium tosylate were purchased from Greenchem ILs, LICP, CAS, China (Lanzhou, China); PEDOT:PSS aqueous solution (solid content 1.0–1.3 wt%, PEDOT to PSS ratio is 1:2.5, OE800) was purchased from Shanghai Ouyi Organic Optoelectronic Materials Co., LTD., Shanghia, China

2.3. Fabrication of Flexible Actuator Based on the CP Ionogel

CP film was fabricated by casting PEDOT:PSS dispersion, drying, and annealing. CP ionogel was then formed by treating the CP film in an ionic liquid. Specifically, ionic additives (BMImOTs, 15 wt%) were added to the PEDOT:PSS solution. After stirring for 1 h and sonication for 15 min, the solution was poured into a Teflon mold and left to stand at room temperature for 24 h. It was then transferred to an electric oven treated at 60 °C for 2 h and 150 °C for 30 min. After cooling to room temperature, the film was peeled off and subsequently immersed in ethanol and DI water. The ionogel was prepared by immersing the swelled film in ionic liquid (EMIMTFSI) and placed at 80 °C for 4 h. The film was then removed and dried in a vacuum oven overnight. Two ionogel electrode films were laminated on a cellulose membrane film that was pre-soaked by ionic liquid (EMIMTFSI). They were then pressed together using two glass slides and placed at 60 °C for 20 min. Afterward, the trilayer was carefully peeled off from the glass slides. All the actuators were cut into long strips with 5 mm width and 26 mm length.

2.4. Characterization and Measurement

The electronic conductivity of the films was tested using a four-point probe station (HPS2523, HELPASS Electronic Technologies Ltd., Changzhou, China). The strain-stress curve was performed on a mechanical tester (QT-6203S, Qiantong Instrument Equipment Co., Ltd., Suzhou, China). Cross-sectional SEM images were shot on a JEOL 7800F (JEOL Ltd., Akishima, Japan). The performance of the actuators was measured using a dual channel source meter (Keithley 2602B,Tektronix Inc., West Chester, OH, USA) for the input power and a laser displacement meter to record the displacement. One end of the actuator was fixed by a Kelvin clamp with two platinum plates on the contact area, and the swing of the other end was recorded by the laser displacement meter. The strain (ε) of the actuator can be calculated by the following equation:

$$\delta = \frac{2d\theta}{L^2 + \delta^2}$$

where δ, d, and L are the tip displacement, the thickness, and the beam length, respectively. The blocking force was measured using a high-precision pressure transducer (Futek LSB200, FUTEK Advanced Sensor Technology, Inc., Irvine, CA, USA) with the probe attached horizontally to the tip of the actuator beam.

2.5. Simulation Method

To further analyze the mechanical behavior of the CP ionogel actuator and identify the key parameters affecting the actuation performance, it is necessary to establish an analytical model containing the necessary physical quantities [16]. We use a Multistage parallel equivalent circuit to simplify the electrical part of the actuator. A single-ended fixed beam model simplifies the statics part, and a pole-zero model is used to couple the electrical and mechanical parts. The parameters of the Pole-zero model are identified using experimental data. Finally, a gray-box model about the control voltage u and the displacement of end position δ with certain physical parameters that can reflect the electrical and mechanical properties of the system is obtained. To establish the analytical model of the actuator, some appropriate assumptions are made as follows:

1. Since the surface resistance of electrodes is very small ($R_s \approx 14\ \Omega$), it is assumed that the resistances in each stage of the RC circuit are equal: $R_n = R_s/n$.

2. Since the thickness of the capacitor layer is very small (0.14 mm), the micro-element capacity of the capacitor in the radial direction is equal.
3. The bending of the actuator is uniform along the whole beam.
4. The influence of bending on the electrical properties of the capacitor layer is negligible.
5. The driving force comes from the ionogel electrode layer with a uniform stress distribution.

3. Results and Discussion

3.1. Driving Characteristics of the Flexible Actuator

The flexible actuator exhibits a sandwich structure consisting of two conducting polymer ion gel electrodes and a quasi-solid ionic liquid electrolyte membrane in the middle (Figure 1a). CP ionogel membranes were prepared from PEDOT:PSS aqueous dispersions with ionic additives to improve membrane quality, including electrical conductivity, electrochemical activity, and mechanical properties, which are all important for driving performance. The introduction of ionic additives into the PEDOT:PSS dispersion can reduce the electrostatic force between PEDOT and PSS, thereby rearranging the molecular chains, thereby improving the interchain interactions during solvent evaporation and recrystallization, forming a molecular chain network structure [17]. After annealing, the PEDOT:PSS film has a layered nano-stacked structure, as shown in the SEM image of Figure 1b, which also facilitates the intercalation of ionic liquids. When immersed in ionic liquids, the films with layered structures can absorb 15 wt% of ionic liquids, forming soft conducting polymer ionogel (CP ionogel, Figure 1c,d).

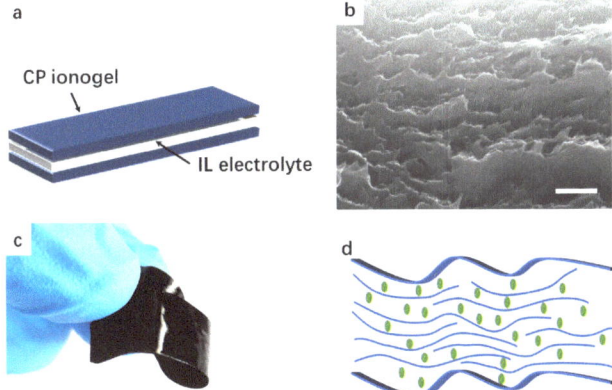

Figure 1. (a) structure of the soft actuator based on CP ionogel; (b) SEM image of the microstructure of CP film (scale bar: 500 nm); (c) photo of the CP ionogel; (d) illustration of the microstructure of the CP ionogel, where green dots represent the ionic liquid.

Ionic liquids in CP ionogels not only effectively reduce Young's modulus of PEDOT:PSS film and render it stretchable but also increase the conductivity. In tensile tests (Figure 2), the dry CP film breaks at 4.6% strain with a high Young's modulus of 700 MPa, while the CP ionogel can be stretched up to 40% with a tensile strength of 20.8 MPa and a low Young's modulus of 217 MPa, which is due to the plasticizing effect of the ionic liquid. Furthermore, the electrical conductivity of the conducting polymer ionogel can reach 500 S/cm, compared with 20 S/cm for dry CP film. The enhanced conductivity can be attributed to the doping effect from the presence of ionic liquid in the ionogels [17].

The soft actuator, consisting of two CP ionogels and a quasi-solid electrolyte membrane, acts like a supercapacitor, storing charge during operation. Previous studies have shown that CP films with polymeric counterions, such as PSS, are mainly driven by cationic motion [18]. Therefore, the introduction of cations on the negative side causes the film to swell, while the removal of cations on the positive side causes the film to shrink (Figure 3a).

The cyclic voltammetric characters of the actuator were examined at different scan rates (Figure 3b). At a slow sweep speed of 10 mV/s, the CV curve is approximately rectangular, indicating that the actuator has a double-layer charge storage characteristic similar to that of a supercapacitor. As the sweep speed increases, the capacitance decreases, and the CV curve gradually becomes conical at the fast sweep speed, indicating that the ion migration rate is the main reason for the capacitance decrease at the fast sweep speed.

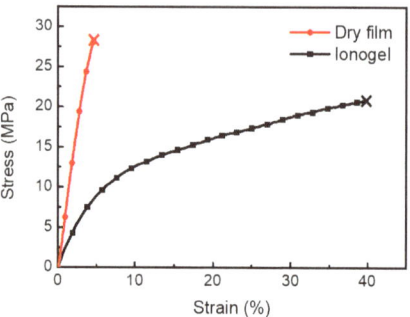

Figure 2. Tensile stress–strain curve of the CP dry and ionogel films.

We tested the actuators operated at varied frequencies from 0.1 Hz to 10 Hz with a fixed amplitude of 1 V (Figure 3c). The maximum strain difference at 0.1 Hz reaches 1.2%. By increasing the applied voltage frequency, the maximum strain decreased, revealing that the slow kinetics of the electrochemical process limited the actuation at a higher frequency. It is worth noting that the strain of CP ionogel actuators is among the highest values from reported works on CP-based actuators (Figure 3d). Furthermore, by comparing with other types of actuators, we found that CP ionogel actuators can provide strain values that other actuators can only achieve at higher driving voltages, which suggests that CP ionogel actuators are promising for flexible wearable devices requiring low energy consumption.

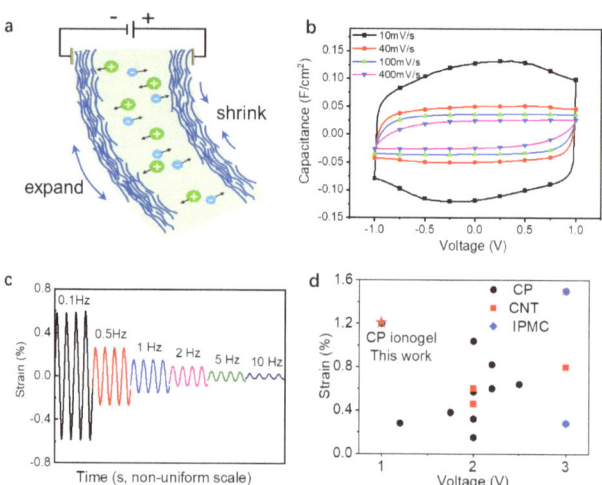

Figure 3. Actuation performance of the CP ionogel actuator. (**a**) illustration of the actuation mechanism of CP ionogel actuator; (**b**) CV curves of the actuator at different scan rates; (**c**) strain of the actuator under different frequencies of triangular voltage between ±1 V; (**d**) comparison of the strain with reported works [19–28].

3.2. Modeling

The physical structure diagram of the ionic actuator is shown in Figure 4. A section of the actuator is clamped by gold electrodes and keeps the whole actuator hanging freely. The specific meaning of each physical quantity is detailed in Table 1.

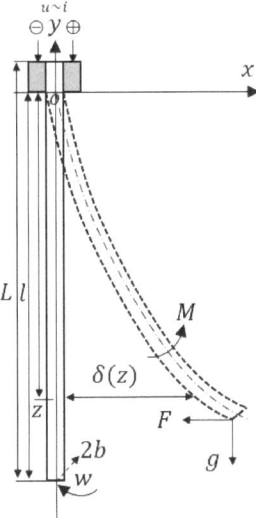

Figure 4. Physical parameters of the actuator for modeling.

Table 1. Model Parameters.

Parameters	Meanings
L	Total Length
l	Free Length
z	Observation Point
F	Blocking Force
$\delta(z)$	Displacement of Observation point
w	Width
b	Half Thickness
u	Control Voltage
i	Current
M	Induced Bending Moment caused by Ionic Migration
g	Gravity

In the electrical part, the electrical model of the actuator is established by simulating the actuator using a simplified multi-level equivalent RC circuit [29] (Figure 5), where Rn/2 is the sheet resistance of each unit in the single-layer ionogel, and Ci is the capacitance of each unit in the electrolyte layer. Through the physical meaning of resistance and capacitance, we can easily obtain

$$R_i = \frac{2\rho_i b}{Lw}, C_i = \frac{\varepsilon_i Lw}{2b} \tag{1}$$

where ρ_i represents the resistivity of each micro-element circuit resistance, and ε_i represents the dielectric constant of the capacitance of each micro-element circuit. The complete state space model of the equivalent circuit is given as follows:

$$\begin{cases} C_1 \dot{u}_1(t) = \dfrac{u(t) - u_1(t)}{R_1} - \dfrac{u_1(t) - u_2(t)}{R_2} \\ C_2 \dot{u}_2(t) = \dfrac{u_1(t) - u_2(t)}{R_2} - \dfrac{u_2(t) - u_3(t)}{R_3} \\ \quad \vdots \\ C_{n-1} \dot{u}_{n-1} = \dfrac{u_{n-2}(t) - u_{n-1}(t)}{R_{n-1}} - \dfrac{u_{n-1}(t) - u_n(t)}{R_n} \\ C_n \dot{u}_n(t) = \dfrac{u_{n-1} - u_n(t)}{R_n} - \dfrac{u_n(t)}{R} \end{cases} \quad (2)$$

where u_n is the voltage on each RC shunt, this state space model is used to describe the charge transfer process in the channels, all the mathematical derivation in this paper take place in Maple (2022 version).

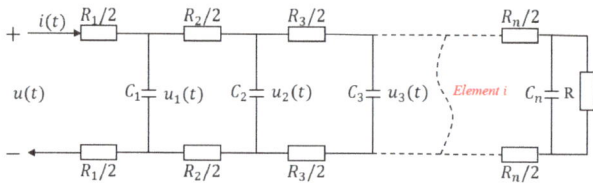

Figure 5. The equivalent electrical circuit model for the actuator.

The number of impedance elements will determine the available degrees of freedom to describe the dynamics of the electrical model [30]. In general, as the number of impedance elements increases, the accuracy of the model will increase. However, this will result in increasingly complex calculations. When the ladder circuit operates at low frequencies, it can be reduced to a suitably low order to achieve a given accuracy [31]. To balance the accuracy and complexity, the second-order link is used as an example for calculation. The transfer function between its charge Q and the input voltage u can be written as follows, where the variable 's' represents the differential operator in the s-domain:

$$G_{Qu}(s) = \frac{RR_1 C_1 C_2 s^2 + (RC_1 + R_2 C_1 + RC_2)s}{RR_1 R_2 C_1 C_2 s^2 + (R_1 R_2 C_1 + RR_1 C_1 + RR_1 C_2 + RR_2 C_2)s + (R_1 + R_2 + R)} \quad (3)$$

The migration of ions driven by the electric field causes the electrode on one side of the actuator to expand and the electrode on the other side to contract. Obviously, if the stress due to volume expansion is assumed to be equal to the stress due to volume reduction, then the entire multilayer model can be simplified to a single layer for the mechanical analysis. Therefore, based on the mechanical model shown in Figure 6, the transfer function of the moment $M(s)$ generated by the induced stress can be written as

$$\begin{aligned} M(s) &= 2 \int_{b_1}^{b} \sigma_c(s) y w \, dy, \\ &= w \sigma_c(s) \left(b^2 - b_1^2 \right), \end{aligned} \quad (4)$$

where $\sigma_c(s)$ is the induced stress caused by the volume change, and w, b, and b_1 are the width of the membrane, half of the overall thickness of the actuator, and the thickness of the electrode layer, respectively. The induced stress arises from the transformation of the volume, so the volume stress is used here to calculate the deformation due to charge transfer.

The electromechanical coupler is introduced to couple the change of charge amount with the induced stress. Here, we choose the pole-zero model, which is shown in function (6).

$$\sigma_c(s) = \frac{Q(s)d(s)}{V} \tag{5}$$

where $V = Lw(b - b_1)$ and L is the overall length of the actuator. $d(s)$ is the electromechanical coupling equation.

$$d(s) = \frac{k \prod_{j=1}^{m}(s + z_j)}{\prod_{i=1}^{n}(s + p_i)} \tag{6}$$

where k, m, n, z_j, and p_i are the parameters to be tuned. Here, k is the gain of the system, m and n are the number of poles and zeros, and z_j, and p_i are the positions of the zeros and poles to be tuned in the frequency domain. Substituting Equation (5) into Equation (4), we obtain

$$M(s) = \frac{(b + b_1)Q(s)d(s)}{L} \tag{7}$$

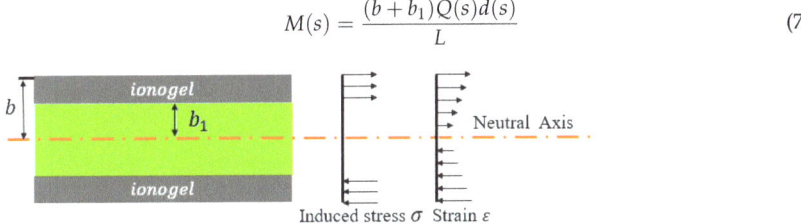

Figure 6. The mechanical model of the actuator.

Through the mechanical balance of the actuator, it is not difficult to obtain the following relationship:

$$\sum_i = M(s) - (F + g)l - \int_b^{-b} \sigma_g(s,x)wxdx = 0 \tag{8}$$

where $\sigma_g(s,x)$ is the internal stress existing against elastic deformation and $(F + g)l$ is the equivalent concentrated stress on the end of the actuator and the moment generated by gravity. The internal stress can be written as:

$$\sigma_g(s,x) = E\kappa(s)x \tag{9}$$

where E is Young's modulus, and κ is the bending curvature of the actuator. Since the weight of the actuator is very light, and there is no blocking force at the end, we set $(F + g)l = 0$ at this time to simplify the calculation. Therefore, we obtain

$$M(s) = \int_b^{-b} \sigma_g(s,x)wxdx$$
$$= \frac{2}{3}Ewb^3\kappa(s) \tag{10}$$

Putting Equation (7) into Equation (10), we obtain

$$\kappa(s) = \frac{3(b + b_1)Q(s)d(s)}{2Ewlb^3} \tag{11}$$

Therefore, the axial displacement can be approximated as follows:

$$\delta(z,s) = \frac{1 - \cos(z\kappa(s))}{\kappa(s)} \tag{12}$$

where z is the observed inflection point.

Here we approximate $\delta(z,s)$ by Taylor series expansions of κ, z at the zero point:

$$\delta(z,s) \approx \frac{1}{2}z^2\kappa(s) \tag{13}$$

Finally, combining Equations (3), (11) and (13), the transfer function $G_{\delta u}$ of the input voltage $u(s)$ to the bending displacement output $\delta(z,s)$ can be written as:

$$G_{\delta u} = \frac{\delta(z,s)}{u(s)} = \frac{3z^2(b+b_1)G_{Qu}(s)d(s)}{4Lwb^3E} \tag{14}$$

With a simple deformation of Equation (8), it is not difficult to obtain the relationship between the blocking force F and the driving voltage u:

$$G_{Fu} = \frac{(b+b_1)G_{Qu}d(s)}{15L} \tag{15}$$

Then, the modeling of the CP ionogel actuator is completed.

3.3. Parameter Identification and Verification

In this section, we demonstrated an actuator for parameter identification. The total length is 26 mm, the width is 5 mm, and the average thickness is 0.18 mm. The free length of the actuator is 20 mm. We use a step signal with a voltage of 1 V for system parameter identification. Based on the assumptions in the previous section, only the parameters of the electromechanical coupling equations need to be identified. The parameters used in the model from experiment data are shown in Table 2. Here, we use the least squares method for parameter identification, and the cost function is as follows:

$$\min \sum (f(x_i) - y_i)^2 \tag{16}$$

where $f(x_i)$ is the system function to be identified, and y is the experimental value.

Table 2. Parameters of Modeling.

Parameters	Values
R_1, R_2	7 Ω
C_1, C_2	0.01 F
R	10,000 Ω
L	20 mm
E	300 MPa
w	4.2 mm
z	18 mm
b	0.09 mm
b_1	0.02 mm

Here, two zeros and two poles are used as the basic parameters of the electromechanical coupling model, and the parameters to be identified are k, z_1, z_2, p_1, and p_2. By performing least squares fitting in MATLAB, all the system parameters are obtained, which are shown in Table 3.

To verify the accuracy of our analytical model, we tested the displacement response of a 20 mm long actuator (total length 26 mm, the electrode clamping part length is 6 mm) at a 1 V step voltage. The experimental data and the theoretical data derived from the model show very high agreement (Figure 7a). Furthermore, when the actuator works at a ramping voltage of 1 V, its displacement response is in good agreement with the simulation results. The deviation range of its maximum displacement is within 10% (Figure 7b). The presence of very small variations in film properties or internal stresses during actuator preparation can introduce disparities in the strain values when the actuator bends in both directions under alternating voltage. At the same time, in the experimental test, the different stiffness of the conductive clamp spring will produce different holding forces, which will also lead to the accumulation of errors in the test. Furthermore, we test the prediction performance of the model under different step voltages (0.1 V and 0.2 V), which are not covered in fitting (Figure 7c,d); the results show that the predicted and experimental data are generally moderate, although the test results showed some deviation when excited by a 0.2 V step voltage, as mentioned earlier, the accumulation of errors during the experimental process can lead to an offset in the test results. However, as shown in the figure, the maximum deviation between the experimental and predicted values is less than 0.2 mm. These results show that our model can well reflect the dynamic response of the actuator and thus make a good prediction of the displacement of the actuator's end position.

Table 3. Parameters of Electromechanical Coupler Function

Parameters	Values
k	86.478
p1	0.036
p2	0.901
z1	2.406
z2	0.048

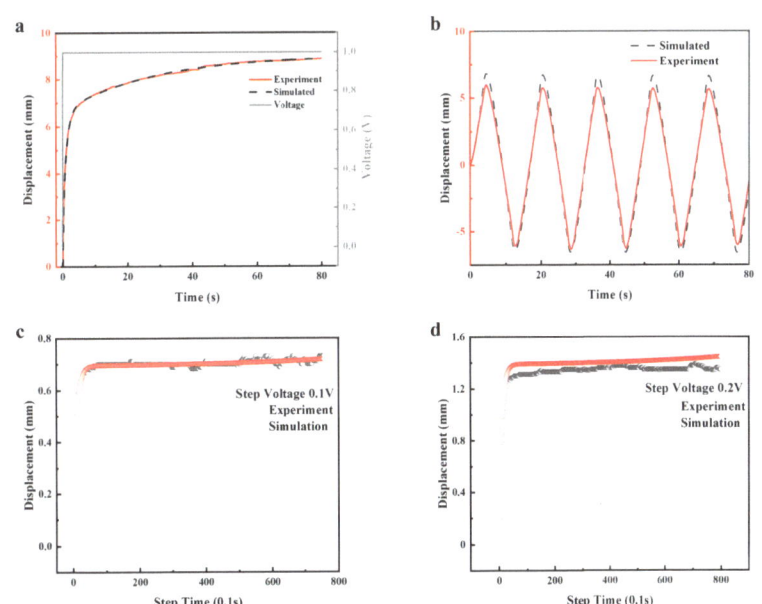

Figure 7. Verification of the model by comparing the experimental and simulated results ((**a**) 1 V step voltage; (**b**) 1 V ramping voltage; (**c**) 0.1 V step voltage; (**d**) 0.2 V step voltage). Difference between simulation results based on analytical model and experimental data. (The structural parameters and electrical parameters of the experiment are consistent with those of the simulation).

3.4. Analysis Result

Based on the established electromechanical coupling model, we further discuss the influence of various parameters of the actuator on the actuation performance. Because the ion migration velocity is reflected in capacitor size, in this section, the concept of equivalent capacitance is introduced. Figure 8a,b shows the end position displacement and blocking force of the actuator as a function of the electrode resistance and capacitance. When the resistance is increased from 10 Ω to 300 Ω, the displacement decreases from 21.9 mm to 21.3 mm, while the blocking force drops from 4.47 mN to 4.35 mN. On the other hand, when the capacitance increases from 0.001 F to 0.02 F, the displacement increases from 1.1 mm to 21.9 mm, while the blocking force improves from 0.2 mN to 4.47 mN. These results show that the capacitance of the electrodes has a large effect on both displacement and blocking force. In contrast, as long as the resistance is within a few hundred orders of magnitude, the resistance has little effect on the actuator. Based on the simulation results, electrodes with low resistive and high capacitive properties can simultaneously generate greater displacement and blocking force, making them ideal candidates for actuator materials.

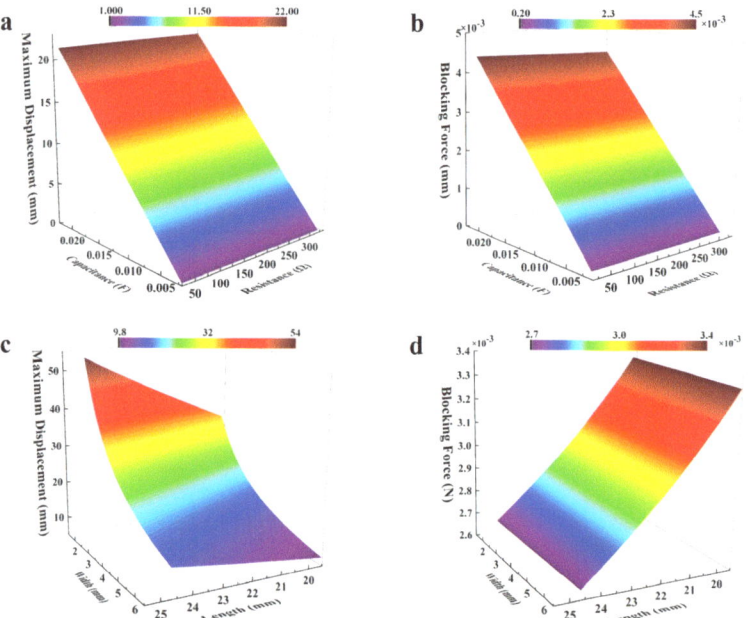

Figure 8. (**a**,**b**) The influence of electrode resistance and capacitance on maximum displacement at the end of the actuator and blocking force; (**c**,**d**) The relationship of actuator's length and width between displacement at the end of the actuator and blocking force.

The mechanical properties of the electrodes also have a very important influence on the performance of the actuator. It can be seen from Figure 9 that although the improvement of Young's modulus can improve the blocking force of the actuator, it also leads to a decrease in displacement. Therefore, it is necessary to select electrode materials with appropriate mechanical properties based on comprehensive consideration of the requirements for strain and gripping force in practical applications.

Optimizing the size of the actuator can also increase the output of its performance. By increasing the length of the actuator, the tip displacement is gradually increased (Figure 8c,d). The displacement of the end of the actuator can be further increased by using a narrower beam. Thus, a larger aspect ratio is beneficial for the unidirectional output of internal stress, resulting in a larger displacement. The blocking force of the actuator decreases with increasing length, while width has a trivial effect on it. This is because the moment generated by the system is only related to the voltage. When the electrical parameters are fixed, the moment generated by the system is also fixed, so a longer length means a longer arm of force, which reduces the blocking force.

Hence, to obtain a high load, large deformation, and fast response, the actuator needs higher capacitance, and Young's modulus should be controlled at around 500 MPa. For the structure of the actuator, reducing the length and increasing the width can achieve greater load capacity, and vice versa can obtain greater displacement of the end position.

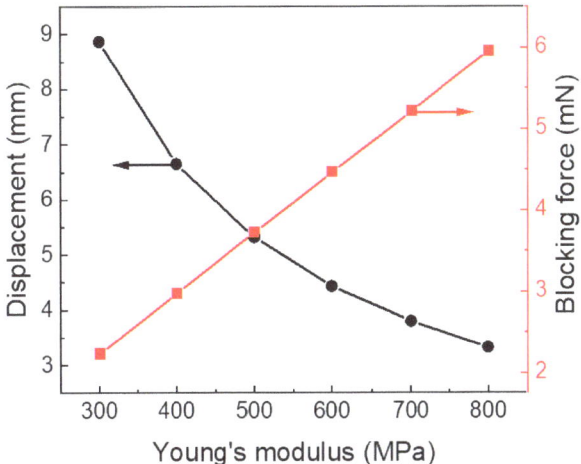

Figure 9. The displacement of end position and blocking force of the actuator based on its Young's Modulus.

4. Conclusions

In this work, a novel electrochemical actuator with large strain and low operating voltage was developed using CP ionogel as the electrode. Due to the high electrical conductivity, flexibility, and excellent electrochemical activity of ionogels, the flexible actuator in this study outperforms most others employing materials like graphene and graphdiyne. Additionally, the preparation process is considerably simpler. An analytical model of the actuator by combining the equivalent electrical and motion processes is established, with the largest disagreement between simulated and the corresponding measured data being less than 10%. In addition, our study systematically scrutinized key parameters influencing actuator performance, including electrode resistance, capacitance, Young's modulus, and actuator size. Simulation results underscore the optimal nature of electrodes with low resistance and high capacitance, contributing to the simultaneous enhancement of bending strain and blocking force. Nevertheless, the increase of Young's modulus, despite augmenting blocking force, compromises bending strain. Moreover, a larger aspect ratio emerges as advantageous for unidirectional stress output, resulting in increased bending strain, while actuator blocking force experiences a reduction with greater length. This thorough investigation elucidates the intricate interplay of these physical factors and their impact on actuator displacement and blocking force. Our work provides valuable practical insights and theoretical guidance for the future development of high-performance electrochemical actuators and the exploration of innovative smart materials.

Author Contributions: Conceptualization, H.H.; Methodology, G.C.; Data curation, S.Z.; Writing—original draft, J.X.; Funding acquisition, J.D. All authors have read and agreed to the published version of the manuscript.

Funding: This work was supported by China Postdoctoral Science Foundation (2022M711372), Postdoctoral Research Program of Jiangsu Province (2021K544C) and General Program of Natural Science Foundation for Higher Education in Jiangsu Province (21KJB510004).

Institutional Review Board Statement: Not applicable.

Data Availability Statement: The experimental data on the results reported in this manuscript are available upon an official request to the corresponding author.

Conflicts of Interest: The authors declare no conflict of interest.

References

1. Li, M.; Pal, A.; Aghakhani, A.; Pena-Francesch, A.; Sitti, M. Soft actuators for real-world applications. *Nat. Rev. Mater.* **2022**, *7*, 235–249. [CrossRef]
2. Apsite, I.; Salehi, S.; Ionov, L. Materials for smart soft actuator systems. *Chem. Rev.* **2021**, *122*, 1349–1415. [CrossRef]
3. Hao, Y.; Zhang, S.; Fang, B.; Sun, F.; Liu, H.; Li, H. A review of smart materials for the boost of soft actuators, soft sensors, and robotics applications. *Chin. J. Mech. Eng.* **2022**, *35*, 37. [CrossRef]
4. Kim, O.; Kim, S.J.; Park, M.J. Low-voltage-driven soft actuators. *Chem. Commun.* **2018**, *54*, 4895–4904. [CrossRef]
5. Wang, H.S.; Cho, J.; Song, D.S.; Jang, J.H.; Jho, J.Y.; Park, J.H. High-performance electroactive polymer actuators based on ultrathick ionic polymer–metal composites with nanodispersed metal electrodes. *ACS Appl. Mater. Interfaces* **2017**, *9*, 21998–22005. [CrossRef]
6. Jo, C.; Pugal, D.; Oh, I.K.; Kim, K.J.; Asaka, K. Recent advances in ionic polymer–metal composite actuators and their modeling and applications. *Prog. Polym. Sci.* **2013**, *38*, 1037–1066. [CrossRef]
7. Fukushima, T.; Asaka, K.; Kosaka, A.; Aida, T. Fully plastic actuator through layer-by-layer casting with ionic-liquid-based bucky gel. *Angew. Chem.* **2005**, *117*, 2462–2465. [CrossRef]
8. Mukai, K.; Asaka, K.; Sugino, T.; Kiyohara, K.; Takeuchi, I.; Terasawa, N.; Futaba, D.N.; Hata, K.; Fukushima, T.; Aida, T. Highly conductive sheets from millimeter-long single-walled carbon nanotubes and ionic liquids: Application to fast-moving, low-voltage electromechanical actuators operable in air. *Adv. Mater.* **2009**, *21*, 1582–1585. [CrossRef]
9. Lu, C.; Yang, Y.; Wang, J.; Fu, R.; Zhao, X.; Zhao, L.; Ming, Y.; Hu, Y.; Lin, H.; Tao, X.; et al. High-performance graphdiyne-based electrochemical actuators. *Nat. Commun.* **2018**, *9*, 752. [CrossRef]
10. Umrao, S.; Tabassian, R.; Kim, J.; Nguyen, V.H.; Zhou, Q.; Nam, S.; Oh, I.K. MXene artificial muscles based on ionically cross-linked Ti3C2T x electrode for kinetic soft robotics. *Sci. Robot.* **2019**, *4*, eaaw7797. [CrossRef]
11. Wu, G.; Hu, Y.; Liu, Y.; Zhao, J.; Chen, X.; Whoehling, V.; Plesse, C.; Nguyen, G.T.; Vidal, F.; Chen, W. Graphitic carbon nitride nanosheet electrode-based high-performance ionic actuator. *Nat. Commun.* **2015**, *6*, 7258. [CrossRef] [PubMed]
12. Spinks, G.M.; Wallace, G.G.; Liu, L.; Zhou, D. Conducting polymers electromechanical actuators and strain sensors. *Macromol. Symp.* **2003**, *192*, 161–170. [CrossRef]
13. Malinauskas, A.; Malinauskiene, J.; Ramanavičius, A. Conducting polymer-based nanostructurized materials: Electrochemical aspects. *Nanotechnology* **2005**, *16*, R51. [CrossRef]
14. Li, L.; Xu, J.; Li, Q.; Jiang, Y.; Dong, X.; Ding, J. Small Multi-Attitude Soft Amphibious Robot. *IEEE Robot. Autom. Lett.* **2023**, 1–8. [CrossRef]
15. Jiang, Y.; Dong, X.; Wang, Q.; Dai, S.; Li, L.; Yuan, N.; Ding, J. A High-Fidelity Preparation Method for Liquid Crystal Elastomer Actuators. *Langmuir* **2022**, *38*, 7190–7197. [CrossRef] [PubMed]
16. Wang, Z.; He, B.; Liu, X.; Wang, Q. Development and modeling of a new ionogel based actuator. *J. Intell. Mater. Syst. Struct.* **2017**, *28*, 2036–2050. [CrossRef]
17. Wang, Y.; Zhu, C.; Pfattner, R.; Yan, H.; Jin, L.; Chen, S.; Molina-Lopez, F.; Lissel, F.; Liu, J.; Rabiah, N.I.; et al. A highly stretchable, transparent, and conductive polymer. *Sci. Adv.* **2017**, *3*, e1602076. [CrossRef]
18. Petroffe, G.; Beouch, L.; Cantin, S.; Aubert, P.H.; Plesse, C.; Dudon, J.P.; Vidal, F.; Chevrot, C. Investigations of ionic liquids on the infrared electroreflective properties of poly (3, 4-ethylenedioxythiophene). *Sol. Energy Mater. Sol. Cells* **2018**, *177*, 23–31. [CrossRef]
19. Lee, J.W.; Yu, S.; Hong, S.M.; Koo, C.M. High-strain air-working soft transducers produced from nanostructured block copolymer ionomer/silicate/ionic liquid nanocomposite membranes. *J. Mater. Chem. C* **2013**, *1*, 3784–3793. [CrossRef]
20. Imaizumi, S.; Kokubo, H.; Watanabe, M. Polymer actuators using ion-gel electrolytes prepared by self-assembly of ABA-triblock copolymers. *Macromolecules* **2012**, *45*, 401–409. [CrossRef]
21. Kim, O.; Kim, S.Y.; Park, B.; Hwang, W.; Park, M.J. Factors affecting electromechanical properties of ionic polymer actuators based on ionic liquid-containing sulfonated block copolymers. *Macromolecules* **2014**, *47*, 4357–4368. [CrossRef]
22. Lee, J.W.; Yoo, Y.T.; Lee, J.Y. Ionic polymer–metal composite actuators based on triple-layered polyelectrolytes composed of individually functionalized layers. *ACS Appl. Mater. Interfaces* **2014**, *6*, 1266–1271. [CrossRef] [PubMed]
23. Rohtlaid, K.; Nguyen, G.T.; Soyer, C.; Cattan, E.; Vidal, F.; Plesse, C. Poly (3, 4-ethylenedioxythiophene): Poly (styrene sulfonate)/polyethylene oxide electrodes with improved electrical and electrochemical properties for soft microactuators and microsensors. *Adv. Electron. Mater.* **2019**, *5*, 1800948. [CrossRef]
24. Terasawa, N.; Asaka, K. High-performance PEDOT: PSS/single-walled carbon nanotube/ionic liquid actuators combining electrostatic double-layer and faradaic capacitors. *Langmuir* **2016**, *32*, 7210–7218. [CrossRef]
25. Terasawa, N.; Asaka, K. Self-standing cellulose nanofiber/poly (3, 4-ethylenedioxythiophene): Poly (4-styrenesulfonate)/ionic liquid actuators with superior performance. *RSC Adv.* **2018**, *8*, 33149–33155. [CrossRef] [PubMed]

26. Terasawa, N.; Asaka, K. Performance enhancement of PEDOT: Poly (4-styrenesulfonate) actuators by using ethylene glycol. *RSC Adv.* **2018**, *8*, 17732–17738. [CrossRef]
27. Li, Y.; Tanigawa, R.; Okuzaki, H. Soft and flexible PEDOT/PSS films for applications to soft actuators. *Smart Mater. Struct.* **2014**, *23*, 074010. [CrossRef]
28. Wang, D.; Lu, C.; Zhao, J.; Han, S.; Wu, M.; Chen, W. High energy conversion efficiency conducting polymer actuators based on PEDOT: PSS/MWCNTs composite electrode. *RSC Adv.* **2017**, *7*, 31264–31271. [CrossRef]
29. Caponetto, R.; Graziani, S.; Pappalardo, F.; Sapuppo, F. Identification of IPMC nonlinear model via single and multi-objective optimization algorithms. *ISA Trans.* **2014**, *53*, 481–488. [CrossRef]
30. Bonomo, C.; Fortuna, L.; Giannone, P.; Graziani, S.; Strazzeri, S. A nonlinear model for ionic polymer metal composites as actuators. *Smart Mater. Struct.* **2006**, *16*, 1. [CrossRef]
31. Dougal, R.; Gao, L.; Liu, S. Ultracapacitor model with automatic order selection and capacity scaling for dynamic system simulation. *J. Power Sources* **2004**, *126*, 250–257. [CrossRef]

Disclaimer/Publisher's Note: The statements, opinions and data contained in all publications are solely those of the individual author(s) and contributor(s) and not of MDPI and/or the editor(s). MDPI and/or the editor(s) disclaim responsibility for any injury to people or property resulting from any ideas, methods, instructions or products referred to in the content.

Article

Investigating the Aging Behavior of High-Density Polyethylene and Polyketone in a Liquid Organic Hydrogen Carrier

Jyothsna Surisetty [1,*], Mohammadhossein Sharifian [2], Thomas Lucyshyn [1] and Clemens Holzer [1]

[1] Polymer Processing, Department of Polymer Engineering and Science, Montanuniversität Leoben, 8700 Leoben, Austria; thomas.lucyshyn@unileoben.ac.at (T.L.); clemens.holzer@unileoben.ac.at (C.H.)
[2] Chemistry of Polymeric Materials, Department of Polymer Engineering and Science, Montanuniversität Leoben, 8700 Leoben, Austria; mohammadhossein.sharifian@unileoben.ac.at
* Correspondence: jyothsna.surisetty@unileoben.ac.at

Abstract: Hydrogen is recognized as a significant potential energy source and energy carrier for the future. On the one hand, storing hydrogen is a challenging task due to its low volumetric density, on the other hand, a particular type of hydrogen in the form of a liquid can be used to store large quantities of hydrogen at ambient conditions in thermoplastic tanks. But storing hydrogen in this form for a long time in polymer tanks affects the physical and chemical properties of the liner. In the current automotive industry, high-density polyethylene (HDPE) has already been used in existing fuel tank applications. However long-term exposure to fuels leads to the permeation of hydrocarbons into the polymers, resulting in a loss of mechanical properties and reducing the efficiency of fuel cells (FC) in automotive applications. Additionally, facing material shortages and a limited supply of resin leads to an increase in the cost of the material. Therefore, an alternative material is being searched for, especially for hydrogen fuel tank applications. In this study, two semi-crystalline thermoplastics, HDPE and polyketone (POK), were compared, which were exposed to a selected liquid organic hydrogen carrier (LOHC) at 25 °C and 60 °C for up to 500 h in an enclosed chamber, to measure their fuel up-take. A short analysis was carried out using differential scanning calorimetry (DSC), thermogravimetric analysis (TGA), Fourier transform infrared spectroscopy (FTIR), and mechanical testing to understand the influence of the LOHC on the polymer over time. Fuel sorption and tensile properties showed a plasticizing effect on HDPE. The material degradation was more pronounced for the aged samples of HDPE in comparison to POK. As expected, thermal aging was increased at 60 °C. The fuel absorption of POK was lower compared to HDPE. A slight increase in crystallinity was observed in POK due to the aging process that led to changes in mechanical properties. Both HDPE and POK samples did not show any chemical changes during the aging process in the oven at 25 °C and 60 °C.

Keywords: LOHC (perhydro-benzyltoluene); POK; aging behavior; fuel uptake; long-term storage

Citation: Surisetty, J.; Sharifian, M.; Lucyshyn, T.; Holzer, C. Investigating the Aging Behavior of High-Density Polyethylene and Polyketone in a Liquid Organic Hydrogen Carrier. *Polymers* **2023**, *15*, 4410. https:// doi.org/10.3390/polym15224410

Academic Editors: Vineet Kumar and Md Najib Alam

Received: 11 October 2023
Revised: 7 November 2023
Accepted: 10 November 2023
Published: 15 November 2023

Copyright: © 2023 by the authors. Licensee MDPI, Basel, Switzerland. This article is an open access article distributed under the terms and conditions of the Creative Commons Attribution (CC BY) license (https:// creativecommons.org/licenses/by/ 4.0/).

1. Introduction

Nowadays, many European countries have set their targets to attain zero emissions by 2050. An increasing awareness of environmental issues and the drawbacks of using fossil fuel have led to a significant progress in the development of hydrogen as an alternative fuel for the transportation sector [1]. During the last two decades, hydrogen has emerged as a promising topic in the current market, and is gaining popularity due to its ability to provide a CO_2-free alternative to gasoline or diesel [1,2]. Moreover, hydrogen can be produced from renewable sources such as solar, wind, and hydroelectric power through the electrolysis of water, making it a potentially sustainable energy source [3,4]. So far, the companies Toyota, Hyundai, General Motors, and BMW have successfully launched hydrogen fuel cell vehicles in the current market, which are approved globally [5–8]. But storing the hydrogen in gaseous and liquid form in automotive applications is challenging due to its low volumetric energy density 5.6 MJ/L (less than gasoline 32 MJ/L), high pressures (700 bar), and low temperatures (−253 °C) which are difficult to handle. High production cost and storage in fuel tanks are challenges that need to be tackled [9,10]. To solve these problems, some researchers suggest storing hydrogen in the form of an organic liquid rather than in its gaseous state. This approach could potentially increase the gravimetric density up to 7.2% and provide a promising solution for safe and efficient storage of hydrogen at room temperatures of 25 °C, using regular polymer tanks. This liquid hydrogen in chemical form can absorb (hydrogenation) and release (dehydrogenation) hydrogen chemically under high temperatures and is known as a liquid organic hydrogen carrier (LOHC) [11–15].

Different types of LOHC are available in the market to store hydrogen such as 9-ethyl-perhydrocarbazole (H-12 NEC), perhydro-benzyltoluene (H12-BT), or methylcyclohexane (MCH) [12,16]. To store these special types of LOHCs, existing infrastructure is sufficient (like gasoline and diesel filling stations) compared to other forms of hydrogen.

In this article, out of the different types of LOHCs, the focus is on storing H12-BT in HDPE and POK. One of the reasons for choosing this benzyltoluene (BT) is a recently introduced, promising approach to convert this BT into an on-board generation of electricity that can be used for mobile applications. For this purpose, a direct LOHC-based fuel cell (LOHC-DIPAFC) is being developed [17,18].

It generally consists of aromatic hydrocarbons with a carbon number of 12 and it exhibits diesel-like properties [19]. It should be considered that LOHC fuels are generally different from each other in terms of molecular structure, which leads to dissimilarity in solubility and diffusion into polymers [20,21]. In the case of petroleum fuels, the most commonly used polymeric material for tanks is HDPE because it can resist chemicals very well and has a low permeability; however, the diffusion coefficient of HDPE is higher due to its non-polar hydrophobic nature [22]. For example, Richaud et al. performed a series of immersion tests on biodiesel fuels to measure the sorption rate in polyethylene. These analyses confirmed that the absorption rate was increased with increasing time and reached a plateau after immersion for 139 h under ambient conditions [23]. Different grades of HDPE were also examined by Böhning et al. A decrease in tensile strength and an increase in strain percentage were observed at 60 °C. As the material is exposed to fuels for a longer time, thermoplastics undergo plasticization effects [21]. Therefore, in order to improve the barrier properties, fluorination (halogen gas) is introduced on the surface of HDPE as a post treatment process, but this alternative method can contribute to global warming because it emits green-house gases. Jianwei Zhao et al. investigated the absorption of different biodiesels in polyamide 6 for 720 h. After 720 h, the polymers did not reach the plateau state and grew linearly. It was also noticed that, due to the longer exposure time in polyamide 6, the material behaved as plasticizer and changes in properties were identified [24]. Michael D. Kass et al. tested immersion tests on different commodity and high-performance polymers for 16 weeks using diesel fuels. Out of those, HDPE and PP also exhibited high swelling percentages (5–15%) when exposed due to higher solubility, resulting in change in the hardness, volume, and mass of the polymers [25]. To develop liquid storage tanks, B.F. Yousif and H. Ku investigated the behavior of aged samples (6 months) in diesel using polyester and coir fibers, and improved the interfacial adhesion and tensile properties, and also the resistance to degradation when compared to glass fibers [26].

From the above research trials by other researchers, there is also the possibility to introduce multi-layers during the production process using co-extrusion; however, the final cost for this manufacturing process is significant [27,28]. For instance, as a substitute for HDPE, POK can be used in fuel delivery systems, which exhibits good barrier properties. It also has excellent impact strength, mechanical and thermal properties, chemical resistance, and a low permeation to gases and liquids [29–32]. POK is polar and has strong inter-molecular interactions between the neighboring polymer chains resulting in low diffusion [33]. These polar and non-polar groups determine the permeation of hydrocarbons into polymers and the changes in thermal and mechanical properties.

Over the last 40 years a lot of research has been focused on polymer interaction with petroleum fuels and the effects of these fuels on polymer properties. However, no papers have been published to address HDPE or POK interaction with BT. In this paper, the physical and chemical properties of POK are compared to HDPE in the presence of hydrogen and BT-containing fuel. Storing this chemically bonded hydrogen in polymeric tanks for a long time can also result in the absorption and diffusion of LOHCs into the plastic over time. Eventually, this could lead to stress cracking, degradation, and a reduction in the mechanical properties of the polymer. Hence, it is important to investigate the aging properties and choose the right material to store LOHCs for a longer duration.

2. Materials and Methods

2.1. Materials

In this study, two different grades of semi-crystalline materials were selected: HDPE (Lupolen 4261 A IM BD) with a density of 0.940 g/cm^3 was provided by LyondellBasell, Frankfurt, Germany. Poketone (M330F) with a density of 1.24 g/cm^3 was obtained from Hyosung, Seoul, Korea. Both are special grades of polymers and have good barrier properties and chemical resistance to fuels in automotive applications. The materials used to investigate are virgin polymers, and no formulations were used during this work.

2.2. Sample Preparation

As shown in Figure 1, in the first step, HDPE and POK pellets were pre-dried in a dryer (DP615, Piovan S.p.A, Venice, Italy) for 3 h before being placed into an injection molding machine. Two different mold plates were used to produce dumbbell and rectangular specimens. Injection-molded cylindrical dumbbell specimens with length = 88 mm and diameter = 5 mm and rectangular plates with length = 138 mm, width = 74 mm, and thickness = 1 mm (as shown in Figure 1) were produced using two different Arburg injection molding machines with an injection speed of 30 and 25 cm^3/s. For HDPE the nozzle temperature was set to 260 °C, and for POK to 250 °C; both mold temperatures were maintained at 60 °C. The rectangular plates were punched with a hydraulic press of 4 mm diameter to obtain circular samples. They were further used to perform gravimetric analysis, DSC, TGA, and FTIR measurements.

Following that, LOHC fuel was put into 100 mL glass tubes, which were entirely filled, and cylindrical dumbbell specimens were dipped to evaluate their mechanical properties. On the other hand, small glass vials were half filled, and circular samples were placed for thermal evaluation. Later on, the samples were aged in a closed oven for up to 500 h at a constant temperature of 25 and 60 °C, respectively. Then, the completely aged samples were removed from the LOHC fuel and wiped with tissues to remove fuel content on the polymer surface to make sure no error values can be recorded during testing. Finally, the samples were sent for immediate evaluation to investigate the test results, which are mentioned in Section 3. In this research, a total of three replicates were tested in each experimental trial.

2.3. Type of Fuel Used

In this study an LOHC with similar properties to diesel was used, with a flash point of 125–180 °C. The advantage of this fuel is that it can be stored under ambient conditions, it is non-flammable, safe, and can be used in automotive fuel tank applications. This LOHC was delivered from Hydrogenious LOHC Technologies GmbH, Erlangen, Germany.

2.4. Ageing Procedure

The dumbbell-shaped specimens of HDPE and POK samples were immersed in 100 mL sealed glass containers which were filled with LOHC fluids at 25 °C and 60 °C for up to 500 h. At regular time intervals, samples were removed from the glass bottles to measure weight changes. Likewise, the circular disk-shaped specimens for DSC and TGA experiments were placed in small glass vials. Then, 10 mL of LOHC was poured into a 50 mL headspace crimp vial and the caps were sealed with tape around them.

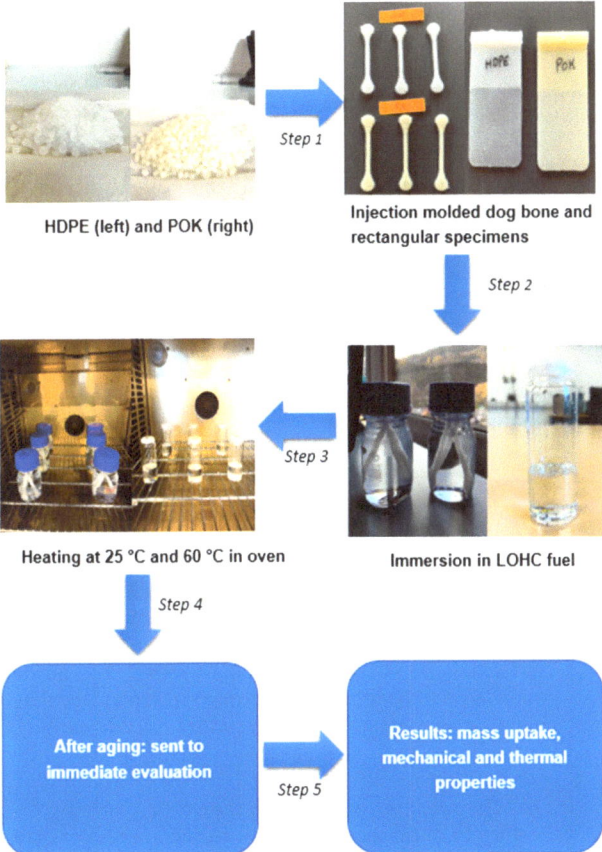

Figure 1. The step-by-step process of injection-molded HDPE and POK samples aged in LOHC fuel in closed chamber.

2.5. Sorption Experiments

The weight gain or weight loss of samples after exposure to LOHC solution were studied using gravimetric analysis. At 60 °C, and also in ambient conditions (25 °C), three replicates of each polymer were weighed in the dry stage and immersed in LOHC solution. Later on, each sample was taken out of the LOHC at regular intervals and weighed. The weight measurement values were recorded on a laboratory electronic microbalance (type M2P, Sartorius AG, Goettingen, Germany). The relative LOHC uptake (M_t) was calculated using Equation (1):

$$M_t = \frac{W_w - W_d}{W_d} \qquad (1)$$

W_w is the weight of the specimen after exposing to LOHC; W_d is the initial weight of the sample before immersion; and M_t is the relative mass uptake with respect to time t.

2.6. Diffusion Coefficient

There are several ways to determine the diffusion of chemicals in polymer materials. Using Fick's law and assuming a single-phase diffusion process is one of the standard methods for describing molecular diffusion in polymers. As time passes, the relative mass of the absorbed phase will grow linearly. The diffusion coefficient (D) is determined using the following equation [34].

$$\frac{M_t}{M_m} = \frac{4}{\sqrt{\pi}} \sqrt{\frac{Dt}{h^2}} \qquad (2)$$

M_t is the LOHC uptake, M_m is the equilibrium concentration of the corresponding curves, t is the immersion time and h is the thickness of the specimen.

2.7. Differential Scanning Calorimetry (DSC)

In order to characterize the changes in the melting behavior of both materials, HDPE and POK were examined using a DSC1 (Mettler-Toledo, Greifensee, Switzerland, Software STARe V16.30a). The samples, weighing 13 mg (HDPE) and 20 mg (POK), were placed in 40 µL aluminum pans. In a nitrogen atmosphere with a gas flow rate of 50 mL min^{-1}, the samples were heated from 20 to 250 °C at a rate of 10 K min^{-1}. Before the initial tests, the samples were first removed from the LOHC, and cleaned with dry tissue paper.

To determine the degree of crystallinity of HDPE and POK, the melting enthalpy is divided by the theoretical melting enthalpy of the 100% crystalline phase (HDPE: 293 J/g [35]; POK: 227 J/g [36]). Therefore, the degree of crystallinity (X_c) was calculated using the following equation [27].

$$X_c = \frac{\Delta H_m}{\Delta H_m^{100\%}} \cdot 100\% \qquad (3)$$

2.8. Thermal Gravimetric Analyzer (TGA)

To determine the fuel uptake of HDPE and POK, the samples were measured using TG/DSC 3 (Mettler-Toledo, Greifensee, Switzerland). Samples weighing 13 mg (HDPE) and 20 mg (POK) were placed in 70 µL alumina crucibles. With a heating rate of 10 K min^{-1} and a nitrogen gas flow rate of 50 mL min^{-1}, TG curves were obtained between 25 and 600 °C under N_2. In addition, further measurement values were also recorded from 600 to 800 °C in an oxygen environment with a gas flow rate of 50 mL min^{-1}.

2.9. Mechanical Testing

The dumbbell specimens for the tensile tests were conditioned by immersion in LOHC fuel up to saturation. The measurements were performed on a Zwick Roell Universal Testing Machine at 25 ± 1 °C with a test speed of 50 mm/min. The gauge length of these specimens was 50 mm.

2.10. Fourier-Transform Infrared Spectroscopy (FTIR)

The infrared spectra were assessed using the Bruker Vertex 70 system (Rosenheim, Germany), which was furnished with an integrated single reflection crystal made of diamond–germanium for ATR measurements. ATR analyses were performed on the surfaces of distinct tablet samples, each exposed to the LOHC for varying durations.

3. Results and Discussions

In the pursuit of exploring the viability of utilizing HDPE and POK as potential candidates for long-term LOHC tank systems, a comprehensive investigation was undertaken. This study involved subjecting both HDPE and POK to prolonged exposure at elevated temperatures within the LOHC environment. The testing regiment encompassed a series of crucial evaluations, including tensile testing, fuel uptake analysis, diffusion studies, melt behavior assessments, FTIR spectroscopy, and measurements of thermal stability. In this section, we present a detailed exposition of the obtained results, which contribute significantly to the broader field of hydrogen storage research and facilitate the development of more efficient and sustainable energy solutions.

3.1. Fuel Uptake
Chemical Effects

Figure 2a,b depict the weight changes in the HDPE and POK samples after immersion at 25 °C and 60 °C with respect to time.

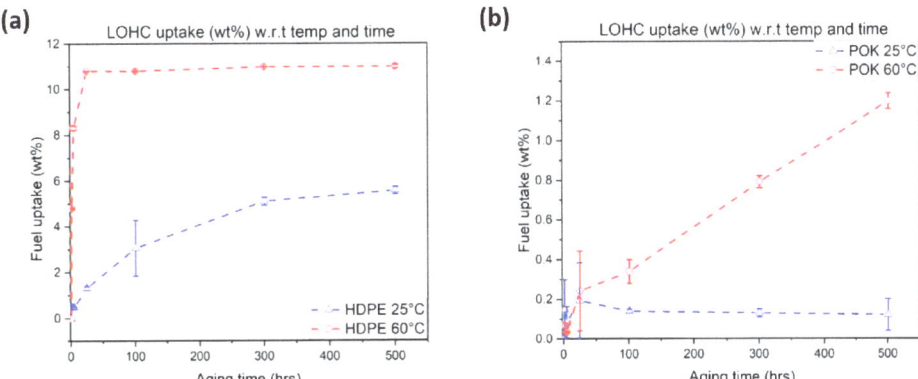

Figure 2. A study of the fuel uptake in injection-molded (**a**) HDPE and (**b**) POK samples which are exposed to air and LOHC fuel in small glass vials. The curves represent mean values along with the standard deviation based on 3 measurements.

After exposure to 500 h in LOHC, HDPE reached a plateau level at 60 °C. It is also noticed that if the temperature increases, the absorption rate of HDPE also increases because, at elevated temperatures, the polymer chains are more mobile, resulting in an increase in the free volume of thermoplastics, which allows molecules to diffuse more easily [37]. At 25 °C the absorption rate is significantly lower because, at low temperatures, atoms are more densely packed in amorphous regions, and thus the absorption rate is very low for chemical fluids into polymers. In addition, the fuel uptake percentage of HDPE can be compared with commercial fuels using the work carried out by Libia et al. [38] because the physical properties of these fuels are similar to those of LOHC [19]. In their work they found that the fuel uptake in HDPE at room temperature (25 °C) was about 5.0–5.3 weight % which is very similar to the results obtained with HDPE aged in LOHC fuel.

Aged samples of POK showed a significantly lower increase in weight with increasing temperature when compared to HDPE. According to Figure 2b, there was an initial drop in mass at the beginning of the exposure time for both 25 and 60 °C. This drop may be attributed to volatilities of residual water or organic additives which might evaporate over time, but there was no degradation caused during these trials. To ensure the material was not dissolved in LOHC, gas chromatography was conducted, but no traces of olefins or carbon monoxide (CO) were found. Additionally, a centrifuge device (Heraeus Labofuge 300) from Thermo Electron (Waltham, MA, USA) was used by inserting small glass tubes in the centrifugal unit and rotating them at 2000 rpm for 300 s. The goal was to investigate the existence of solid particles. However, no polymer particles were found during this process.

POK is generally polar and contains carbonyl groups which are electronegative. These groups attract electrons away from neighboring carbon atoms, enhancing the intermolecular attraction of the polymer chains [29]. As a result, POK exhibits low absorption of liquids and gases compared to HDPE. At 60 °C these samples did not reach the equilibrium state and grew linearly with time. Therefore, POK could be compared with the work conducted by Zhao et al. [24]. In this work, they investigated the absorption rate of diesel and biodiesels in two different grades of polyamides (which was 1.6 and 2.0 wt%). At 700 h none of these fuels reached an equilibrium state because polyamides contain amide groups and are also polar in nature and this makes the diffusion process slow during the ageing process.

Moreover, H12-BT also contains saturated aromatic hydrocarbons like diesel, which are non-polar in nature and that tend to diffuse into polymers.

Comparing both graphs, the fuel uptake in HDPE increased rapidly i.e., ~5.5 wt.% and ~11 wt.% under isothermal conditions (25 °C and 60 °C) contrary to POK which was significantly lower at 0.2 wt.% and 1.1 wt.%. at 500 h.

The diffusion coefficient is calculated from Equation (2) using the initial stages of the sorption phase. It is clear that HDPE is more sensitive to LOHC fuel. As the aging temperature increased, the rate of diffusion also increased from 1.2×10^{-7} to 1.64×10^{-7} cm^2 s^{-1}, which further obeys the Fickian diffusion model. In the case of POK, diffusion is less at 25 °C with 5.6×10^{-8} cm^2 s^{-1}, showing high resistance to LOHC fuel. As it did not reach an equilibrium state at 60 °C, this follows the non-Fickian diffusion model, and no diffusion coefficient could be determined for POK exposed to 60 °C. Table 1 shows the measured values of the diffusion coefficients.

Table 1. Diffusion coefficient D and equilibrium concentration M_m of LOHC fuel in HDPE and POK.

Temperature (°C)	HDPE (D in (cm^2 s^{-1}))	POK (D in (cm^2 s^{-1}))	M_m (HDPE/POK) (-)
25	1.25×10^{-7}	5.6×10^{-8}	0.0557/0.0014
60	1.64×10^{-7}	-	0.1098/-

3.2. Annealing Effects and Crystallinity

The melting behavior of the samples was determined using DSC before and after exposure to LOHC fuel. The melting peak temperature (T_m) and enthalpy (ΔH_m) values of the first heating scan are shown in Table 2. A single melting peak at 132.6 °C (HDPE) and 220.9 °C (POK) was recorded in unaged samples (shown in Figure 3). Note that the ΔH_m of HDPE increased in aged samples compared to unaged. Likewise, enthalpy also increased in POK aging samples which were aged for 100 h to 500 h.

Table 2. Melting temperature (T_m) and melting enthalpy (ΔH_m) of HDPE and POK samples aged in LOHC fuel for 100 and 500 h at 25 and 60 °C.

Sample	T_m (°C) 1st Heating	ΔH_m (J/g) 1st Heating
Unaged @HDPE	132.6 ± 1.4	135.6 ± 1.6
HDPE @25 °C/100 h	131.9 ± 0.3	144.9 ± 2.5
HDPE @25 °C/500 h	131.4 ± 0.1	142.4 ± 3.1
HDPE @60 °C/100 h	130.6 ± 0.7	122.1 ± 1.2
HDPE @60 °C/500 h	130.5 ± 1.2	137.2 ± 1.0
Unaged @POK	220.9 ± 0.7	74.9 ± 0.6
POK @25 °C/100 h	221.8 ± 0.1	82.6 ± 1.3
POK @25 °C/500 h	221.8 ± 0.2	81.9 ± 1.4
POK @60 °C/100 h	222.3 ± 0.6	83.6 ± 0.8
POK @60 °C/500 h	222.6 ± 0.4	84.0 ± 2.0

Figure 3a–d depicts the first heating cycle of HDPE and POK thermograms. For these measurements, all the samples underwent chemical oxidation in the oven. As a result of aging at 60 °C, the melting endotherms of HDPE samples showed a slight decrease in melting temperature with an average peak value of 130.5 °C (after aging for 500 h) when compared to samples aged at 25 °C, which showed an average peak value of 131.4 °C. Furthermore, as the aging temperature increased, (60 °C) the melting temperature of POK also increased with a melting peak of 222.6 °C at 500 h. The samples aged at 25 °C were almost similar to samples that were aged at 60 °C with a melting temperature of 221.8 °C.

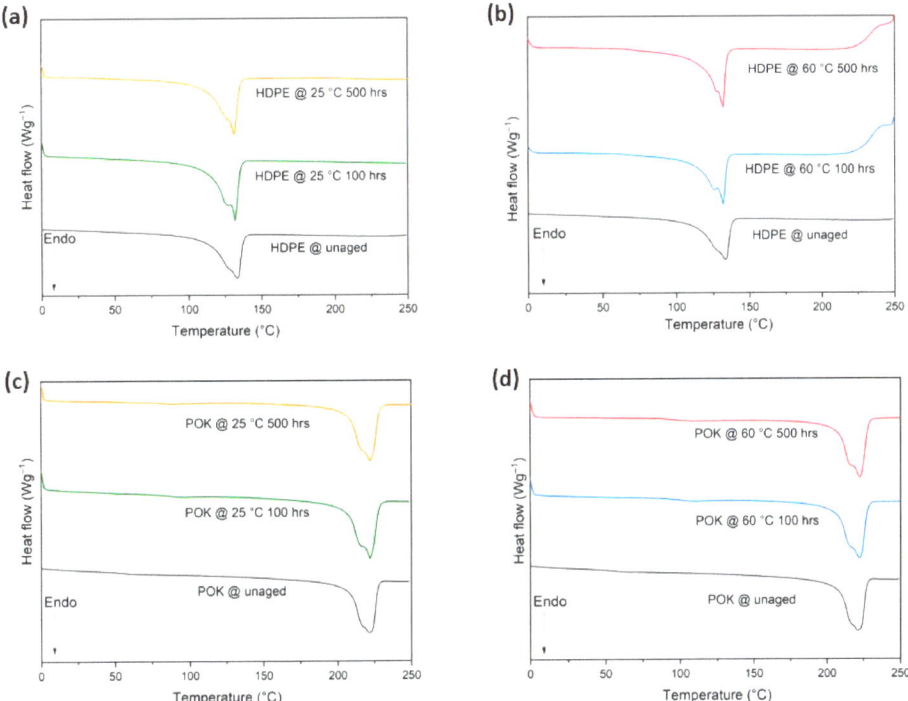

Figure 3. (**a**,**b**) DSC first heating curves for unaged and aged samples of HDPE in LOHC fuel at 25 °C and 60 °C measured for 100 h and 500 h; (**c**,**d**) DSC first heating curves for POK samples on same aging conditions as HDPE (all curves are based on mean values of three measurements).

As shown in Table 2, the rise in enthalpy ΔH_m from the first scan in POK from 74.9 J/g (before immersion) to 81.9 J/g and 84 J/g at 25 °C and 60 °C for 500 h (after immersion) in LOHC fuel is due to the strong intermolecular bonding (spaces reduce between polar groups), which leads to an increase in melting point [39]. Thus, from the below DSC graphs, it can be concluded that with increasing aging time and aging temperature the melting temperature of HDPE decreased slightly, whereas, in the case of POK, the melting temperature increased after aging in LOHC fuel.

From Table 3 it is observed that HDPE crystallinity percentage decreased during the aging process at 60 °C and POK crystallinity percentage (X_c) increased slightly with increasing aging temperature.

Table 3. Crystallinity (X_c) of aged and unaged samples of HDPE and POK obtained from DSC measurements as a function of aging times that are obtained using Equation (3).

Aging Time (h)	Material 1	Crystallinity (%) (1st Heating)	Material 2	Crystallinity (%) (1st Heating)
unaged		46.3 ± 0.3		33.0 ± 0.2
100	HDPE @25 °C	49.5 ± 0.8	POK @25 °C	36.4 ± 0.6
500		48.6 ± 0.9		36.1 ± 0.6
100	HDPE @60 °C	41.7 ± 0.4	POK @60 °C	36.8 ± 0.4
500		46.8 ± 0.3		37.0 ± 1.0

In HDPE, at 25 °C (500 h) the degree of crystallinity is reduced a little. This phenomenon occurs due to the fact that fuels like H12-BT can penetrate into the polymers in the amorphous region and disrupt the intermolecular forces in polymer chains, resulting in a reduction in crystallinity percentage over time from 100 to 500 h [40]. In addition to this, HDPE also exhibited a higher absorption rate compared to POK. This absorbed LOHC fuel is trapped in between the polymer chains and acts as a plasticizer that could lead to reduction in the crystallinity of the material [38].

In the case of POK, a minor increase in crystallinity was observed, which was ~5% after aging for longer exposure times at 25 °C and 60 °C when compared to the unaged samples. This could be due to the interaction of polar carbonyl groups within the LOHC fuel and the rearrangement of polymer chains at elevated temperatures. In addition to this, for all the unaged and aged samples a small shoulder appears before the melting peak which is caused by annealing effects at high temperatures.

3.3. Thermogravimetric Analysis (TGA)

According to Figure 4a,b, HDPE exhibited mass losses between 100 and 300 °C before decomposition, which occurred at approximately 500 °C. This initial mass loss in the aged samples can be attributed to the evaporation of absorbed fuel, which reflects the fuel uptake measurements. As HDPE is exposed to LOHC fuel for 500 h under ambient conditions (25 °C), degradation is slightly faster than the sample aged at 100 h which can be seen in Figure 4a. In addition to this, the degradation caused by the mass loss of samples aged in LOHC fuel at 60 °C for 500 h is much faster than the samples aged for 500 h at 25 °C.

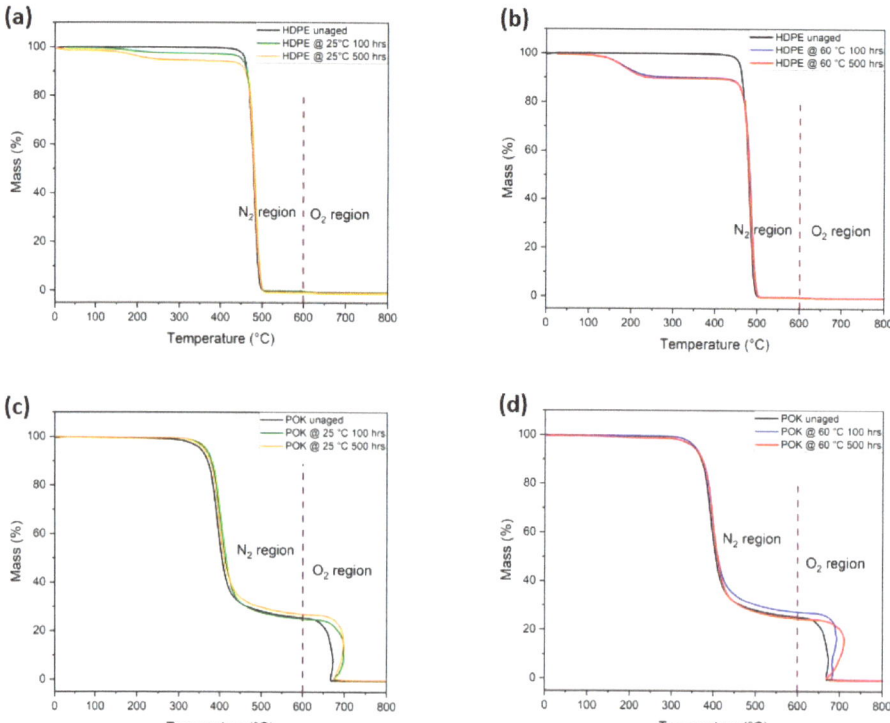

Figure 4. (**a**,**b**) TGA thermograms of unaged and aged injection-molded HDPE samples at 25 and 60 °C; (**c**,**d**) are the TGA thermograms of unaged and aged POK samples at 25 and 60 °C (all curves are based on mean values of three measurements).

Figure 4c,d shows POK thermograms from the TGA, indicating a two-step process before complete degradation occurs. As the absorption rate is very low in POK, no mass loss occurred below 200 °C. All the samples started to lose mass in between 200 °C and 400 °C. In addition to that, only 75% of the polymer broke down under a nitrogen atmosphere, where the ketone groups being eliminated during the first step of the process. The remaining 25% of the degradation occurred in the second step under the oxygen atmosphere. This particular step plays a crucial role in the analytical process, allowing us to discern and quantify the presence of non-hydrocarbon materials within the samples. It is also noticed that the rate of degradation of POK in the first step is faster than the second step which was also observed by Al-Muaikel et al. [41]. Similar to HDPE, POK also lost mass slightly faster for samples aged at 60 °C when compared to samples aged at 25 °C.

Overall, it is worth noting that, based on the weight loss from volatile compounds, POK demonstrates greater inertness to LOHC fuel compared to HDPE. However, it is shown that the POK's drawback lies in its lower thermal stability, which needs to be enhanced through the use of stabilizers.

3.4. Mechanical Testing

Figure 5a illustrates the stress–strain curves of injection molded HDPE samples immersed in LOHC fuel which are aged in a closed oven at 25 °C and 60 °C. It shows that the maximum stress (σ) of all the aged polymers decreased during exposure to LOHC with respect to time when compared to unaged samples, and this is due to the decrease in crystallinity. On the other hand, the strain at break is increased by immersing samples for a longer time in the LOHC fuel. A distinct difference in the elongation at break was also observed between aged and unaged samples. The strain percentage of the aged sample increased from 79% to 84% at 25 °C over time; at 60 °C, the sample increased from 101% to 103%. But there is a huge difference between the aged samples at 60 °C and the unaged samples, where the unaged sample remained at 77%. This increased strain percentage was due to the higher fuel concentration in the polymers which leads to a plasticizing effect that can soften the material over time.

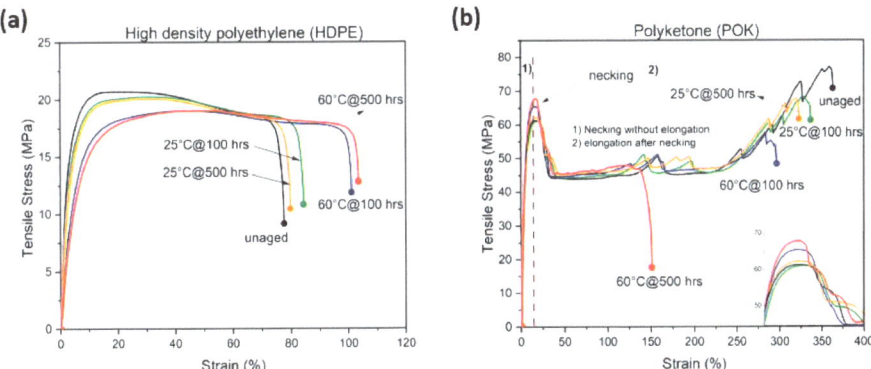

Figure 5. The stress–strain curves of aged and unaged samples of (**a**) HDPE and (**b**) POK in a closed oven are measured after immersion for up to 500 h in LOHC fuel. The inset in (**b**) shows the magnified detail of the stress peaks in the necking region (all curves are based on mean values of three measurements).

According to Figure 5b, POK shows a rise in the maximum stress (σ) in the initial region of the curve after the aging process. At higher strains, the tensile stress of the POK samples decreases gradually due to strain hardening. It is also noticed that the strain at break of aged samples almost remained constant at 25 °C over time, whereas at the elevated temperature of 60 °C, a sharp decline in strain at break was observed from about 300% to 150% which could be due to the increased crystallinity and a reduction in the mobility of the polymer chains that makes the material less flexible and decreases the strain at break.

In addition, due to the aging at high temperatures for 500 h, the intermolecular forces in POK may improve and strengthen the material, leading to a slight increase in tensile stress from 60 MPa to 67 MPa. In Table 4, the E-modules of HDPE and POK are listed along with the strain at break percentage.

Table 4. Youngs modulus E and strain at break percentage of HDPE and POK are defined below.

Aging Time (h)	Oven Temperature (°C)	E-Modulus (GPa) (HDPE)	E-Modulus (GPa) (POK)	Strain (%) (HDPE)	Strain (%) (POK)
Unaged	25	0.61 ± 0.01	1.96 ± 0.20	78 ± 3.9	363 ± 29
Aged @ 100	25	0.55 ± 0.01	1.70 ± 0.01	84 ± 7.7	336 ± 43
Aged @ 500	25	0.53 ± 0.01	1.72 ± 0.04	80 ± 3.1	322 ± 40
Aged @ 100	60	0.30 ± 0.01	1.56 ± 0.09	101 ± 2	296 ± 25
Aged @ 500	60	0.27 ± 0.01	1.63 ± 0.01	103 ± 8.4	151 ± 33

The above results indicate that the exposure temperature of the LOHC strongly influences the tensile behavior of HDPE and POK. The POK aged in LOHC showed higher tensile properties and a drop in elongation, whereas, in the case of HDPE, the elongation at break percentage increased but tensile properties decreased due to the plasticization effect.

3.5. FTIR Spectrum

Figure 6a,b contrasts the FTIR spectra of HDPE and POK that are displayed for before and after being immersed in LOHC until 500 h at both 25 and 60 °C. All the spectra exhibit characteristic IR absorbance peaks commonly associated with HDPE and POK. For HDPE, the peaks at 2913 and 2847 cm^{-1} correspond to asymmetric and symmetric C-H bond stretching, respectively. Additionally, a peak at 1461 and 1463 cm^{-1} indicates the presence of CH_3 umbrella bending mode, while the peak at 719 and 720 cm^{-1} signifies a rocking movement.

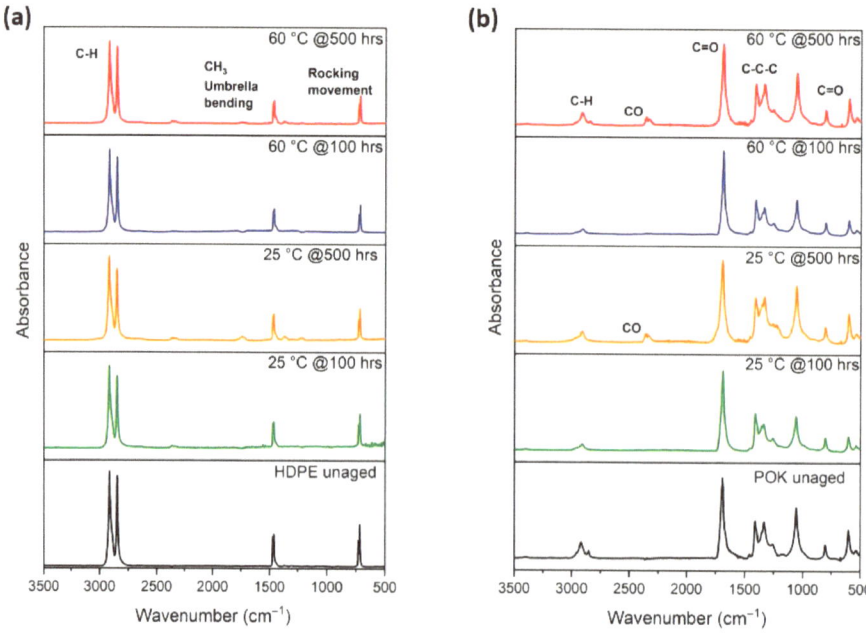

Figure 6. (**a**,**b**) are the FTIR spectra of the unaged and aged (**a**) HDPE and (**b**) POK at 25 and 60 °C for 100 and 500 h.

In the case of POK, a distinct peak at 1687 cm^{-1} indicates the presence of carbonyl groups. The methylene groups in polyketones contribute to a broad peak in the range of 2800–3000 cm^{-1}, and there are also additional peaks around 1300 cm^{-1} corresponding to stretching vibrations of carbon–carbon–carbon (C-C-C) bonds. Peaks at 600 and 800 cm^{-1} represent C=O bond stretching. By comparing the immersed samples to the non-immersed one, it can be concluded that no noticeable chemical changes occurred as a result of LOHC absorption in the polymers. Despite the increase in temperature, there is also no discernible difference in the FTIR spectra of HDPE and POK.

4. Conclusions

This study compared the aging behavior of HDPE and POK that are exposed to LOHCs (at ambient conditions and at 60 °C). HDPE, as a non-polar polymer, showed a greater fuel uptake compared to POK after 3 weeks. However, an initial mass loss was observed in POK below 100 h at 60 °C due to the evaporation of volatiles. Nevertheless, the POK samples did not reach a plateau level like HDPE, and therefore, more immersion tests need to be performed. From mass uptake measurements, the diffusion of chemicals into polymers was calculated. The diffusion coefficient increased from 1.25×10^{-7} at 25 °C to 1.64×10^{-7} cm^2 s^{-1} in HDPE at 60 °C due to the weak bonds in the polymer chain, whereas, in POK, the rate of diffusion was less with 5.62×10^{-8} cm^2 s^{-1} at 25 °C.

Fuel uptake, plasticization effect, and crystallinity led to differences in tensile strengths of HDPE and POK. Due to the presence of non-polar groups the absorption rate of HDPE samples immersed in LOHC fuel was increased. The LOHC fuel used in this work contains hydrocarbons which act as plasticizers. Fuels like these can penetrate and break down the polymer chains and increase the free volume between them, which results in a decrease in crystallinity and tensile strength and increased flexibility. Thus, HDPE lost more strength and the elongation at break percentage increased due to the plasticization effect.

In the case of POK, LOHC fuel exposure may cause polymer chain restructuring. The interaction with the fuel can result in the creation of highly ordered and crystalline regions within the polymer matrix, improving its overall crystallinity and tensile strength and making the material more rigid. Therefore, the elongation at break percentage was reduced in POK as the aging temperature increased from 25 °C to 60 °C, but the tensile strength increased gradually after aging for 500 h. This could be due to the formation or enhancement of crosslinking within the polymer structure.

Furthermore, the FTIR results indicated no significant chemical changes in both materials. After the long experimental trials neither of the wet specimens developed discoloration or cracks.

Therefore, in order to store fuels like LOHC for a longer time, not only at room temperature (25 °C) but even in high temperatures (60 °C), POK could be considered as an alternative choice to store and transport hydrogen due to its excellent barrier properties and mechanical properties when compared to HDPE, because materials like HDPE are less suitable for applications that require a high mechanical strength and dimensional stability.

Author Contributions: Conceptualization, J.S.; methodology, J.S. and T.L.; validation, J.S.; formal analysis, J.S.; investigation, J.S. and M.S.; resources, J.S.; data curation, J.S.; writing—original draft preparation, J.S.; writing—review and editing, J.S., M.S., T.L. and C.H.; visualization, J.S.; supervision, T.L. and C.H.; project administration, T.L. All authors have read and agreed to the published version of the manuscript.

Funding: This research received no external funding.

Institutional Review Board Statement: Not applicable.

Data Availability Statement: The data presented in this study are available on request from the corresponding author.

Acknowledgments: We would like to thank LyondellBasell for supplying the materials and the Polymer Competence Center Leoben GmbH (PCCL) for providing closed oven chambers to investigate experimental results.

Conflicts of Interest: The authors declare no conflict of interest.

References

1. European Commission. Communication from the Commission to the European Parliamen, the Council, the European Economic and Social Committee and the Committee of the Regions: A hydrogen Strategy for a Climate-Neutral Europe. Available online: https://ec.europa.eu/energy/sites/ener/files/hydrogen_strategy.pdf (accessed on 17 October 2023).
2. Erbach, G.; Jensen, L. *Hydrogen as an Energy Carrier for a Climate-Neutral Economy: EU Hydrogen Policy*; European Parliamentary Research Servic: Brussels, Belgium, 2021.
3. Ishaq, H.; Dincer, I.; Crawford, C. A review on hydrogen production and utilization: Challenges and opportunities. *Int. J. Hydrogen Energy* **2022**, *47*, 26238–26264. [CrossRef]
4. Owusu, P.A.; Asumadu-Sarkodie, S. A review of renewable energy sources, sustainability issues and climate change mitigation. *Cogent Eng.* **2016**, *3*, 1167990. [CrossRef]
5. Relying on Hydrogen Energy to Run 1360 Kilometers! The New Generation Toyota Mirai Sets the Guinness World Record. *News Directory 3 [Online]*, 10 September 2021. Available online: https://www.newsdirectory3.com/relying-on-hydrogen-energy-to-run-1360-kilometersthe-new-generation-toyota-mirai-sets-the-guinness-world-record/ (accessed on 27 July 2023).
6. Hyundai Nexo: The SUV That Could Make the Market for Hydrogen-Powered Cars Really Take Off. *The Telegraph [Online]*. Available online: https://www.telegraph.co.uk/cars/hyundai/hyundai-nexo-suv-could-make-market-hydrogen-powered-cars-really/ (accessed on 27 July 2023).
7. Eberle, U.; Müller, B.; von Helmolt, R. Fuel cell electric vehicles and hydrogen infrastructure: Status 2012. *Energy Environ. Sci.* **2012**, *5*, 8780. [CrossRef]
8. Bruce, G. Road Testing BMW's Hydrogen 7. Available online: https://web.archive.org/web/20140429204449/http://archive.wired.com/cars/energy/news/2006/11/72100?currentPage=all (accessed on 27 July 2023).
9. Ozsaban, M.; Midilli, A. A parametric study on exergetic sustainability aspects of high-pressure hydrogen gas compression. *Int. J. Hydrogen Energy* **2016**, *41*, 5321–5334. [CrossRef]
10. Møller, K.T.; Jensen, T.R.; Akiba, E.; Li, H. Hydrogen—A sustainable energy carrier. *Prog. Nat. Sci. Mater. Int.* **2017**, *27*, 34–40. [CrossRef]
11. Daisuke, K. *Introduction of Liquid Organic Hydrogen Carrier and the Global Hydrogen Supply Chain Project*; Chiyoda Corporation: Jokohama, Japan, 2018.
12. Jorschick, H.; Geißelbrecht, M.; Eßl, M.; Preuster, P.; Bösmann, A.; Wasserscheid, P. Benzyltoluene/dibenzyltoluene-based mixtures as suitable liquid organic hydrogen carrier systems for low temperature applications. *Int. J. Hydrogen Energy* **2020**, *45*, 14897–14906. [CrossRef]
13. Zhang, Y.-Q.; Stolte, S.; Alptekin, G.; Rother, A.; Diedenhofen, M.; Filser, J.; Markiewicz, M. Mobility and adsorption of liquid organic hydrogen carriers (LOHCs) in soils–environmental hazard perspective. *Green Chem.* **2020**, *22*, 6519–6530. [CrossRef]
14. Wunsch, A.; Berg, T.; Pfeifer, P. Hydrogen Production from the LOHC Perhydro-Dibenzyl-Toluene and Purification Using a 5 μm PdAg-Membrane in a Coupled Microstructured System. *Materials* **2020**, *13*, 277. [CrossRef]
15. Rivard, E.; Trudeau, M.; Zaghib, K. Hydrogen Storage for Mobility: A Review. *Materials* **2019**, *12*, 1973. [CrossRef]
16. Hurskainen, M.; Ihonen, J. Techno-economic feasibility of road transport of hydrogen using liquid organic hydrogen carriers. *Int. J. Hydrogen Energy* **2020**, *45*, 32098–32112. [CrossRef]
17. Sievi, G.; Geburtig, D.; Skeledzic, T.; Bösmann, A.; Preuster, P.; Brummel, O.; Waidhas, F.; Montero, M.A.; Khanipour, P.; Katsounaros, I.; et al. Towards an efficient liquid organic hydrogen carrier fuel cell concept. *Energy Environ. Sci.* **2019**, *12*, 2305–2314. [CrossRef]
18. Hauenstein, P.; Seeberger, D.; Wasserscheid, P.; Thiele, S. High performance direct organic fuel cell using the acetone/isopropanol liquid organic hydrogen carrier system. *Electrochem. Commun.* **2020**, *118*, 106786. [CrossRef]
19. Daniel, T. Hydrogen storage and distribution via liquid organic carriers. In Proceedings of the Bridging Renewable Electricity with Transportation Fuels Workshop, Denver, CO, USA, 27–28 August 2015.
20. Böhning, M.; Niebergall, U.; Adam, A.; Stark, W. Impact of biodiesel sorption on mechanical properties of polyethylene. *Polym. Test.* **2014**, *34*, 17–24. [CrossRef]
21. Böhning, M.; Niebergall, U.; Zanotto, M.; Wachtendorf, V. Impact of biodiesel sorption on tensile properties of PE-HD for container applications. *Polym. Test.* **2016**, *50*, 315–324. [CrossRef]
22. Selke, S.; Hernandez, R. *Packaging: Polymers for Containers*; Elsevier: Amsterdam, The Netherlands, 2001; ISBN 0-08-0431526.
23. Richaud, E.; Flaconnèche, B.; Verdu, J. Biodiesel permeability in polyethylene. *Polym. Test.* **2012**, *31*, 1070–1076. [CrossRef]
24. Zhao, J.; Mallick, P.K. Effect of biodiesel on polyamide-6-based polymers. *Polym. Eng. Sci.* **2019**, *59*, 1445–1454. [CrossRef]
25. Kass, M.D.; Janke, C.J.; Connatser, R.M.; Lewis Sr, S.A.; Keiser, J.R.; Gaston, K. Compatibility Assessment of Fuel System Infrastructure Plastics with Bio-oil and Diesel Fuel. *Energy Fuels* **2017**, *32*, 542–553. [CrossRef]

26. Yousif, B.F.; Ku, H. Suitability of using coir fiber/polymeric composite for the design of liquid storage tanks. *Mater. Des. (1980–2015)* **2012**, *36*, 847–853. [CrossRef]
27. Fillot, L.-A.; Ghiringhelli, S.; Prebet, C.; Rossi, S. Biofuels Barrier Properties of Polyamide 6 and High Density Polyethylene. *Oil Gas Sci. Technol.—Rev. IFP Energ. Nouv.* **2015**, *70*, 335–351. [CrossRef]
28. Yang, H.; Yang, S.; Wu, C.; Fei, Y.; Wei, X. The Applications of Direct Fluorinated HDPE in Oil & Gas storage and Transportation. *Adv. Materails Res.* **2011**, *328–330*, 2436–2439.
29. Smaardijk, A.; Drent, E. Polyketones: Aliphatic. *Sci. Technol.* **2001**, 7194–7197. [CrossRef]
30. Nam, J.U.; Choi, E.Y.; Park, H.J.; Kim, C.K. Fabrication of polyketone-grafted multi-walled carbon nanotubes using Grignard reagent and their composites with polyketone. *Compos. Sci. Technol.* **2018**, *167*, 199–205. [CrossRef]
31. You, J.; Choi, H.-H.; Kim, T.A.; Park, M.; Ha, J.S.; Lee, S.-S.; Park, J.H. High-performance polyketone nanocomposites achieved via plasma-assisted mechanochemistry. *Compos. Sci. Technol.* **2019**, *183*, 107800. [CrossRef]
32. Belov, G.P.; Novikova, E.V. Polyketones as alternating copolymers of carbon monoxide. *Russ. Chem. Rev.* **2004**, *73*, 267–291. [CrossRef]
33. He, Y.; Luo, R.; Li, Z.; Lv, R.; Zhou, D.; Lim, S.; Ren, X.; Gao, H.; Hu, W. Comparing Crystallization Kinetics between Polyamide 6 and Polyketone via Chip-Calorimeter Measurement. *Macromol. Chem. Phys.* **2018**, *219*, 1700385. [CrossRef]
34. Crank, J. *The Mathematics of Diffusion*, 2nd ed.; Clarendon Press: Oxford, UK, 1975; ISBN 0198533446.
35. Lee, Y.H.; Kuboki, T.; Park, C.B.; Sain, M.; Kontopoulou, M. The effects of clay dispersion on the mechanical, physical, and flame-retarding properties of wood fiber/polyethylene/clay nanocomposites. *J. Appl. Polym. Sci.* **2010**, *118*, 452–461. [CrossRef]
36. Sim, M.-J.; Shim, J.; Lee, J.-C.L.; Cha, S.-H. Preparation of a novel phosphorus–nitrogen flame retardant and its effects on the flame retardancy and physical properties of polyketone. *J. Appl. Polym. Sci.* **2020**, *137*, 49199. [CrossRef]
37. Lei, W.; Xiuling, C.; Zhiguang, Z.; Shan, X.; Canghai, M.; Nanwen, L. Enhanced molecular selectivity and plasticization resistance in ring-opened Trger's base polymer membranes. *J. Membr. Sci.* **2021**, *634*, 119399.
38. Baena, L.M.; Zuleta, E.C.; Calderón, J.A. Evaluation of the Stability of Polymeric Materials Exposed to Palm Biodiesel and Biodiesel–Organic Acid Blends. *Polymers* **2018**, *10*, 511. [CrossRef]
39. Schweitzer, P.A. *Corrosion of Polymers and Elastomers: Engineering & Technology, Physical Sciences*, 1st ed.; CRC Press: Boca Raton, FL, USA, 2006; ISBN 9780429129391.
40. Sabard, M.; Gouanvé, F.; Espuche, E.; Fulchiron, R.; Seytre, G.; Fillot, L.-A.; Trouillet-Fonti, L. Influence of film processing conditions on the morphology of polyamide 6: Consequences on water and ethanol sorption properties. *J. Membr. Sci.* **2012**, *415–416*, 670–680. [CrossRef]
41. Al-Muaikel, N.S.; Aly, K.I. Friedel–Crafts Polyketones: Synthesis, Thermal and Antimicrobial Properties. *Int. J. Polym. Mater.* **2009**, *59*, 45–59. [CrossRef]

Disclaimer/Publisher's Note: The statements, opinions and data contained in all publications are solely those of the individual author(s) and contributor(s) and not of MDPI and/or the editor(s). MDPI and/or the editor(s) disclaim responsibility for any injury to people or property resulting from any ideas, methods, instructions or products referred to in the content.

Article

Effects of Nd$_2$O$_3$ Nanoparticles on the Structural Characteristics and Dielectric Properties of PVA Polymeric Films

Khulaif Alshammari [1,*], Thamer Alashgai [1], Alhulw H. Alshammari [1], Mostufa M. Abdelhamied [2], Satam Alotibi [3] and Ali Atta [1]

1. Physics Department, College of Science, Jouf University, P.O. Box 2014, Sakaka 72388, Saudi Arabia; 421100043@ju.edu.sa (T.A.); ahalshammari@ju.edu.sa (A.H.A.); aamahmad@ju.edu.sa (A.A.)
2. Charged Particles Lab., Radiation Physics Department, National Center for Radiation Research and Technology (NCRRT), Egyptian Atomic Energy Authority (EAEA), Cairo 13759, Egypt; m_elbana52@yahoo.com
3. Department of Physics, College of Science and Humanities in Al-Kharj, Prince Sattam bin Abdulaziz University, Al-Kharj 11942, Saudi Arabia; sf.alotibi@psau.edu.sa
* Correspondence: knnalshammari@ju.edu.sa

Abstract: Polyvinyl alcohol (PVA) and Neodymium (III) oxide (Nd$_2$O$_3$) were combined to synthesized flexible innovative PVA/Nd$_2$O$_3$ polymer composite samples utilizing a solution casting approach for use in dielectric devices. The XRD, FTIR, and SEM methods are all investigated to characterize the composite films. In a frequency of 50 Hz to 5 MHz, the effects of additive Nd$_2$O$_3$ on the dielectric behavior of PVA were recorded. The PVA/Nd$_2$O$_3$ composite films were successfully fabricated, as shown by XRD and infrared spectroscopy. The scanning microscopy pictures showed that the Nd$_2$O$_3$ was loaded and distributed uniformly throughout the PVA. After the incorporation of Nd$_2$O$_3$, the composite PVA/Nd$_2$O$_3$ has a conductivity of 6.82×10^{-9} S·cm^{-1}, while the PVA has a conductivity of 0.82×10^{-9} S·cm^{-1}. Another improvement is the decrease in the relaxation time from 14.2×10^{-5} s for PVA to 6.35×10^{-5} s for PVA/Nd$_2$O$_3$, and an increase in the dielectric constant of 0.237 for PVA to 0.484 at a frequency of 100 Hz. The results showed that the composite samples have considerable changes as flexible films in different applications, including batteries and electronic circuits.

Keywords: PVA/Nd$_2$O$_3$; structural investigation; dielectric properties; energy applications

Citation: Alshammari, K.; Alashgai, T.; Alshammari, A.H.; Abdelhamied, M.M.; Alotibi, S.; Atta, A. Effects of Nd$_2$O$_3$ Nanoparticles on the Structural Characteristics and Dielectric Properties of PVA Polymeric Films. *Polymers* **2023**, *15*, 4084. https://doi.org/10.3390/polym15204084

Academic Editors: Vineet Kumar and Md Najib Alam

Received: 15 August 2023
Revised: 1 October 2023
Accepted: 9 October 2023
Published: 14 October 2023

Copyright: © 2023 by the authors. Licensee MDPI, Basel, Switzerland. This article is an open access article distributed under the terms and conditions of the Creative Commons Attribution (CC BY) license (https://creativecommons.org/licenses/by/4.0/).

1. Introduction

The use of different polymers, i.e., Polyvinyl chloride (PVC), Polyvinylpyrrolidone (PVP), Polyvinyl alcohol (PVA), etc., has recently increased due to their distinctive properties such as being lightweight and transparent [1]. Their mechanical, optical, or electrical behavior can be influenced by the addition of nanoparticles, resulting in polymer nanocomposites (PNs) [2]. The polymer nanocomposites are launched on the polymer or the polymer matrix which includes the nanofillers within suitable ratios. The weight of the nanofillers is usually far less than the weight of the polymers because some nanofillers are expensive, and they may also have a negative impact on the transparency of polymer films. The efficiency of the polymer nanocomposites' material in inhibiting electromagnetic interference (EMI) depends on a number of factors, including its dielectric permittivity and electrical conductivity. Dielectric loss, dielectric constant, and their frequency dependencies [3] are all measures of an EMI binding material's ability to absorb or scatter EM waves. Polymer composites may be a viable alternative to metals and ceramics as a shielding material due to their thermal stability, mechanical performance, and cost [4].

Polymeric materials have become a well-known hosting matrix for rare earth ions because of their attractive advantages, including the ease of fabrication, in addition to their good transparency to the visible and infrared regions [5]. PVA is a polymer that features a carbon chain and hydroxyl group. The PVA polymer has outstanding characteristics such as humidity, water absorption, and being easily produced, which has motivated further studies with this substance [6]. In recent decades, the MOS (metal oxide semiconductor) structure has gained popularity in the field of solid-state electronics. It is simple to examine the dielectric behavior of the insulating oxide layer [7] and the material's potential compatibility by microelectronic applications [8] using MOS capacitor designs. The ability to modify the properties of nanocomposites has been made possible by incorporating inorganic nanofillers into the polymer. Polymer chains can be altered at their very core thanks to the nanofillers conducting connections within the polymer [9]. Nanoparticle fillers are more efficient than micron-sized fillers; therefore, their integration into polymers results in better polymer characteristics. Variations in the polymer chains' dielectric properties can be observed as a direct result of the differences in their structural properties [10]. Polymer nanocomposites (PNs) are applied to various promising devices such as nanotechnology and supercapacitors [11]. The doping polymer matrix with suitable nanoparticles will generate PNs with distinctive properties, including mechanical, optical, or electrical properties. Dopants can be rare earth elements such as metal oxide or neodymium oxide (Nd_2O_3), which are characterized by their excellent properties.

The most attention has been paid to rare earth oxides because they have many promising applications. In addition, among the lanthanide series, neodymium oxide (Nd_2O_3) is regarded as the most important rare earth oxide. Nd_2O_3 is a rare earth element that has been used in sophisticated applications including magnetic devices and protective coatings because of its desirable property [6]. Because of its high dielectric constant, Nd_2O_3 is essentially electrically and thermally stable. In addition, its conduction band is well suited for use in microelectronics [12]. These features make Nd_2O_3 a superior gate dielectric for semiconductor devices compared to silicon dioxide (SiO_2). In addition, Nd_2O_3 has been extensively applied in optoelectronics because of its narrower gap energy [13]. Several research efforts have focused on studying the alterations in the oxidation properties of rare earth metals through the introduction of various photonics materials, including optical crystals (e.g., Al), glasses, and semiconductors (e.g., Si, SiGe) [14]. Neodymium oxide has shown promising uses in a variety of fields in recent years. This is because neodymium, unlike most rare earth elements, is abundant in the Earth's crust and rapidly forms its oxides. Crystals of Nd_2O_3, a compound with a light grayish blue color, are employed in the production of solid-state lasers. Sunglasses, welding goggles, and other types of glass benefit from Nd_2O_3 dichroic characteristics [15].

Moreover, neodymium oxide (Nd_2O_3) nanoparticles have garnered a lot of attention among the rare earth elements due to their superior dielectric characteristics. In comparison to excited electrons, the UV- excitation produces more effective light radiation [16]. In recent times, there has been a significant focus on Nd_2O_3 nanoparticles as a dielectric filler due to their outstanding properties that have potential applications in various fields such as thermo-luminescence, protective coatings, and dielectric devices [17]. The nanoparticles of Nd_2O_3 have several potential applications because of their size impact. Therefore, Nd_2O_3 nanoparticle synthesis has received considerable interest in recent years [18]. Keikhaei et al. have investigated the incorporation of Nd_2O_3 on the properties of the PVA polymer [19]. Various characterizations, namely structural, electrical, and optical, were conducted on the PVA/PVP blend doped with Nd_2O_3 [6]. AlAbdulaal et al. [20] studied the optical behavior for PVA doped by various quantities of Nd_2O_3, which were processed by the casting method. The novelty of this research is to improve the electrical characteristics of the PVA polymer doped by Nd_2O_3 nanoparticles, which have hardly been studied by other researchers. Therefore, the present research modified the dielectric characteristics of PVA for applications in energy devices. The XRD, SEM, and FTIR methods were used to investigate the PVA/Nd_2O_3 characteristics. In a frequency spanning from 100 Hz to 5 MHz, the permittivity, impedance, and dielectric efficiency of PVA/Nd_2O_3 films were measured.

2. Materials and Methods

The Nd_2O_3 powder (purity of 99.95% and average size of 25–40 nm) was purchased from Nanografi Nano Technology Company (Darmstadt, Germany). The PVA powder (molecular weight: 29,000–69,000 g/mol, and 87.90% hydrolyzed) was given by Sigma-Aldrich Company, Darmstadt, Germany. The solution casting method [21,22] was elected for making polymer films because it involves a simple preparation, is inexpensive, allows for the precise management of composite composition, and produces highly uniform nanocomposite films. Firstly, 1.0 g of PVA was dissolved in 85 mL of deionized water with magnetic stirring for 1.5 h. After stirring the PVA polymer for 1.5 h, different amounts of Nd_2O_3 (0.05, 0.10, 0.15, and 0.20 wt%) were added and mixed in PVA. The PVA/Nd_2O_3 solution was then poured into a glass Petri dish and allowed to air dry at room temperature. The sheets were then removed from the glass plates with an average thickness of 170 µm.

The chemical changes were studied with FTIR (Shimadzu FTIR, Tracer 100, Kyoto, Japan) in the wavenumber of 410–3900 cm^{-1}, while XRD (Shimadzu XRD 7000, Kyoto, Japan) was used to examine the structural characteristics at 2θ range (10–80)°. In order to determine the morphology and geometric shape of the films, FESEM (Thermo Fisher Scientific, Waltham, MA, USA) was used. Typically, a gold film is sputtered onto the surface of the film before the sample is placed under an electron microscope. The LCR meter (RS-232C interface, Hioki, Japan) is used to record the dielectric parameters at room temperature at a frequency of 50 Hz to 5 MHz.

3. Results and Discussion

The XRD was utilized for studying the structure of the PVA film and PVA/Nd_2O_3 nanocomposite, as shown in Figure 1. The pattern of the PVA displayed a peak located at $2\theta = 19.6°$, which refers to the (101) plane. A significant peak is observed at approximately $2\theta = 40°$, indicating that the structure of (PVA) exhibits semi-crystalline characteristics [23]. The XRD peaks observed of Nd_2O_3 were found to be consistent with the standard reference card for Nd_2O_3 with JCPDS Card No: 74–1147. The distinct diffraction peak observed in the XRD pattern of Nd_2O_3 indicates the presence of an orthorhombic lattice structure. Furthermore, it has been observed that the peak at 19.4° exhibits a decrease by the incorporation of Nd_2O_3. The inclusion of Nd_2O_3 causes modifications to the amorphous structure of the PVA, resulting in a change in crystallinity [24]. In addition, the number of planes of the peaks' angles is modified as Nd_2O_3 concentration increases, indicating that the PVA and Nd_2O_3 nanoparticles interacted [25]. The addition of Nd_2O_3 to PVA results in higher crystallinity in comparison to pure PVA. X-ray diffraction (XRD) analysis revealed the formation of a complex between the Nd_2O_3 additives and the polyvinyl alcohol (PVA) polymer chain.

The crystallite size of Nd_2O_3 is determined by using the Scherer formula [26].

$$D = \frac{0.9\lambda}{\beta \cos\theta} \quad (1)$$

The crystallite size is D, β is the full width, and λ is the wavelength. With increasing concentrations of PVA, the crystallite size of Nd_2O_3 grows from 13.5 nm for PVA/0.1Nd_2O_3 to 15 nm and 18.2 nm, respectively, for PVA/0.15Nd_2O_3 and PVA/0.2Nd_2O_3, demonstrating that Nd_2O_3 is formed at the nanoscale. Because of its low concentration (0.05 wt%), PVA/0.05Nd_2O_3 does not exhibit any peaks characteristic of Nd_2O_3 in XRD spectra. In turn, this makes it easier for Nd_2O_3 to be distributed uniformly throughout the PVA.

An additional investigation through the utilization of FTIR analysis was conducted to prove the changes occurred in the chemical characteristics of the composite after the incorporation of Nd_2O_3. Furthermore, FT-IR was utilized to analyze the functional groups present in the composite films of PVA doped with the Nd_2O_3 nanocomposite. The FT-IR measurements were obtained from 4000 to 500 cm^{-1} in wavenumber, as shown in Figure 2. The FTIR of PVA showed a band observed at a wavenumber of 3285 cm^{-1}, with the stretching O–H resulting from the hydrogen bonds between and inside molecules, as reported in previous studies [27]. Moreover, for the C–H band, asymmetric stretching vibrations are found at a wavenumber of 2920 cm^{-1}. The bands of 1720 cm^{-1} and 1080 cm^{-1} correspond to carbonyl stretching (C=O) and C-O bond stretching from the acetate groups. C-H bending appeared in the infrared (IR) region at 1430 cm^{-1}. The wavenumber at 840 cm^{-1} provided evidence of the specific formation of CH$_2$ bending [28]. The observed reduction in the maximum intensity of PVA due to the addition of Nd$_2$O$_3$ indicates the presence of intermolecular bands, which can be attributed to the hydroxyl group present in PVA. The Nd$_2$O$_3$ nanoparticle incorporation changed the electron-hole recombination rate because of the defects created into the PVA polymeric structure, leading to a corresponding change in the peak positions. At the same time, the ability of Nd$_2$O$_3$ to dampen the brightness of PVA/Nd$_2$O$_3$ indicates that there are too many electrons present, leading to enhanced performance in terms of the materials' electrical properties.

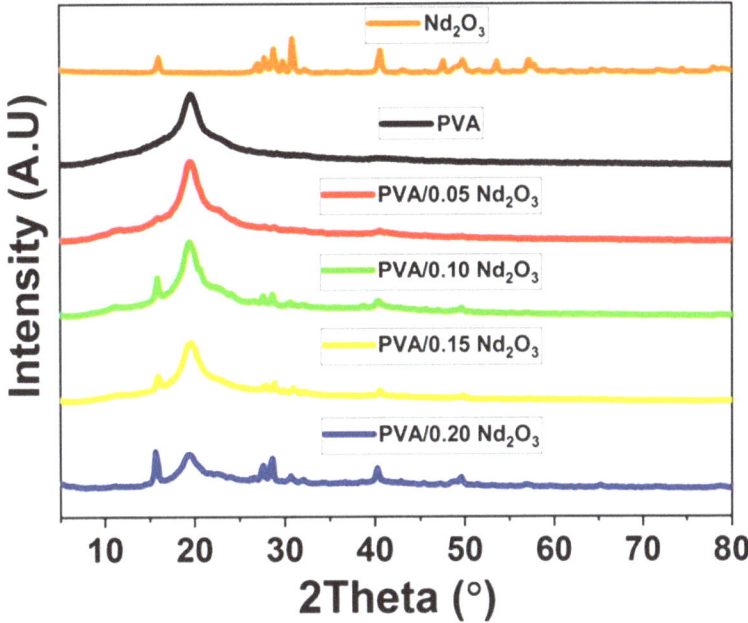

Figure 1. XRD diffraction for the polymer PVA, Nd$_2$O$_3$, and the composite PVA/Nd$_2$O$_3$.

The morphology of PVA/Nd$_2$O$_3$ is shown in Figure 3a–d. A micrograph of PVA film, as shown in Figure 3a, has a uniform and smooth surface. Figure 3b–d depict PVA/Nd$_2$O$_3$ films with 0.05Nd$_2$O$_3$, 0.10Nd$_2$O$_3$, and 0.20Nd$_2$O$_3$. The SEM pictures of the composites reveal no discernible agglomeration, cracks, or breaks, suggesting that the cross-linker is well dispersed in the PVA chains by increasing Nd$_2$O$_3$ [28]. The increase in Nd$_2$O$_3$ content (from 5% to 20%) causes the dispersion to become highly apparent, as clearly shown in Figure 3b–d. The presence of the cross-linkers created in the composite is attributed to the formation of strong hydrogen bonds between PVA and Nd$_2$O$_3$ or the increase in viscosity during the composite preparation. This phenomenon has been documented in previous investigations [5,20,28]. It is possible that the observed morphological alterations are due to Nd$_2$O$_3$ and the preparation method conditions. The concentration of Nd$_2$O$_3$ in PVA also affects its dispersion and the interfacial interaction of PVA chains and Nd$_2$O$_3$ [5].

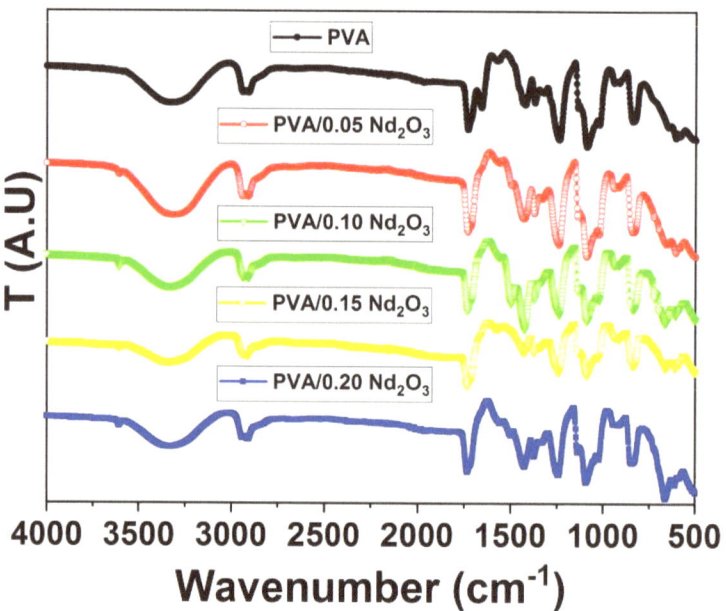

Figure 2. FTIR spectrum for polymer nanocomposite PVA/Nd$_2$O$_3$ and different concentrations of Nd$_2$O$_3$.

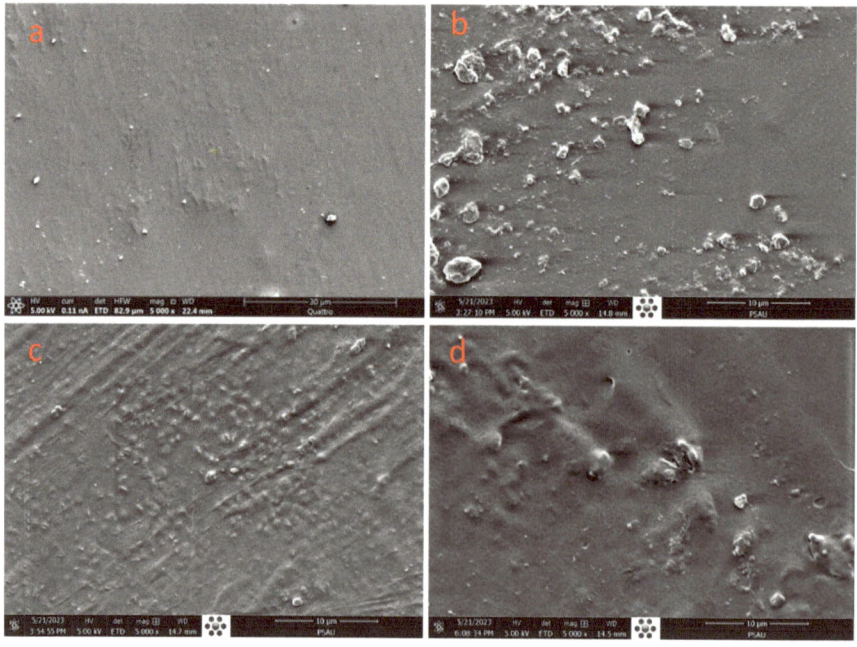

Figure 3. SEM of (**a**) PVA, (**b**) PVA/0.05 Nd$_2$O$_3$, (**c**) PVA/0.10Nd$_2$O$_3$, and (**d**) PVA/0.20Nd$_2$O$_3$.

The dielectric real ε' and imaginary ε'' value, which are related to complex dielectric permittivity (ε^*), are given by [29]:

$$\varepsilon^* = \varepsilon' - i\,\varepsilon'' \tag{2}$$

The real ε' dielectric constant ε' is given by [30]:

$$\varepsilon' = \frac{c \cdot d}{\varepsilon_o \cdot A} \tag{3}$$

c is the capacitance, the film thickness is t, and A is the area. Figure 4 displays the ε' with frequency for the PVA, PVA/0.05Nd$_2$O$_3$, PVA/0.10Nd$_2$O$_3$, PVA/0.15Nd$_2$O$_3$, and PVA/0.20Nd$_2$O$_3$ films. At a lower frequency, the frequency-dependent attenuation of the ε' is observed for all samples. This occurs because the dipoles have enough time to encounter polarization between themselves at smaller frequencies. Then, the ε' remains nearly constant at higher frequencies, which may be because the dipoles do not have more time to align with the orientation of the external fields [31]. In addition, the high frequency values of the applied external fields are too rapid for the interface states to rearrange.

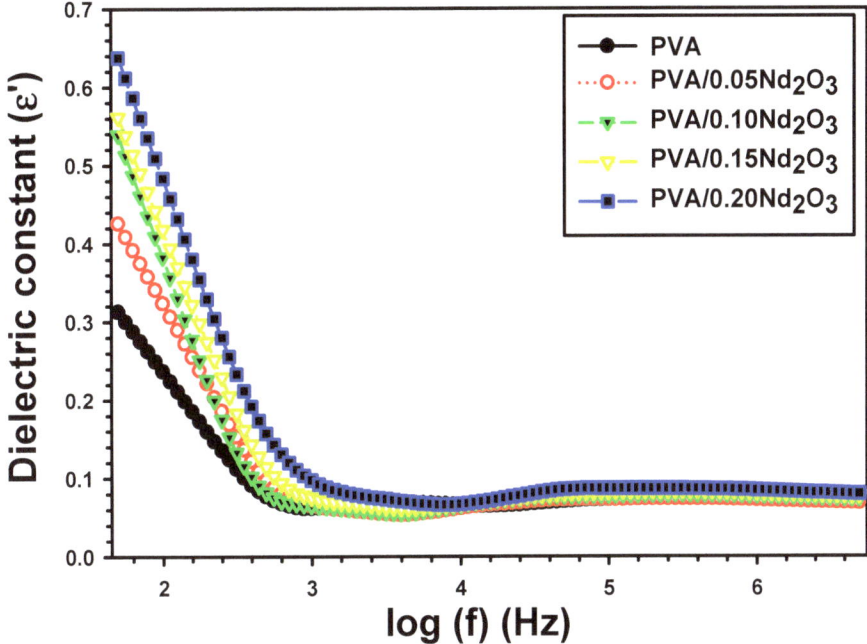

Figure 4. The dielectric constant (ε') with frequency for PVA and different concentrations of Nd$_2$O$_3$ nanocomposite.

Incorporating Nd$_2$O$_3$ increases the ε' from 0.236 for pure PVA at 100 Hz to 0.323, 0.383, 0.418, and 0.483 for PVA/0.05 Nd$_2$O$_3$, PVA/0.10 Nd$_2$O$_3$, PVA/0.15 Nd$_2$O$_3$, and PVA/0.20 Nd$_2$O$_3$ (Table 1). The rapid growth due to Nd$_2$O$_3$ also increases dipole density and hence, enhances polarizability at interfaces. Thus, the increased polarization and dielectric characteristics leads to an increased number of dipoles caused by the flaws in the PVA polymer matrix. Atta [32] also showed this phenomenon when CuNPs were deposited on PTFE and PET polymeric films. He showed that the ε' increased from 0.63 for pure PET to 1.18 after 25 min of deposition Cu on PET, and from 1.95 for pure PTFE to 2.25 for

Cu on PTFE. The improvement in the dielectric constant supports the use of these samples in energy storage applications with a high ability to store charge.

Table 1. Values of ε', ε'', σ_{ac}, M', M'', and U of PVA and PVA/Nd$_2$O$_3$.

	ε'	ε''	M'	M''	σ_{ac} (S/cm)	U (J/m^3)
PVA	0.237	0.312	4.21	2.31	0.82×10^{-9}	1.05×10^{-6}
PVA/0.05Nd$_2$O$_3$	0.324	0.539	2.48	1.49	1.67×10^{-9}	1.43×10^{-6}
PVA/0.10Nd$_2$O$_3$	0.383	0.592	2.31	1.25	2.01×10^{-9}	1.70×10^{-6}
PVA/0.15Nd$_2$O$_3$	0.428	0.585	1.46	1.01	4.32×10^{-9}	$1.85 10^{-6}$
PVA/0.20Nd$_2$O$_3$	0.483	0.669	0.488	0.84	6.82×10^{-9}	2.14×10^{-6}

The imaginary dielectric loss ε'' is given by [33]:

$$\varepsilon'' = \varepsilon' \tan \delta \qquad (4)$$

Figure 5 illustrates the frequency-dependent dielectric loss ε'' for PVA, PVA/0.5Nd$_2$O$_3$, PVA/0.10Nd$_2$O$_3$, PVA/15Nd$_2$O$_3$, and PVA/0.20Nd$_2$O$_3$. It has been shown that the enhancement of the frequency significantly reduces the dielectric loss at a lower frequency [34,35]. This shift in ε'' is also associated with the low-frequency active characteristic trait of dipolar relaxation. Charge carriers is another explanation for the varying of ε'' with frequency. Furthermore, lowering the frequency triggers a charge transfer across the interface. It has been established that PVA/0.05Nd$_2$O$_3$ and PVA/0.20Nd$_2$O$_3$ both increase the ε'' value from 0.313 for PVA to 0.539 and 0.669, respectively. The improved PVA chain and Nd$_2$O$_3$ links are responsible for the increased ε'' with the addition of Nd$_2$O$_3$ resulting in a more robust coupling across the grain boundary [35]. The mini-capacitor networks formed in PVA/Nd$_2$O$_3$ films are also responsible for the enhancement in the dielectric value.

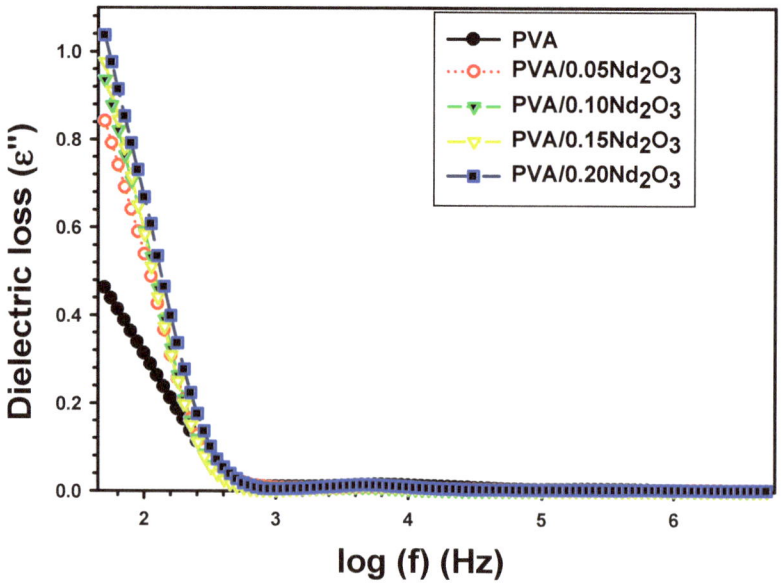

Figure 5. The dielectric loss (ε'') with frequency for PVA and different concentrations of Nd$_2$O$_3$ nanocomposite.

The electrical modulus (M*) is estimated by [35]:

$$M^* = \frac{1}{\varepsilon^*} = M' + i M'' \tag{5}$$

The M′ and M″ are, respectively, the real and imaginary electric modulus, and are recorded by [36]:

$$M' = \frac{\varepsilon'}{\varepsilon'^2 + \varepsilon''^2} \tag{6}$$

$$M'' = \varepsilon'' / \left(\varepsilon'^2 + \varepsilon''^2\right) \tag{7}$$

Figure 6 displays the M′ versus frequency with different concentrations of Nd_2O_3. M′ increases exponentially with frequency, and at a higher frequency, it becomes straight and constant. The polymer electrical dipoles have a considerable self-orientation propensity at low frequencies, while the orientation produced by an external electric field is minimal at high frequencies [37]. The M′ is also affected by the electrode polarization [38]. As can be shown in Figure 7, the incorporation of Nd_2O_3 causes a decrease in M′. The value of M′ of PVA is 4.21 at 100 Hz, and it dropped to 2.488, 2.318, 1.469, and 0.487 when doped with varying concentrations of Nd_2O_3. The permittivity for the electric modulus reduces due to the dipolar contribution. The dispersion of charges in the oxide may also play a role in this attenuation.

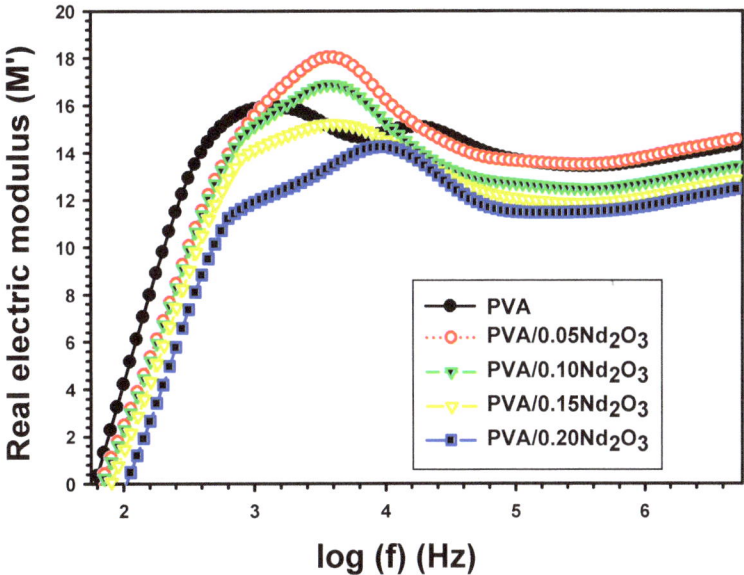

Figure 6. The real electric modulus (M′) with frequency for PVA and PVA/%Nd_2O_3 films.

Figure 7 displays the imaginary modulus M″ for PVA, PVA/0.05Nd_2O_3, PVA/0.10Nd_2O_3, PVA/0.15Nd_2O_3, and PVA/0.20Nd_2O_3. A peak in M″ refers to the presence of a relaxation process (as depicted in the picture). Higher permittivity is the result of an oriented M″ value at a low frequency, which indicates that the ions are transferred by hopping from one location to another. With an increase in frequency, the orientation of more dipolar groups becomes more challenging. The M″ of PVA is 2.318 at 100 Hz, but it drops to 1.493, 1.255, 1.01, and 0.848 when doped with 0.05Nd_2O_3, 0.10Nd_2O_3, 0.15Nd_2O_3, and 0.20Nd$_2$O$_3$. This is because the charged interface region is formed of PVA and Nd$_2$O$_3$. The time of relaxation (τ_r) is estimated by the following relation [39]:

$$\tau_r = \frac{1}{\omega_r} \quad (8)$$

where τ_r is the angular frequency that corresponds to the crest of the M″ peak. The relaxation time, measured in seconds, drops from 14.2×10^{-5} s for pure PVA film to 12.67×10^{-5} s for PVA/0.05Nd$_2$O$_3$, to 11.29×10^{-5} s for PVA/0.10Nd$_2$O$_3$, and finally to 8.97×10^{-5} s and 6.35×10^{-5} s for PVA/0.15Nd$_2$O$_3$ and PVA/0.20Nd$_2$O$_3$. This trend arises because the relaxation duration varies between the acceptor- and donor-of-interface states. Abdelhamied et al. [40] found that the relaxation time for PVA/PANI/Ag is changed by frequency. They discovered that the PVA film had a relaxation time of 3.3×10^{-5} s, but this reduced to 3.25×10^{-5} s and 1.9×10^{-5} s, respectively, for PVA/PANI and PVA/PANI/Ag. The Nd$_2$O$_3$ shortens the time needed for dipole orientation. The hopping process directly enhanced dipole mobility, which in turn decreased M″ [41].

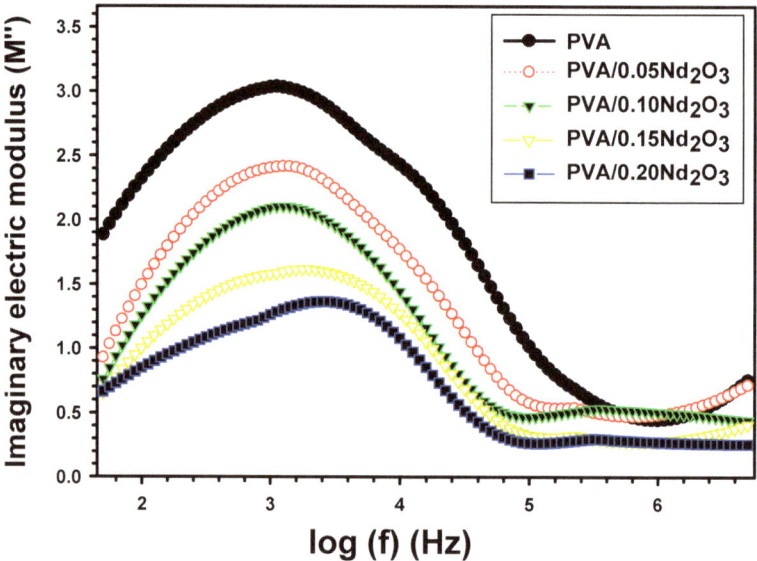

Figure 7. The imaginary electric modulus (M″) versus frequency for PVA and PVA/%Nd$_2$O$_3$ films.

The complex impedance of PVA and PVA/Nd$_2$O$_3$ is given by [42]:

$$Z^* = Z' + iZ'' \quad (9)$$

Z^* stands for the complex impedance, Z' for the real impedance, and Z'' for the imaginary impedance. Figure 8 shows that the real Z' impedance of pure PVA and PVA/0.20Nd$_2$O$_3$ films varies as a function of frequency. In the lower frequency, the Z' of all samples decreases significantly with frequency and is often fixed in the high frequency range. This behavior is explained by the fact that an enhancement in the number of free charges causes the conductivity to rise and the impedance to remain stable [43]. The addition of Nd$_2$O$_3$ has the opposite effect, progressively lowering the Z' value. The inclusion of Nd$_2$O$_3$ enhanced conductivity and directly decreased the films' impedance, leading to a decrease in Z'. Notably, the occurrence of peaks in PVA/Nd$_2$O$_3$ films is confirmed to be an improvement in the composite films, making them more suitable for devices.

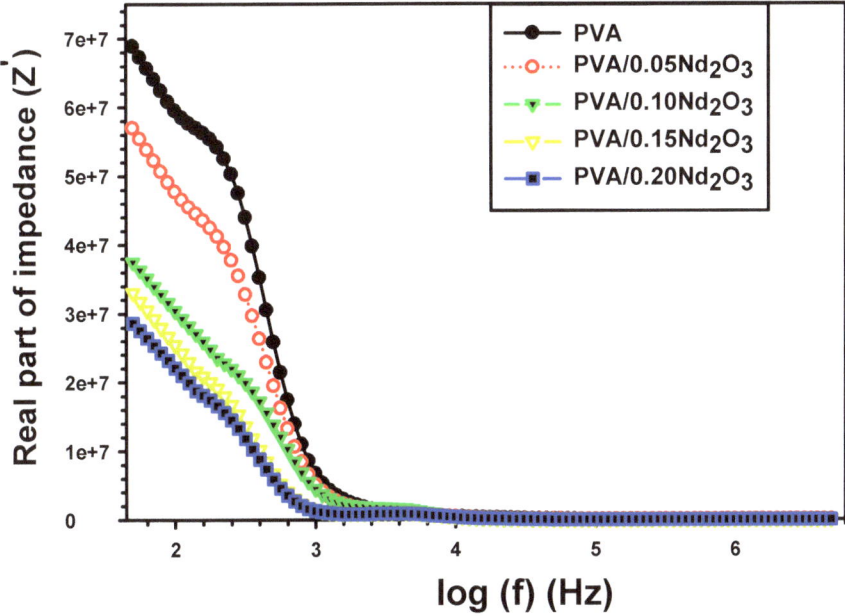

Figure 8. The real impedance (Z') with frequency for PVA and PVA/%Nd$_2$O$_3$ films.

Figure 9 displays the relationship between the imaginary impedance component Z'' and frequency for films of pure PVA, PVA/0.05Nd$_2$O$_3$, PVA/0.10Nd$_2$O$_3$, PVA/0.15Nd$_2$O$_3$, and PVA/0.20Nd$_2$O$_3$. The Z'' drops by frequency. Incorporating more Nd$_2$O$_3$ into PVA decreases the Z'' values, resulting in a faster charge transfer compared to pure PVA [44].

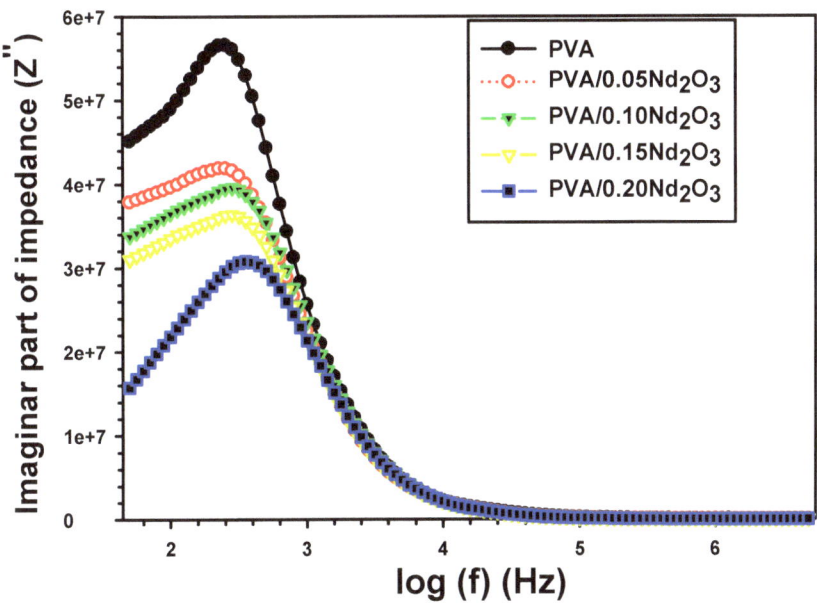

Figure 9. The imaginary impedance (Z'') with frequency for PVA and PVA/%Nd$_2$O$_3$ films.

The energy density (U) as a function with ε' is given by [45]:

$$U = \frac{1}{2} \varepsilon' \varepsilon_O E^2 \qquad (10)$$

E is the electric field, which is assumed to be 1 V/mm, and ε_O is the electric permittivity, which is equal to 8.85×10^{-11} NVm. The energy density at various frequencies for pure PVA, PVA/0.05Nd$_2$O$_3$, PVA/0.10Nd$_2$O$_3$, PVA/0.15Nd$_2$O$_3$, and PVA/0.20Nd$_2$O$_3$ films are displayed in Figure 10. Interface states that vary in strength with frequency are one possible explanation for this pattern. Since the contents of Nd$_2$O$_3$ increase, PVA had a higher power density. Table 1 shows that the addition of varying amounts of neodymium oxide into PVA results in an enhancement of its energy density. Particularly, the energy density values increase from 1.05×10^{-6} J/m^3 (the value for pure PVA) to 1.43×10^{-6} J/m^3, 1.70×10^{-6} J/m^3, 1.85×10^{-6} J/m^3, and 2.14×10^6 J/m^3, respectively. One potential explanation for this phenomenon could be the presence of border traps in close proximity to the interface between Nd$_2$O$_3$ and PVA. These findings provide further evidence of the effective incorporation of Nd$_2$O$_3$ within PVA.

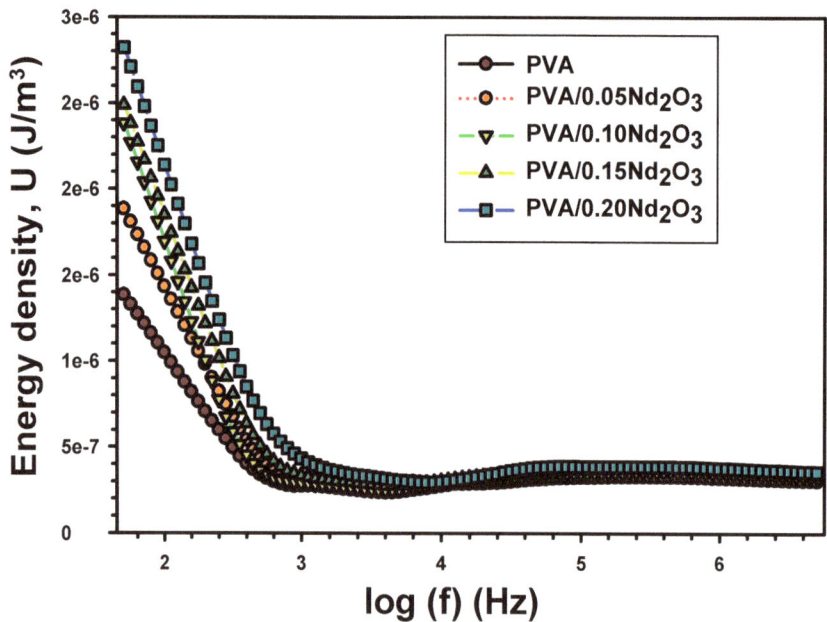

Figure 10. The energy density (U) with frequency for PVA and PVA/%Nd$_2$O$_3$ films.

The conductivity (σ_{ac}) as a function of the dielectric constant is given by [46]:

$$\sigma_{ac} = \varepsilon_o \varepsilon'' \omega \qquad (11)$$

ω is the angular frequency and ε_o is permittivity [47]. Films made of pure PVA, PVA/0.05Nd$_2$O$_3$, PVA/0.10Nd$_2$O$_3$, PVA/0.15Nd$_2$O$_3$, and PVA/0.20Nd$_2$O$_3$ show different changes in electrical conductivity as the frequency increases from 100 Hz to 5 MHz (Figure 11). In addition, the activation of trapped charges results in a shift in σ_{ac} conductivity [48]. According to Table 1, the electrical conductivity of PVA at 100 Hz ranges from 0.82×10^{-9} S/cm for pure PVA to 1.67×10^{-9} S/cm for PVA/0.05Nd$_2$O$_3$ and 6.82×10^{-9} for PVA/0.20Nd$_2$O$_3$.

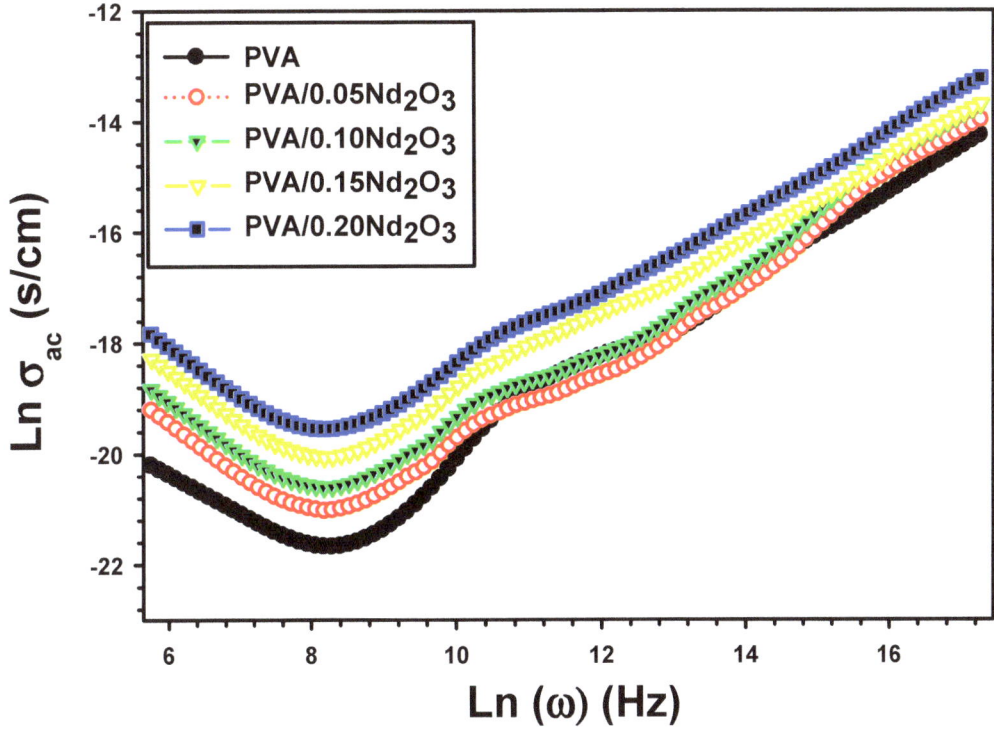

Figure 11. The conductivity (σ_{ac}) with frequency for PVA and PVA/%Nd$_2$O$_3$ films.

Moreover, the addition of Nd$_2$O$_3$ causes the greatest improvement in electrical conductivity because chain scissoring causes an increase in the speed of transported ions [49]. The conductivity of the PVA/MWCNT composite is investigated by Atta et al. [50]. The σ_{ac} of PVA is 0.89 × 10^{-4} S/cm at 1 MHz, while the σ_{ac} of PVA/MWCNT increased to 2.6 × 10^{-6} S/cm. This is because the MWCNT-induced increase in the carrier's charges led to the coalescence of grain boundaries, which improved conductivity.

The potential barrier height W_m is estimated by [51]:

$$W_m = \frac{-4k_BT}{m} \quad (12)$$

k_B is the Boltzmann constant and m is given by the slope of $\ln\varepsilon_2$ and $\ln\omega$, as investigated in Figure 12 by [52]:

$$\varepsilon'' = A\omega^m \quad (13)$$

where ε'' is the dielectric loss and ω is the angular frequency. As the percentage of Nd$_2$O$_3$ increased in the formula, the predicted potential barrier W_m is modified. The W_m energy of PVA with various concentrations of neodymium oxide varies from 0.264 eV to 0.248 eV and 0.255 eV for pure PVA, and to 0.284 eV and 0.252 eV for PVA/0.05Nd$_2$O$_3$, PVA/0.10Nd$_2$O$_3$, PVA/0.15Nd$_2$O$_3$, and PVA/0.20Nd$_2$O$_3$. The XRD measurements show that the inclusion of Nd$_2$O$_3$ alters the crystal structure and increases the defect density, leading to a lower potential barrier. Therefore, this evidence of a lower energy barrier for charge carrier hopping demonstrates that increasing the Nd$_2$O$_3$ content in PVA improves its σ_{ac} electrical conductivity.

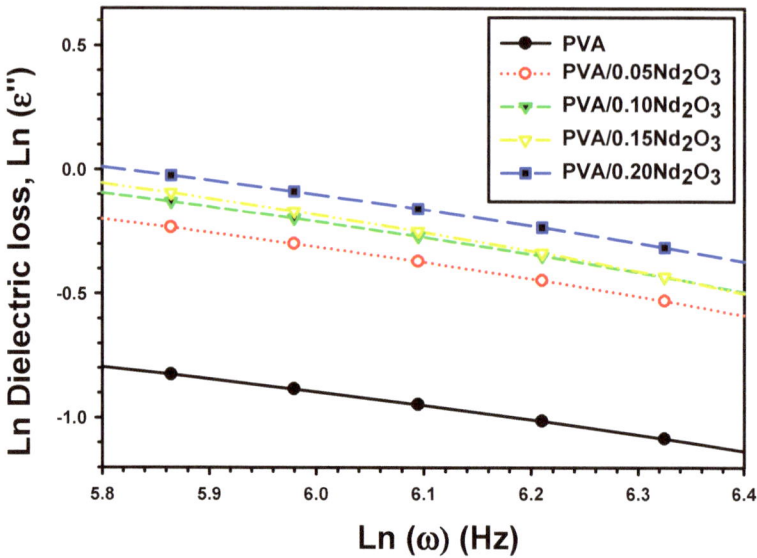

Figure 12. Variation of Ln (σ_{ac}) versus Ln (ω) for PVA and PVA/%Nd$_2$O$_3$ films.

4. Conclusions

The casting reduction method was used to successfully create PVA/Nd$_2$O$_3$ composite films, which were then characterized using a variety of methods. The XRD and FTIR methods verify that PVA/Nd$_2$O$_3$ was successfully synthesized. The inclusion of Nd$_2$O$_3$ causes modifications to the amorphous structure of the PVA, resulting in a change in PVA crystallinity. Moreover, SEM analysis shows that Nd$_2$O$_3$ is well incorporated into the PVA. It is observed that the morphology is altered due to Nd$_2$O$_3$. The concentration of Nd$_2$O$_3$ in PVA also affects its dispersion and the interfacial interaction of PVA chains and Nd$_2$O$_3$. The inclusion of Nd$_2$O$_3$ has also been observed to improve electrical conductivity and dielectric characteristics. The addition of Nd$_2$O$_3$ causes the greatest improvement in electrical conductivity because chain scissoring causes an increase in the speed of transported ions. In addition, there was a correlation between the amount of Nd$_2$O$_3$ and the changes in the electric modulus and complex impedance. Dielectric dispersion qualities were enhanced as Nd$_2$O$_3$ concentrations were raised from 0.05 wt% to 0.20 wt%, thanks to the activity of charge carriers. The electric modulus increases exponentially in a low frequency region, and at a higher frequency it becomes nearly constant. This is because the polymer electrical dipoles have a considerable self-orientation propensity at low frequencies, while the orientation produced by an external electric field is minimal at high frequencies. Based on the electrical results, the produced results have paved the way for using flexible PVA/Nd$_2$O$_3$ samples in a wide variety of possible applications, including batteries, super-capacitors, and microelectronic devices.

Author Contributions: Methodology, S.A.; Investigation, M.M.A.; Data curation, T.A.; Writing—original draft, A.H.A. and A.A.; Writing—review & editing, K.A. All the authors have contribution in the work, and they are known with the submission. All authors have read and agreed to the published version of the manuscript.

Funding: This research was funded by Deanship of Scientific Research at Jouf University through research Grant No. (DSR2022-NF-22).

Institutional Review Board Statement: Not applicable.

Data Availability Statement: Data will be made available upon reasonable request.

Acknowledgments: The authors extend their appreciation to the Deanship of Scientific Research at Jouf University for funding this work through research Grant No. (DSR2022-NF-22).

Conflicts of Interest: The authors declare that they have no known competing financial interest or personal relationships that could have appeared to influence the work reported in this paper.

References

1. Suhailath, K.; Thomas, M.; Ramesan, M.T. Studies on mechanical properties, dielectric behavior and DC conductivity of neodymium oxide/poly (butyl methacrylate) nanocomposites. *Polym. Polym. Compos.* **2021**, *29*, 1200–1211. [CrossRef]
2. Benabid, F.Z.; Kharchi, N.; Zouai, F.; Mourad AH, I.; Benachour, D. Impact of co-mixing technique and surface modification of ZnO nanoparticles using stearic acid on their dispersion into HDPE to produce HDPE/ZnO nanocomposites. *Polym. Polym. Compos.* **2019**, *27*, 389–399. [CrossRef]
3. Rani, P.; Ahamed, M.B.; Deshmukh, K. Structural, dielectric and EMI shielding properties of polyvinyl alcohol/chitosan blend nanocomposites integrated with graphite oxide and nickel oxide nanofillers. *J. Mater. Sci. Mater. Electron.* **2021**, *32*, 764–779. [CrossRef]
4. Sadasivuni, K.K.; Cabibihan, J.J.; Deshmukh, K.; Goutham, S.; Abubasha, M.K.; Gogoi, J.P.; Klemenoks, I.; Sekar, B.S.; Sakale, G.; Knite, M.; et al. A review on porous polymer composite materials for multifunctional electronic applications. *Polym.-Plast. Technol. Mater.* **2019**, *58*, 1253–1294. [CrossRef]
5. Alshammari, A.H.; Alshammari, M.; Alshammari, K.; Allam, N.K.; Taha, T.A. PVC/PVP/SrTiO$_3$ polymer blend nanocomposites as potential materials for optoelectronic applications. *Results Phys.* **2023**, *44*, 106173. [CrossRef]
6. Zyoud, S.H.; Almoadi, A.; AlAbdulaal, T.H.; Alqahtani, M.S.; Harraz, F.A.; Al-Assiri, M.S.; Yahira, I.S.; Zahran, H.Y.; Mohammed, M.I.; Abdel-Wahab, M.S. Structural, Optical, and Electrical Investigations of Nd$_2$O$_3$-Doped PVA/PVP Polymeric Composites for Electronic and Optoelectronic Applications. *Polymers* **2023**, *15*, 1351. [CrossRef]
7. Büyükbaş Uluşan, A.; Tataroğlu, A.; Azizian-Kalandaragh, Y.; Altındal, Ş. On the conduction mechanisms of Au/(Cu 2 O–CuO–PVA)/n-Si (MPS) Schottky barrier diodes (SBDs) using current–voltage–temperature (I–V–T) characteristics. *J. Mater. Sci. Mater. Electron.* **2018**, *29*, 159–170. [CrossRef]
8. Yerişkin, S.A.; Demir, G.E.; Yücedag, İ. On the frequency-voltage dependence profile of complex dielectric, complex electric modulus and electrical conductivity in Al/ZnO/p-GaAs type structure at room temperature. *J. Nanoelectron. Optoelectron.* **2019**, *14*, 1126–1132. [CrossRef]
9. Atta, A.; Negm, H.; Abdeltwab, E.; Rabia, M.; Abdelhamied, M.M. Facile fabrication of polypyrrole/NiOx core-shell nanocomposites for hydrogen production from wastewater. *Polym. Adv. Technol.* **2023**, *34*, 1633–1641. [CrossRef]
10. Alshammari, A.H.; Alshammari, K.; Alshammari, M.; Taha, T.A.M. Structural and optical characterization of g-C3N4 nanosheet integrated PVC/PVP polymer nanocomposites. *Polymers* **2023**, *15*, 871. [CrossRef]
11. Menazea, A.A.; El-Newehy, M.H.; Thamer, B.M.; El-Naggar, M.E. Preparation of antibacterial film-based biopolymer embedded with vanadium oxide nanoparticles using one-pot laser ablation. *J. Mol. Struct.* **2021**, *1225*, 129163. [CrossRef]
12. Lok, R.; Budak, E.; Yilmaz, E. Structural characterization of Nd 2 O 3 by sol–gel method and electrical properties. *J. Mater. Sci. Mater. Electron.* **2020**, *31*, 3111–3118. [CrossRef]
13. Huang, B.; Huang, C.; Chen, J.; Sun, X. Size-controlled synthesis and morphology evolution of Nd2O3 nano-powders using ionic liquid surfactant templates. *J. Alloys Compd.* **2017**, *712*, 164–171. [CrossRef]
14. Saxena, S.K. Polyvinyl Alcohol (PVA) Chemical and Technical Assessment. *Solids* **2004**, *1*, 3–5.
15. Wahid, A.; Asiri, A.M.; Rahman, M.M. One-step facile synthesis of Nd2O3/ZnO nanostructures for an efficient selective 2, 4-dinitrophenol sensor probe. *Appl. Surf. Sci.* **2019**, *487*, 1253–1261. [CrossRef]
16. Vendrame, Z.B.; Gonçalves, R.S. Electro chemical evidences of the inhibitory action of propargy lalcohol on the electrooxidation of nickel in sulfuric acid. *J. Braz. Chem. Soc.* **1998**, *9*, 441–448. [CrossRef]
17. Campbell, J.H.; Suratwala, T.I. Nd-doped phosphate glasses for high-energy/high-peak-power lasers. *J. Non-Cryst. Solids* **2000**, *263*, 318–341. [CrossRef]
18. Zhaorigetu, B.; Ridi, G.; Min, L. Preparation of Nd2O3 nanoparticles by tartrate route. *J. Alloys Compd.* **2007**, *427*, 235–237. [CrossRef]
19. Keikhaei, M.; Motevalizadeh, L.; Attaran-Kakhki, E. Optical properties of neodymium oxide nanoparticle-doped polyvinyl alcohol film. *Int. J. Nanosci.* **2016**, *15*, 1650012. [CrossRef]
20. AlAbdulaal, T.H.; Ali, H.E.; Ganesh, V.; Aboraia, A.M.; Khairy, Y.; Hegazy, H.H.; Soldatov, V.A.V.; Zahran, H.Y.; Abdel-Wahab, M.S.; Yahia, I.S. Investigating the structural morphology, linear/nonlinear optical characteristics of Nd2O3 doped PVA polymeric composite films: Kramers-Kroning approach. *Phys. Scr.* **2021**, *96*, 125831. [CrossRef]
21. Alshammari, A.H.; Alshammari, M.; Ibrahim, M.; Alshammari, K.; Taha, T.A.M. New hybrid PVC/PVP polymer blend modified with Er$_2$O$_3$ nanoparticles for optoelectronic applications. *Polymers* **2023**, *15*, 684. [CrossRef] [PubMed]
22. Alshammari, A.H.; Alshammari, M.; Ibrahim, M.; Alshammari, K.; Taha, T.A.M. Processing polymer film nanocomposites of polyvinyl chloride–Polyvinylpyrrolidone and MoO3 for optoelectronic applications. *Opt. Laser Technol.* **2024**, *168*, 109833. [CrossRef]

23. Abdelhamied, M.M.; Atta, A.; Abdelreheem, A.M.; Farag, A.T.M.; El Okr, M.M. Synthesis and optical properties of PVA/PANI/Ag nanocomposite films. *J. Mater. Sci. Mater. Electron.* **2020**, *31*, 22629–22641. [CrossRef]
24. Hodge, R.M.; Edward, G.H.; Simon, G.P. Water absorption and states of water in semicrystalline poly (vinyl alcohol) films. *Polymer* **1996**, *37*, 1371–1376. [CrossRef]
25. Ahmad Fauzi, A.A.; Osman, A.F.; Alrashdi, A.A.; Mustafa, Z.; Abdul Halim, K.A. On the Use of Dolomite as a Mineral Filler and Co-Filler in the Field of Polymer Composites: A Review. *Polymers* **2022**, *14*, 2843. [CrossRef] [PubMed]
26. Atta, A.; Abdelhamied, M.M.; Essam, D.; Shaban, M.; Alshammari, A.H.; Rabia, M. Structural and physical properties of polyaniline/silver oxide/silver nanocomposite electrode for supercapacitor applications. *Int. J. Energy Res.* **2021**, *46*, 6702–6710. [CrossRef]
27. Ali, F.; Kershi, R.; Sayed, M.; AbouDeif, Y. Evaluation of structural and optical properties of Ce^{3+} ions doped (PVA/PVP) composite films for new organic semiconductors. *Phys. B Condens. Matter* **2018**, *538*, 160–166. [CrossRef]
28. Kayış, A.; Kavgacı, M.; Yaykaşlı, H.; Kerli, S.; Eskalen, H. Investigation of structural, morphological, mechanical, thermal and optical properties of PVA-ZnO nanocomposites. *Glass Phys. Chem.* **2021**, *47*, 451–461. [CrossRef]
29. Althubiti, N.A.; Al-Harbi, N.; Sendi, R.K.; Atta, A.; Henaish, A.M. Surface Characterization and Electrical Properties of Low Energy Irradiated PANI/PbS Polymeric Nanocomposite Materials. *Inorganics* **2023**, *11*, 74. [CrossRef]
30. Nahar, K.; Aziz, S.; Bashar, M.S.; Haque, A. Synthesis and characterization of Silver nanoparticles from Cinnamomum tamala leaf extract and its antibacterial potential. *Int. J. Nano Dimens.* **2020**, *11*, 88–98.
31. Ishaq, S.; Kanwal, F.; Atiq, S.; Moussa, M.; Azhar, U.; Gul, I.; Losic, D. Dielectric and impedance spectroscopic studies of three phase graphene/titania/poly (vinyl alcohol) nanocomposite films. *Results Phys.* **2018**, *11*, 540–548. [CrossRef]
32. Abutalib, M.M.; Rajeh, A. Preparation and characterization of polyaniline/sodium alginate-doped TiO_2 nanoparticles with promising mechanical and electrical properties and antimicrobial activity for food packaging applications. *J. Mater. Sci. Mater. Electron.* **2020**, *31*, 9430–9442. [CrossRef]
33. Atta, A. Enhanced dielectric properties of flexible Cu/polymer nanocomposite films. *Surf. Innov.* **2020**, *9*, 17–24. [CrossRef]
34. Agrawal, A.; Satapathy, A. Thermal and dielectric behaviour of polypropylene composites reinforced with ceramic fillers. *J. Mater. Sci. Mater. Electron.* **2015**, *26*, 103–112. [CrossRef]
35. Mahendia, S.; Tomar, A.K.; Kumar, S. Electrical conductivity and dielectric spectroscopic studies of PVA–Ag nanocomposite films. *J. Alloys Compd.* **2010**, *508*, 406–411. [CrossRef]
36. Lotfy, S.; Atta, A.; Abdeltwab, E. Comparative study of gamma and ion beam irradiation of polymeric nanocomposite on electrical conductivity. *J. Appl. Polym. Sci.* **2018**, *135*, 46146. [CrossRef]
37. Latif, I.; AL-Abodi, E.E.; Badri, D.H.; Al Khafagi, J. Preparation, characterization and electrical study of (carboxymethylated polyvinyl alcohol/ZnO) nanocomposites. *Am. J. Polym. Sci.* **2012**, *2*, 135–140. [CrossRef]
38. Amin GA, M.; Abd-El Salam, M.H. Optical, dielectric and electrical properties of PVA doped with Sn nanoparticles. *Mater. Res. Express* **2014**, *1*, 025024. [CrossRef]
39. Dhatarwal, P.; Choudhary, S.; Sengwa, R.J. Dielectric and optical properties of alumina and silica nanoparticles dispersed poly (methyl methacrylate) matrix-based nanocomposites for advanced polymer technologies. *J. Polym. Res.* **2021**, *28*, 63. [CrossRef]
40. Thangarasu, R.; Senthilkumar, N.; Babu, B.; Mohanraj, K.; Chandrasekaran, J.; Balasundaram, O.N. Structural, optical, morphological and electrical properties of v2o5 nanorods and Its Ag/n-V2O5/p-Si/Ag diode application. *J. Adv. Phys.* **2018**, *7*, 312–318. [CrossRef]
41. Abdelhamied, M.M.; Abdelreheem, A.M.; Atta, A. Influence of ion beam and silver nanoparticles on dielectric properties of flexible PVA/PANI polymer composite films. *Plast. Rubber Compos.* **2022**, *51*, 1–12. [CrossRef]
42. Gupta, R.; Kumar, R. Influence of low energy ion beam implantation on Cu nanowires synthesized using scaffold-based electrodeposition. *Nano-Struct. Nano-Objects* **2019**, *18*, 100318. [CrossRef]
43. Sahu, G.; Das, M.; Yadav, M.; Sahoo, B.P.; Tripathy, J. Dielectric relaxation behavior of silver nanoparticles and graphene oxide embedded poly (vinyl alcohol) nanocomposite film: An effect of ionic liquid and temperature. *Polymers* **2020**, *12*, 374. [CrossRef] [PubMed]
44. Atiq, S.; Majeed, M.; Ahmad, A.; Abbas, S.K.; Saleem, M.; Riaz, S.; Naseem, S. Synthesis and investigation of structural, morphological, magnetic, dielectric and impedance spectroscopic characteristics of Ni-Zn ferrite nanoparticles. *Ceram. Int.* **2017**, *43*, 2486–2494. [CrossRef]
45. Abdel Reheem, A.M.; Atta, A.; Afify, T.A. Optical and electrical properties of argon ion beam irradiated PVA/Ag nanocomposites. *Surf. Rev. Lett.* **2017**, *24*, 1750038. [CrossRef]
46. Jilani, W.; Fourati, N.; Zerrouki, C.; Gallot-Lavallée, O.; Guermazi, H. Optical, dielectric properties and energy storage efficiency of ZnO/epoxy nanocomposites. *J. Inorg. Organomet. Polym. Mater.* **2019**, *29*, 456–464. [CrossRef]
47. Ahmed, H.; Hashim, A. Fabrication of PVA/NiO/SiC nanocomposites and studying their dielectric properties for antibacterial applications. *Egypt. J. Chem.* **2020**, *63*, 805–811. [CrossRef]
48. Bouaamlat, H.; Hadi, N.; Belghiti, N.; Sadki, H.; Bennani, M.N.; Abdi, F.; Lamcharfi, T.-D.; Bouachrine, M.; Abarkan, M. Dielectric Properties, AC Conductivity, and Electric Modulus Analysis of Bulk Ethylcarbazole-Terphenyl. *Adv. Mater. Sci. Eng.* **2020**, *2020*, 8689150. [CrossRef]

49. Elamin, N.Y.; Modwi, A.; Abd El-Fattah, W.; Rajeh, A. Synthesis and structural of Fe3O4 magnetic nanoparticles and its effect on the structural optical, and magnetic properties of novel Poly (methyl methacrylate)/Polyaniline composite for electromagnetic and optical applications. *Opt. Mater.* **2023**, *135*, 113323. [CrossRef]
50. Raghu, S.; Archana, K.; Sharanappa, C.; Ganesh, S.; Devendrappa, H. Electron beam and gamma ray irradiated polymer electrolyte films: Dielectric properties. *J. Radiat. Res. Appl. Sci.* **2016**, *9*, 117–124. [CrossRef]
51. Atta, A.; Lotfy, S.; Abdeltwab, E. Dielectric properties of irradiated polymer/multiwalled carbon nanotube and its amino functionalized form. *J. Appl. Polym. Sci.* **2018**, *135*, 46647. [CrossRef]
52. Ebrahim, S.; Kashyout, A.H.; Soliman, M. Ac and Dc conductivities of polyaniline/poly vinyl formal blend films. *Curr. Appl. Phys.* **2009**, *9*, 448–454. [CrossRef]

Disclaimer/Publisher's Note: The statements, opinions and data contained in all publications are solely those of the individual author(s) and contributor(s) and not of MDPI and/or the editor(s). MDPI and/or the editor(s) disclaim responsibility for any injury to people or property resulting from any ideas, methods, instructions or products referred to in the content.

Article

Electrochemical Comparison of 2D-Flexible Solid-State Supercapacitors Based on a Matrix of PVA/H$_3$PO$_4$

Bianca K. Muñoz *, Andrés González-Banciella, Daniel Ureña, María Sánchez * and Alejandro Ureña

Material Science and Engineering Area, Universidad Rey Juan Carlos, ESCET, C/Tulipán s/n. Móstoles, 28933 Madrid, Spain; andres.banciella@urjc.es (A.G.-B.); duremed.ingenieroenergia@gmail.com (D.U.); alejandro.urena@urjc.es (A.U.)
* Correspondence: bianca.munoz@urjc.es (B.K.M.); maria.sanchez@urjc.es (M.S.); Tel.: +34-9148-87141 (B.K.M.); +34-9148-88156 (M.S.)

Abstract: Different modifications of woven carbon fiber (WCF) based on carbon aerogel (CAG), copper oxide nanoparticles (CuO-NPs), and lignin (LIG) has been tested and used to study their effect on the fabrication and performance of a flexible supercapacitor. New symmetric flexible supercapacitors (SFSCs) were fabricated using different separators. According to the electrochemical results, the device fabricated using CAG and woven glass fiber (WGF) in a sandwich type configuration CAG/WGF/CAG embedded in H$_3$PO$_4$/PVA exhibited the best performance (1.4 F/g, 0.961 W/kg, 0.161 Wh/kg). A proof of concept based on a LED powered on and a bending test was done, and the capacitor demonstrated excellent electrochemical values even during and after bending. The new device was able to recover 96.12% of its capacitance when returned to its original unbent position. The manufacturing process was critical, as the fibers or layers must be completely embedded in the gel electrolyte to function effectively. A double flexible supercapacitor connected in parallel was fabricated and it showed higher stability, in the same voltage window, yielding 311 mF/cm^2 of areal capacitance.

Keywords: flexible supercapacitors; woven carbon fiber; polyvinyl alcohol-based electrolyte; energy storage

Citation: Muñoz, B.K.; González-Banciella, A.; Ureña, D.; Sánchez, M.; Ureña, A. Electrochemical Comparison of 2D-Flexible Solid-State Supercapacitors Based on a Matrix of PVA/H$_3$PO$_4$. *Polymers* **2023**, *15*, 4036. https://doi.org/10.3390/polym15204036

Academic Editors: Md Najib Alam and Vineet Kumar

Received: 15 September 2023
Revised: 29 September 2023
Accepted: 8 October 2023
Published: 10 October 2023

Copyright: © 2023 by the authors. Licensee MDPI, Basel, Switzerland. This article is an open access article distributed under the terms and conditions of the Creative Commons Attribution (CC BY) license (https://creativecommons.org/licenses/by/4.0/).

1. Introduction

In the last decade, the number of wearable electronic devices including multiple applications, such as watches, suitcases, sports gadgets, clothes, etc., has increased [1,2]. The increasing demand for them in the market has encouraged researchers to study the best method of energy storage by designing flexible batteries (FBs) and supercapacitors (FSCs) that can be modulated, bended, stretched, twisted, or folded without losing their electrochemical performance [3]. It is well-known that supercapacitors are considered a good alternative for energy storage due to the high-power density delivered, which refers to how quickly a device can discharge its energy and is a property usually needed for mobility applications. Fast charge/discharge cycles and relatively easy assembly also make them more attractive [4].

Different strategies to design FSCs have appeared, such as 1D (coaxial fibers) [5], 2D (thin films or sandwich type) [6], and 3D (hierarchical frameworks) [7] and they have been applied to different materials such as carbonous fibers [8], textile fibers (polyester) [9], flexible papers and foams [10,11], and woven fibers [12]. Carbon cloth (CC) is a very convenient textile due to its high conductivity performance, price, flexibility, and feasibility of surface modification to enhance its behavior as electrodes [13]. Electrical double layer capacitors (EDLCs) are highly dependent on the surface-interface character of the electrodes. For this reason, carbon fiber surface modifications by oxidation treatments [14] or by the incorporation of carbon nanostructures such as carbon nanotubes (CNTs) [11] or graphene nanoplatelets (GNPs) [15] have previously been reported. Moreover, the modification by

3D metal oxide arrays on the carbon cloth has also been reported as well as combinations of carbon nanostructures and metal oxides [14]. Other interesting modifications include some wastes from biomass as lignocellulosic biomass-derived carbon [16], which are used directly as electrodes due to the high porosity and hierarchical structures. In other cases, the use of lignin-modified carbon fiber was reported to improve the interfacial adhesion in carbon fiber-reinforced polymers [17].

By exfoliation and oxidation of the surface of CC (modified Hummer's method) using $KMnO_4$, H_2SO_4, and HNO_3 solution, a symmetric FSC fabricated with sulfuric acid and poly(vinyl alcohol) (PVA) gel electrolyte showed a capacitance of 31 mFcm^{-2} (0.0015 Fg^{-1}) at a scan speed of 10 mVs^{-1} [18]. Liu and Zhou included a carbon nanotube 3D network on CC for a FSC assembled using KOH/PVA, obtaining a capacitance of 106.1 Fg^{-1}, an areal capacitance of 38.75 mFcm^{-2}, and an ultralong cycle life of 100,000 times (capacitance retention: 99%) [19]. Symmetric FSCs using carbon fabric modified with metal oxides as electrodes have also been efficiently developed. MnO_2 has been extensively used to modify the surface in symmetric FSCs, an areal capacitance of 42.4 mF/cm^{-2} at 5 mVs^{-1} has been reported. Moreover, a FSC using manganese dioxide nanorod arrays on carbon fiber was fabricated using PVA/H_3PO_4 as electrolyte [20]. The modification of the fiber showed a very good performance (678 Fg^{-1} at a current density of 0.3 Ag^{-1}). For heterostructure nanoarrays with $CuCo_2O_4$@MnO_2 nanowire-based electrodes, a high cell-area-specific capacitance of 714 mFcm^{-2} at 1 mAcm^{-2} was achieved for the device using (PVA)/KOH polymer electrolyte [21]. The Co_3O_4 nanoparticles on vertically aligned graphene nanosheets (VAGNs) were also tested for FSC (PVA/KOH as electrolyte), delivering high capacitance 580 Fg^{-1} and high energy density (80 Wh/kg) and power density (20 kW/kg) [14]. $CoMoO_4$@NiCo-layered double hydroxide nanowire arrays have also been used to modify CC and led to the fabrication of asymmetric FSC with PVA/KOH as electrolyte, exhibiting a maximum energy density of 59 Whkg^{-1} at a power density of 800 Wkg^{-1} [22]. CuO-based nanomaterials have been found to be very attractive for supercapacitor applications due to their low cost, facile and reproducible preparation methods, and electrochemical responses [23]. It is well known that light-weight carbon aerogel (CAG) has some characteristics that make them ideal for SC electrodes, such as large surface area, low mass density, and high porosity [24,25].

The separator also plays an important role in the supercapacitor performance and flexibility. In the flexible supercapacitors approach, some authors suppress the use of them because the polymer electrolyte can act as a dielectric. In other cases, authors prefer to use separators based on woven glass fiber, woven glass fiber (WGF), filter paper mat (FP), and, for structural applications, Kevlar® fiber fabric [26].

As it has been previously mentioned above, one of the most interesting approaches for the development of flexible supercapacitors is the use of woven carbon fiber as electrodes. The carbon cloth has some characteristics that make it attractive to be used as an electrode from an economic and environmental point of view. An appropriate combination of electrodes based on carbon fabric with PVA/H_3PO_4 matrix electrolyte can make the development of new supercapacitors for smart applications possible, being scalable and supporting their increasing demand. In general, the electrodes developed so far showed outstanding properties. However, the modified electrodes have complicated fabrication processes and usually they are very expensive, not to mention that in some cases, they are still far from being used for immediate applications. In this sense, modifying carbon fabric in an easy and cheap manner represents a more challenging goal. Moreover, despite the great number of studies in FSCs using CC as substrate electrode, their performance is not easy to compare and unify due to different ways of fabrication (assembling, separator, and electrolyte) and characterization.

In this work, we present the modification and characterization of woven carbon fiber (WCF) including carbon aerogel (CAG), copper oxide nanoparticles (CuO NPs), and lignin (LIG), as well as their application as electrodes for flexible supercapacitors using PVA/H_3PO_4 as gel polymer electrolyte. We tested their performance before, during, and after bending tests to demonstrate their flexibility and ability to recover the electrochemical response. The electrochemical performance of the devices is discussed, and a proof of concept based on a red LED powered on was done with the system WCF/WGF/WCF.

2. Materials and Methods

2.1. Materials

Woven carbon fabric HexForce® 48,193 plain 12 K, manufactured by Hexcel (Stamford, CT, USA), was used in this work. The separator used was of filter paper (FP) or woven glass fiber (WGF). The glass fiber separator (E-Fiberglass Woven Roving) was supplied by Castro Composites® (Pontevedra, Spain). Glass Filter paper Whatman, Lignin alkaline (low sulfonate content), resorcinol, formaldehyde (37 wt% in H_2O), copper acetate dihydrate, copper chloride, (L)-ascorbic acid (L-AA), PVP K90, K_2CO_3, KOH, H_3PO_4 ($\geq$85 wt% in H_2O), and Poly(vinyl alcohol) (PVA, Mw 89,000–98,000, 99+% hydrolyzed) were purchased from Sigma-Aldrich (St. Louis, MO, USA). Liquid sizing agent (SICIZYL) was purchased from Nanocyl S.A. (Sambreville, Belgium). The CuO seed on carbon cloth was done following the procedure previously reported in the literature [27]. The carbon aerogel modification was prepared following the procedure reported [28].

2.2. Preparation and Characterization of Electrodes

In all cases, fabric was washed with ethanol before any particle deposition and dried in the oven at 60 °C.

2.2.1. Dip-Coating (DC) of Lignin

A water suspension of lignin (10 wt%) in distilled water and sizing solution (5 wt%) was stirred at room temperature for 30 min. Then, the carbon fiber fabric was soaked for 10 min and dried in the oven at 80 °C for 10 more minutes until solvent evaporation. This procedure was repeated twice.

2.2.2. Hydrothermal Growth of CuO Nanoparticles

The hydrothermal method used here is a modification of a procedure previously reported [29], using an optimized mixture of $CuCl_2$/NaCl/L-AA of 1:3.5:15 in DIW (375 mL) containing 1 wt.% of PVP K90. A solution of $CuCl_2$ was stirred in 150 of DIW for 20 min. In a different vessel, L-AA was also dissolved in 200 mL of DIW under sonication for 30 min, then mixed with NaCl and PVP-K90 and stirred. The copper chloride solution was added to AA solution under stirring and the pH was adjusted using NaOH. The seeded WCFs were immersed in the growth solution into the stainless-steel autoclave in an oven at 80 °C. The desired growth temperature for the CuO/WCF hybrid at a fixed growth time of 2 h. After the hydrothermal processing, the CuO/WCF hybrid was rinsed with DIW to stop further growth of CuO NPs and dried at room temperature for 1 day.

2.2.3. Surface Morphology and Area Characterization

The specific surface area of the modified carbon fiber fabric was calculated by the Brunauer, Emmett, and Teller (BET) method based on N_2 adsorption–desorption isotherms recorded with a Micrometrics ASAP 2020 analyzer. Fabric surface characterization was also performed evaluating the images from Scanning Electron Microscopy (SEM, S-3400 N from Hitachi, Tokyo, Japan) and Field Emission Gun SEM (FEG-SEM, Nova NanoSEM FEI 230 from Philips, Amsterdam, The Netherlands). Samples were previously coated by a 6 nm layer of gold for proper characterization.

2.3. Electrochemical Characterization of Electrodes

Cyclic voltammetry (CV) tests on modified carbon fibers were performed at room temperature with a three-electrode cell, using an Ag/AgCl reference electrode and a platinum counter electrode in 3 M KCl aqueous electrolyte. The working electrode consisted of a single tow partly immersed into the electrolyte and electrically contacted at the dry end. Tows were measured and weighed before the test and the active mass (*ma*) of the working electrode was determined from the immersed length.

Experiments were conducted using an Autolab PGSTAT302N system. Different scan rates (from 10 to 100 mV/s) were tested on a representative specimen from each condition. The specific capacitance (C_{sp}) was calculated with Equation (1).

$$Csp = \frac{\int_{v}^{vf} I dV}{s \cdot \Delta V \cdot ma} \quad (1)$$

where *I* is the current (A), *dV* is the differential voltage corresponding to both charge and discharge processes between final voltage (*vf*) and initial voltage (*v*), *ma* (g) is the mass of the electrodes, *s* (V s^{-1}) is the scan rate, and ΔV is the potential window of the CV. Representative capacitance results were calculated using a potential window (ΔV) from -0.1 to 0.1 V and a scan rate (*s*) of 10 mV/s.

2.4. FSC Fabrication

The gel polymer electrolyte (GPE) was prepared by mixing H_3PO_4 (3 g) and PVA (3 g) in distilled water (30 mL) [30] under stirring at 85 °C for 30 min or until the solution became clear. The electrodes (previously modified and attached to a copper current collector wire glued using silver ink) and separator (glass fiber or filter paper) were immersed in a bath containing the GPE mixture for a few minutes to favor the impregnation. Then, the device was sandwich-type assembled following the configuration electrode/separator/electrode over a metallic mold and left to cast a 60 °C in the oven for 8 h.

2.5. Electrochemical Characterization of FSCs

The electrochemical characterization of the symmetric SC device was measured in a two-electrode setup. Cyclic voltammetry (CV) tests on FSCs were carried out by using a potentiostat AUTOLAB PGSTAT302N with a software Nova 2.1. Different scan rates (from 10 to 100 mV/s) were tested on a representative specimen from each condition. Each sample was subjected to 5 consecutive voltammetry cycles to determine its specific capacitance using Equation (1). Representative capacitance results were calculated using a potential window (ΔV) from -0.5 V to 0.5 V and a scan rate (*s*) of 10 mV/s. Galvanostatic charge-discharge tests (GCD) and electrochemical impedance spectroscopy (EIS) tests were carried out to characterize in depth their capacitor capabilities. From GCD analysis, specific capacitance (C, F/g) can be derived from Equation (2), while energy density (E) and power density can be obtained according to Equations (3) and (4) [22]. EIS was performed in samples in a frequency range of 10^5–0.1 Hz. Both tests were also performed in an AUTOLAB PGSTAT302N module with a software Nova 2.1.

$$C = \frac{I \times \Delta t}{m \times \Delta V} \quad (2)$$

$$E = \frac{1}{2} C \times \Delta V^2 \quad (3)$$

$$P = \frac{E}{\Delta t} = \frac{1}{2} \frac{I \times \Delta V}{m} \quad (4)$$

where *I* is the current (A), Δt (s) the discharge time, *m* (g) the electrode mass, and ΔV the potential window.

3. Results and Discussion

3.1. Electrodes Modification and Characterization

The surface morphology of the new modified carbon fibers was investigated by FEGSEM (Figure 1).

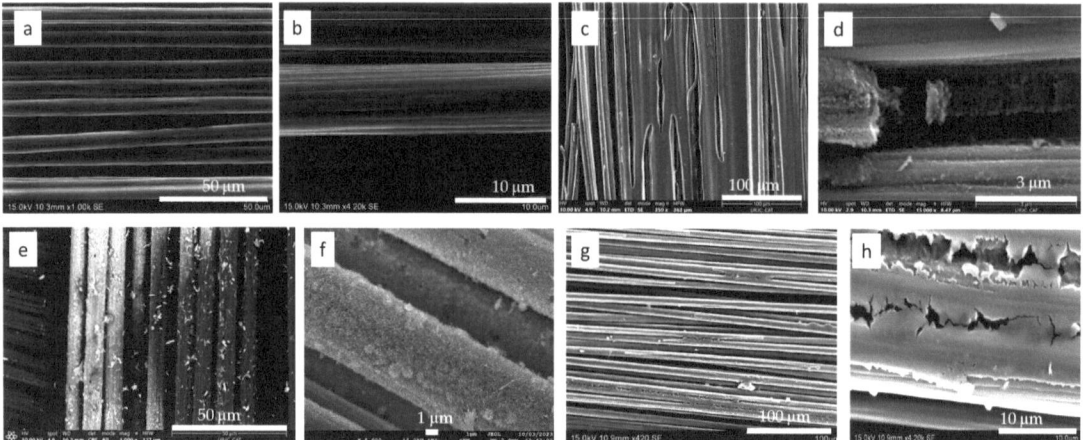

Figure 1. SEM images of modified carbon fibers. (**a,b**) Two magnifications of bare carbon fiber, (**c,d**) two magnifications of carbon fiber modified with carbon aerogel, (**e,f**) two magnifications of carbon fiber modified with CuP NPs, (**g,h**) two magnifications of carbon fiber modified with lignin.

In this image, the plain WCF appears as a corrugated non-porous surface. The images of modified fibers containing CAG, CuO, and lignin are shown in Figure 1c–h. For the sample modified with CAG (Figure 1c,d), the covering looks like a thick layer; however, in the closer picture, it is possible to observe a porous surface in this layer. The modification with CAG is done from carbonization RF polymer directly cured over the CC; then, the CAG layer thickness depends on the pressing step during the cured. Even though the CAG layer is highly porous as can be demonstrated from the specific surface area (SSA) of the electrode reported in Table 1 (BET area), the CuO NPs growth over the carbon fiber shows a more homogeneous layer (Figure 1e,f). The SSA is not as high as those values obtained for CAG but is higher than the pristine WCF (0.244 m^2 g^{-1}). The SEM images of fiber modified using lignin by dip-coating (Figure 1g,h) show a brittle organic layer corresponding to lignin and sizing (commercial organic sizing). The SSA obtained for this sample is even smaller than the pristine WCF, meaning the lignin by itself is not porous, and is not conferring any extra surface area.

Table 1. T BET areas and specific capacitances of carbon fibers.

Entry	Sample	S_{BET} (m^2/g$_{C+cover}$)	C_{sp} (F/g$_{C+cover}$)
1	WCF	0.244	0.19
2	CuO	0.853	5.42
3	CAG	63.47	1.7
4	LIG	0.278	0.18

The electrodes were analyzed by cyclic voltammetry at different scan rates and the specific capacitance was determined at 10 mVs^{-1}. The specific capacitances of the electrodes are shown in Table 1.

As could be expected, the specific capacitance for the pristine WCF is low (0.19 Fg^{-1}). As the BET surface area increased for the electrodes modified with CuO-NPs or CAG (entries 2 and 3), it is evident that the C_{sp} increased, reaching values between 1.5 and 5 Fg^{-1}. The CAG on WCF was previously studied by different authors using EMIMTFSI as an electrolyte and the specific capacitance reported is in the same order of magnitude (2.4 F/g) [31]. In the case of WCF-CuO, it has been previously reported that depending on the structure of the NPs, they can provide either large BET area and behave as an EDLC [32] or they can present certain pseudocapacitance as a dominant energy storage mechanism. In the last case, the electrodes can record high capacitances through a Faradaic process. For transition metal oxide nanostructures, the effect of the morphology on the electrochemical behavior is quite remarkable [23]. A schematic illustration of both EDLC and pesudocapacitor charge storage mechanism are depicted in Scheme 1. Pseudocapacitors usually exhibit better capacitance than double layer capacitors [33].

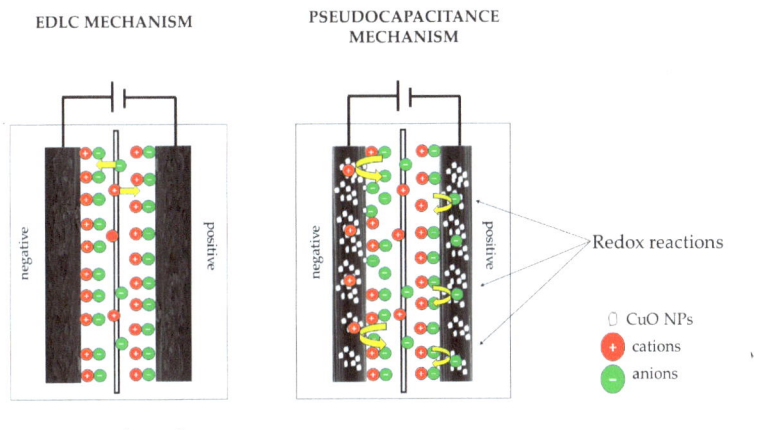

Scheme 1. Schematic illustration of EDLC (**left**) and pseudocapacitance (**right**) mechanisms.

The sample modified with lignin showed a very low specific capacitance, even lower than the pristine WCF. This value can be attributed to the isolating nature of the sizing used as a binder, which is an organic oligomer. This effect when polymeric binders or surfactants are used has been previously reported [34].

With these electrodes, symmetric flexible supercapacitors were fabricated using PVA/H_3PO_4 as gel polymer electrolyte and using a separator of filter paper (FP).

3.2. Flexible Supercapacitors' Electrochemical Characterization

The modified carbon fiber cloths were used as electrodes to fabricate the flexible supercapacitors. The electrolyte was prepared, and the electrodes and separator were dipped in for a few minutes. Then, they were piled up using a sandwich type configuration by placing the layers in the following order: electrode/separator/electrode. The FSCs obtained were initially characterized by CV and EIS (Figures 2 and S1).

The CV of FSC with the bare WCF and those modified with CuO NPs or CAG have the same scale, while the CV of FSC fabricated using the electrodes modified with lignin is shown in a different scale. The CV curves show how the surface modifications have affected the electrochemical performance of the devices. The FSC fabricated using plain WCF exhibits good electrochemical stability at different potentials. In this case, the area inside the CV curve is small but the shape indicates typical capacitive non-Faradaic behavior.

The specific capacitances were normalized by the electrode masses due to the different additional weights corresponding to the electrolyte (as they do not contain the same amount and not all the gel electrolyte is in contact with the electrodes). The FSCs modified either with CAG or CuO NPs show higher specific capacitance since the area inside the curve is higher. The electrochemical values obtained from these curves are shown in Table 2.

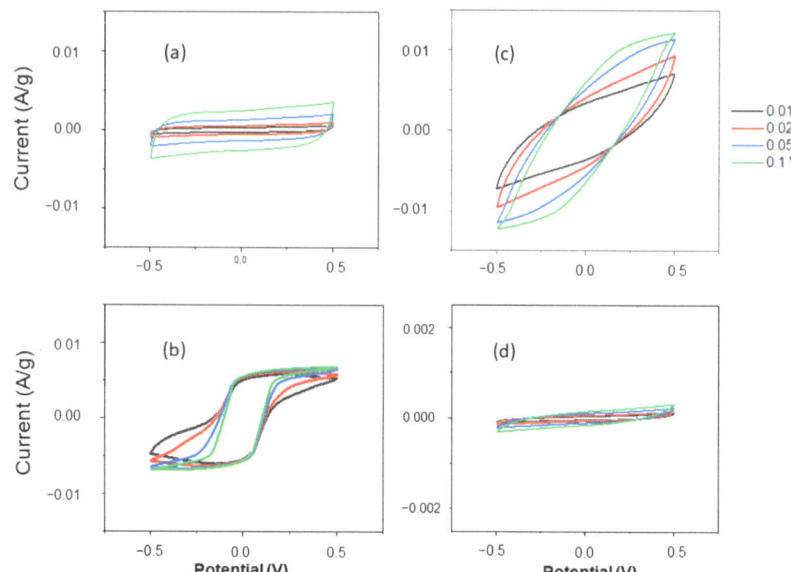

Figure 2. CV of the FSCs fabricated from the modified electrodes. (**a**) Bare carbon fiber electrodes, (**b**) CuO NPs modified electrodes, (**c**) CAG modified electrodes, (**d**) lignin modified electrodes.

Table 2. Electrochemical data from CV and EIS analysis of FSCs.

SCF	C_{sp} (mF/cm^2)	C_{sp} (F/g)	σ (mS cm^{-1})	R_s (Ω)	$R_s + R_{ct}$ (Ω)
FSC-WCF	-	0.023	-	-	-
FSC-CuO	12.1	0.328	2.08	2.95	10.61
FSC-CAG	9.6	0.259	1.44	8.26	69.10
FSC-LIG	0.14	0.004	0.034	164.9	414.68

Both FSCs, either with CuO NPs or CAG, present similar specific capacitances (from 0.26 to 0.33 F/g at 10 mV/s). The areal capacitances are comparable to systems modified with Mn oxides using KOH in PVA as electrolytes [35]. Regarding the electrochemical stability, the FSC using electrodes modified with CuO seems to be less stable in the same potential window compared to CAG, which shows less prominent vertical changes. It has been reported that some zero-dimensional nanoparticles of pristine copper oxide could show poor electrochemical stability and high tendency towards rapid capacity decay [23]. In Figure 2b, at the lower scan rate (0.01 Vs^{-1}), it is possible to detect some deviations in the CV, suggesting some Faradaic processes; these small humps disappear at faster scan rates. However, more studies are currently in progress to obtain more information about the plausible energy storage mechanism followed by these CuO nanoparticles. In the case of FSC modified with WCF-LIG, the specific capacitance of the devices are two orders of magnitude smaller than those obtained for CuO and CAG electrodes, consistent with the results obtained for plain electrodes analyzed in KCl. Concerning the results extracted from the EIS characterization (Figure S1), the EIS curves show a depressed semicircle at high frequencies associated with the ion transport processes, followed by a spike at low frequencies. The semicircle for the FSC-CuO is smaller, meaning that there is less resistance due to the best surface interface contact between the electrode and the matrix as it has been reported for similar CuO-WCF electrodes using structural epoxy matrices. The FSC-CAG shows a larger spike that is translated into a more capacitive behavior, typical for porous carbon electrodes. As a counterpart, the FSC-LIG exhibits high resistance values. The R_s corresponds to the ionic transport through the electrolyte and electrode. That means that the values of R_s can be directly related to the ionic conductivity capabilities. The isolating sizing layer present in the LIG electrodes led to an increase in the resistance values, diminishing the ionic movement between the electrolyte/electrode and affording an extremely low specific capacitance value. The conductivity of the electrolyte in the flexible device modified with CuO is slightly higher than the obtained for the electrode modified with CAG, due to a better interface interaction that exhibits less resistance to the ion transport. The value corresponding to the FSC fabricated with the bare WCF is shown in Table 2 for comparison.

3.3. Bending Tests

To demonstrate their application as flexibles devices, their flexibility was evaluated by bending and analyzing the FSC by CV. The results are shown in Figure 3.

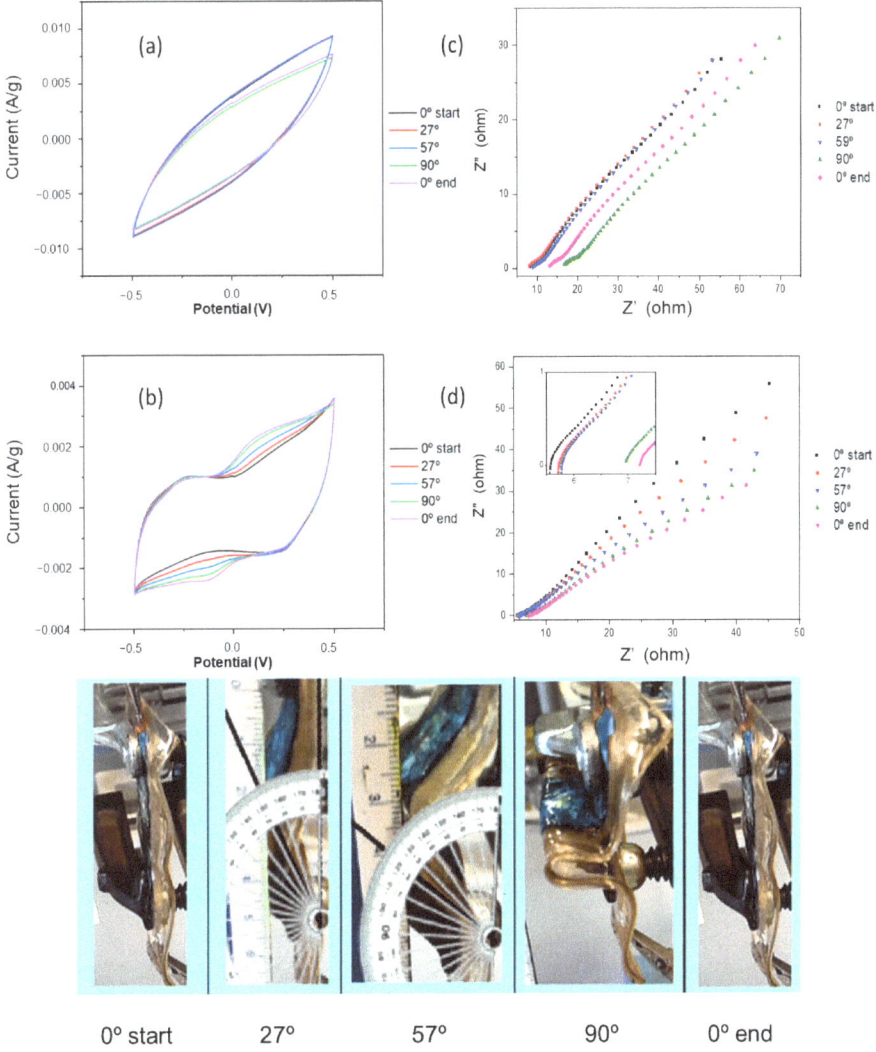

Figure 3. CV and EIS analysis of the FSCs during the bending test. (**a,c**) for CAG/CC and (**b,d**) for CuO/CC. The picture shown of the bending angle corresponds to the sample modified with CAG.

The absence of significant distortions in the shape of the CV curves at the different bending angles for both FSCs reveal high flexibility and certain stability of both devices. The resistances are more affected instead, revealing a higher resistance after the bending. According to the internal area of the CV curves, it was possible to calculate the capacitance retention (%) during and after bending (Figure S2). For FSC-CuO and FSC-CAG, the behavior was similar. At 27°, both devices retained >95% of the capacitance; at 57°, CAG-FSC retained 96% and CuO-FSC 93%; and at 90°, the capacitance dropped to 80% in both devices, but they recovered their capacitance over 85% when returned to the initial position. The picture at the bottom of Figure 3 shows the dramatic bending at 90°. For the sample FSC-LIG, the capacitance retention fell to 69% at 90° and in general, it exhibited a worst performance.

A new device was fabricated using a different separator, woven glass fiber (WGF), to compare its performance with the filter paper (FP). The CV at 10 mVs^{-1} and EIS analysis shown in Figure 4 were used to characterize the device before, during, and after bending. The FSC-CAG fabricated using WGF as separator showed better electrochemical values, such as less resistance and a higher specific capacitance. The capacitance retention (%) at different bending angles for the flexible supercapacitors fabricated with WGF or FP as separator indicated that there were not big differences in using both separators. However, after the bending, the device using WGF recovered 96% of the initial capacitance, showing that it is more flexible. The GCD tests were also measured for both samples (Figures S3 and S4) and the ratio for the device modified using the FP separator had a $t_{discharge}/t_{charge}$ ratio of 0.62 compared to the 0.47 value obtain for the sample separated using WCF. The specific capacitance and other electrochemical parameters calculated from EIS and GCD tests (at 2 mA g^{-1}) are listed in Table 3.

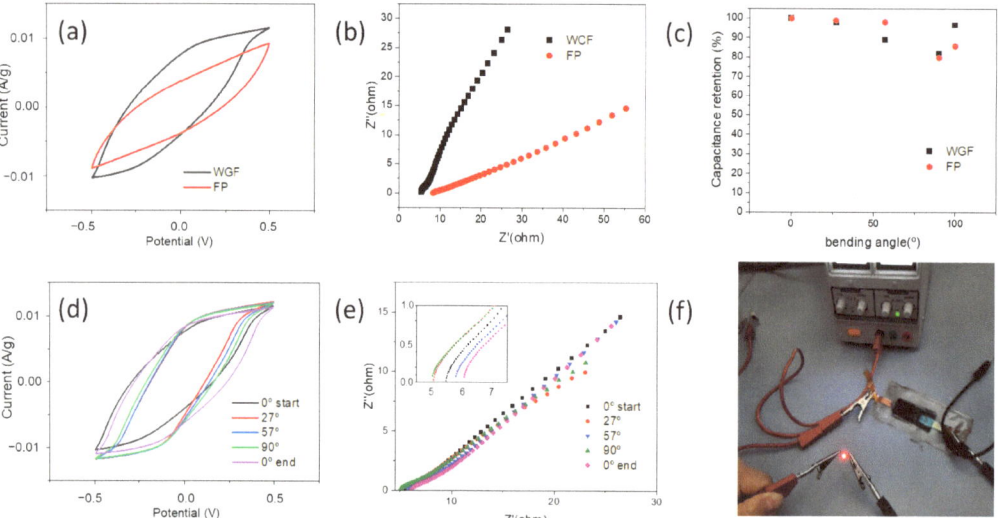

Figure 4. Characterization of the new device (CAG/separator/CAG) using a different separator (WGF). (**a**–**c**) Electrochemical performance comparison of FSCs. (**d**,**e**) CV and EIS of the device fabricated using woven glass fiber separator. (**f**) Proof of concept of red LED switched on.

Table 3. Electrochemical performance of FSCs with different separators.

Separator	$R_s/R_s + R_{ct}$ (Ω)	C_{sp} (F/g)	P (W/kg)	E (Wh/kg)
FSC-FP	8.26/69.1	0.563	0.815	0.048
FSC-WGF	5.48/8.48	1.369	0.962	0.161

3.4. Proof of Concept

The modified device using CAG electrodes and WGF separator was charged for 30 min at 2 V and was able to switch on a 1.6 V red LED, but only for a few minutes (1 min of bright light) (Figure 4f). This short time agrees with the fast discharge previously observed during the CDG. Despite the short time of the proof, this device kept 83.34% capacitance at 90° of bending and recovered 96.12% when returned to the initial unbending position. A new device was prepared following the same procedure but coupling two devices connected in parallel. For this fabrication, the modified CAG electrodes corresponded to the same batch to avoid the introduction of new variables. The CV and EIS curves obtained from the electrochemical characterization of the double and singles devices are depicted in Figure 5.

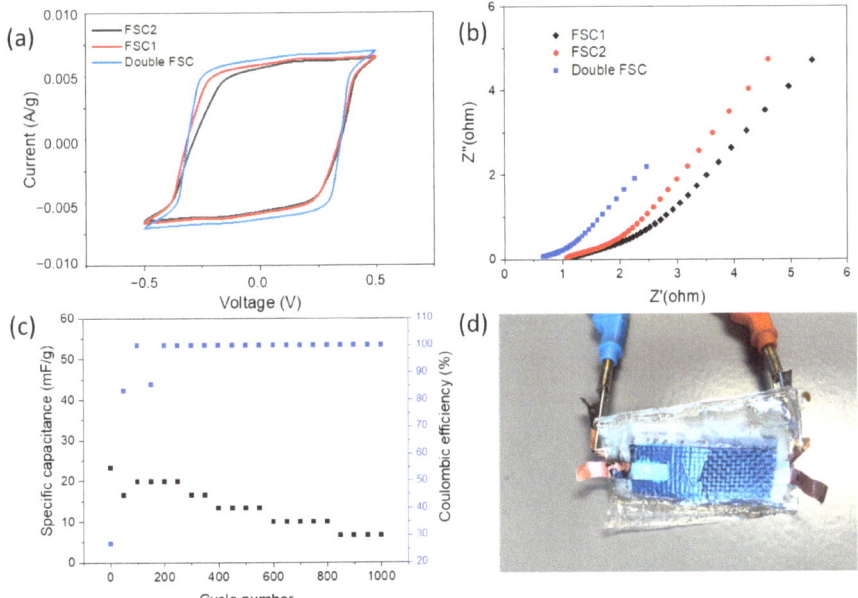

Figure 5. Electrochemical characterization of the double FSC-CAG connected in parallel, and each device separated (FSC1 and FSC2). (**a**) CV curves; (**b**) EIS analysis; (**c**) CDG results; and (**d**) double FSC device.

As can be observed, the square shape of the three CV curves is consistent with an ideal non-Faradaic capacitive behavior, and it shows higher stability in the voltage window analyzed. The results obtained from the CDG and EIS test are listed in Table 4.

Table 4. Electrochemical data of flexible supercapacitors connected in parallel.

Device	Capacitance (F)	Areal C (F/cm^2)	R_s/R_{ct} (W)
FSC1	2.14	0.143	1.02/1.14
FSC2	2.00	0.134	1.13/1.36
Double FSC	4.66	0.311	0.6/1.13

The areal capacitance for this new device was 0.311 F/cm^2, much higher than the results obtained for the FSC-CuO, and the resistances observed from the EIS analysis (R_s and R_{ct}) were exceptionally low compared to the single devices previously fabricated. The better electrochemical values obtained for the double FSC-CAG can be attributed to a better impregnation of the electrodes with the gel electrolyte during the fabrication, which led to the electrodes being better embedded in the matrix. This better encapsulation could favor the ion migration and transport, increasing the ionic conductivity and specific capacitance.

Two important aspects about the fabrication method must be considered: (1) The fabrication method of these devices can be a limitation, because depending on how good it is, there may be a different performance from one batch to another. (2) There are too many factors affecting the fabrication such as covering layer thickness, surface area and porosity, cured process, electrolyte impregnation, and casting process, which must be controlled to improve reproducibility and stability.

Even though these results are far from being scaled up, they demonstrate that the FSCs fabricated in this study exhibit better performance and capacitance than other FSCs made from carbon aerogel on woven carbon fiber. Of various sources of carbon that can be used to create potential electrodes for flexible applications, carbon fabric remains an ideal substrate despite its low specific surface area. The carbon aerogel synthesized from RF resin produces a highly porous covering that can modify the carbon cloth in a less expensive manner compared to other different carbon nanostructure approaches. Recent studies have focused on the use of CAG on CF to study their performance in structural supercapacitors [36].

Some results obtained for CAG/WCF electrodes using different electrolytes are depicted in Table 5 for comparison with the results obtained in this study. These results are also compared with those for bare carbon fiber, activated carbon fiber (etched and grafted), and other relevant carbonaceous coverings such as CNTs or GNP.

Table 5. Electrochemical data from CV analysis of FSCs.

Composites	Electrolyte	Electrodes (F/g)	Capacitor	Reference
WCF	3 M KCl	0.02	-	This work
WCF	3 M KCl	0.06	-	[25]
a-WCF	3 M	2.63	-	[25]
a-CFC	6 M KCl	197 (0.1 A/g)	0.5 Fcm^{-2}	[37]
ACF	PVA-H$_3$PO$_4$	18.6	300 mFcm^{-2}	[36]
WCF-CNT	EMIMTFSI	13.26	0.010	[38,39]
WCF-CNT	3 M KCl	0.015		[28]
WCF-GNP	3 M KCl	0.405		[28]
WCF/CAG	EMIMTFSI	2.4	-	[31]
WCF/CAG	EMIMTFSI	11.4	1.74 Fg^{-1}; 135 mFcm^{-2}	[40]
WCF/CAG	Epoxy resin/EMIMTFSI	-	0.212 Fg^{-1}	[40]
WCF/CAG	3 M KCl	5.9	-	[25]
CF/CAG	EMIMTFSI	-	1.49 Fg^{-1}	[25]
CF/CAG	EMIMTFSI	-	0.75 Fg^{-1}	[41]
WCF/CAG	3 M KCl	4.35	1.4 Fg^{-1}	This work
WCF/CuO	3 M KCl	5.42	0.33 Fg^{-1}	This work

The FSCs prepared in this study are competitive, highly flexible, and low cost, which is an important remark. It is expected that the materials developed in this work will have practical value and interest. Further studies involving more efficient fabrication methods are currently underway.

4. Conclusions

This study explores the effect of modifications made to woven carbon fiber on the performance of flexible supercapacitors. The electrodes were modified by the incorporation of copper oxide nanoparticles (CuO-NPs), carbon aerogel (CAG), and lignin covering (LIG) as coverings. Reinforced carbon fiber electrodes were employed to fabricate 2D-flexible supercapacitors using a sandwich configuration of CF-electrode/separator/CF-electrode, all coated with H$_3$PO$_4$/PVA gel electrolyte. The FSC produced a specific capacitance of 1.4 Fg^{-1}, which is comparable to other supercapacitors outlined in the existing literature.

This translated into energy and power densities of 0.161 Wh kg^{-1} and 0.961 W kg^{-1}, respectively. The CAG/WGF/CAG configuration showed a R_s of 5.48 ohm and demonstrated excellent electrochemical values, even during and after bending. The capacitor was able to recover 96.12% of its capacitance when returned to its original unbent position. In this study, the use of woven glass fiber (WGF) as a separator was found to improve the electrochemical performance. Finally, the fabrication process is critical, as the fibers must be embedded thoroughly in the gel electrolyte to perform effectively.

Supplementary Materials: The following supporting information can be downloaded at: https://www.mdpi.com/article/10.3390/polym15204036/s1, Figure S1: EIS analyses of the FSCs. CuO (Blue), CAG (Black), and LIG (Green); Figure S2: % Capacitance retention before, during and after bending for the FSCs analyzed; Figure S3: GCD tests for the samples FSC-CAG/FP/CAG (a and c) and FSC-CAG/WGF/CAG (b and d) at 0.002 A g^{-1}; Figure S4: GCD tests for the doubled FSC CAG/WGF/CAG/CAG/WCF/CAG at different scan rates.

Author Contributions: B.K.M.: Investigation, Methodology, Conceptualization, Formal analysis, Visualization, Writing—original draft. A.G.-B.: Investigation, Methodology. D.U.: Investigation, Methodology, Formal analysis. M.S.: Writing—review and editing, Funding acquisition. A.U.: Writing—review and editing, Funding acquisition. All authors have read and agreed to the published version of the manuscript.

Funding: This work was supported by the Agencia Estatal de Investigación of Spanish Government Project MULTISENS PID2022-136636OB-I00.

Data Availability Statement: The data presented in this study are available on request from the corresponding author.

Conflicts of Interest: The authors declare no conflict of interest.

References

1. Mao, M.; Hu, J.; Liu, H. Graphene-based materials for flexible electrochemical energy storage. *Int. J. Energy Res.* **2015**, *39*, 727–740. [CrossRef]
2. Keum, K.; Kim, J.W.; Hong, S.Y.; Son, J.G.; Lee, S.-S.; Ha, J.S. Flexible/Stretchable Supercapacitors with Novel Functionality for Wearable Electronics. *Adv. Mater.* **2020**, *32*, 2002180. [CrossRef] [PubMed]
3. Jiang, S.; Shi, T.; Zhan, X.; Long, H.; Xi, S.; Hu, H.; Tang, Z. High-performance all-solid-state flexible supercapacitors based on two-step activated carbon cloth. *J. Power Sources* **2014**, *272*, 16–23. [CrossRef]
4. Xue, Q.; Sun, J.; Huang, Y.; Zhu, M.; Pei, Z.; Li, H.; Wang, Y.; Li, N.; Zhang, H.; Zhi, C. Recent progress on flexible and wearable supercapacitors. *Small* **2017**, *13*, 1701827. [CrossRef] [PubMed]
5. Yu, D.; Qian, Q.; Wei, L.; Jiang, W.; Goh, K.; Wei, J.; Chen, Y. Emergence of fiber supercapacitors. *Chem. Soc. Rev.* **2015**, *44*, 647–662. [CrossRef] [PubMed]
6. Zhai, S.; Karahan, H.E.; Wei, L.; Qian, Q.; Harris, A.T.; Minett, A.I.; Chen, Y. Textile energy storage: Structural design concepts, material selection and future perspectives. *Energy Storage Mater.* **2016**, *3*, 123–139. [CrossRef]
7. Lv, Z.; Tang, Y.; Zhu, Z.; Wei, J.; Li, W.; Xia, H.; Jiang, Y.; Liu, Z.; Luo, Y.; Ge, X.; et al. Honeycomb-lantern-inspired 3D stretchable supercapacitors with enhanced specific areal capacitance. *Adv. Mater.* **2018**, *30*, 1805468. [CrossRef]
8. Xie, P.; Yuan, W.; Liu, X.; Peng, Y.; Yin, Y.; Li, Y.; Wu, Z. Advanced carbon nanomaterials for state-of-the-art flexible supercapacitors. *Energy Storage Mater.* **2021**, *36*, 56–76. [CrossRef]
9. Yu, G.; Hu, L.; Vosgueritchian, M.; Wang, H.; Xie, X.; McDonough, J.R.; Bao, Z. Solution-Processed Graphene/MnO$_2$ Nanostructured Textiles for High-Performance Electrochemical Capacitors. *Nano Lett.* **2011**, *11*, 2905–2911. [CrossRef]
10. Liu, W.; Lu, C.; Li, H.; Tay, R.Y.; Sun, L.; Wang, X.; Yu, A. Based all-solid-state flexible micro-supercapacitors with ultra-high rate and rapid frequency response capabilities. *J. Mater. Chem. A* **2016**, *4*, 3754–3764. [CrossRef]
11. Liu, Z.; Wu, Z.S.; Yang, S.; Dong, R.; Feng, X.; Müllen, K. Ultraflexible in-plane micro-supercapacitors by direct printing of solution-processable electrochemically exfoliated graphene. *Adv. Mater.* **2016**, *28*, 2217–2222. [CrossRef] [PubMed]
12. Wang, Y.; Tang, S.; Vongehr, S.; Ali Syed, J.; Wang, X.; Meng, X. High-performance flexible solid-state carbon cloth supercapacitors based on highly processible N-graphene doped polyacrylic acid/polyaniline composites. *Sci. Rep.* **2016**, *6*, 12883. [CrossRef] [PubMed]
13. Dubal, D.P.; Chodankar, N.R.; Kim, D.H.; Gomez-Romero, P. Towards flexible solid-state supercapacitors for smart and wearable electronics. *Chem. Soc. Rev.* **2018**, *47*, 2065–2129. [CrossRef] [PubMed]

14. Liao, Q.; Li, N.; Jin, S.; Yang, G.; Wang, C. All-Solid-State Symmetric Supercapacitor Based on Co$_3$O$_4$ Nanoparticles on Vertically Aligned Graphene. *ACS Nano* **2015**, *9*, 5310–5317. [CrossRef]
15. Javaid, A.; Zafrullah, M.B.; Khan, F.; Bhatti, G.M. Improving the multifunctionality of structural supercapacitors by interleaving graphene nanoplatelets between carbon fibers and solid polymer electrolyte. *J. Compos. Mater.* **2018**, *53*, 1401–1409. [CrossRef]
16. Hu, W.; Xiang, R.; Lin, J.; Cheng, Y.; Lu, C. Lignocellulosic Biomass Derived-Carbon Electrodes fro Flexible Supercapacitor: An Overview. *Materials* **2021**, *14*, 4571. [CrossRef]
17. Szabó, L.; Imanishi, S.; Tetsuo, F.; Hirose, D.; Ueda, H.; Tsukegi, T.; Ninomiya, K.; Takahashi, K. Lignin as a functional Green Coating on Carbon Fiber Surface to Improve Interfacial Adhesion in Carbon Fiber Reinforced Polymers. *Materials* **2019**, *12*, 159. [CrossRef]
18. Wang, G.; Wang, H.; Lu, X.; Ling, Y.; Yu, M.; Zhai, T.; Li, Y. Solid-state supercapacitor based on activated carbon cloths exhibits excellent rate capability. *Adv. Mater.* **2014**, *26*, 2676–2682. [CrossRef]
19. Zhou, C.; Liu, J. Carbon nanotube network film directly grown on carbon cloth for high-performance solid-state flexible supercapacitors. *Nanotechnology* **2014**, *25*, 035402. [CrossRef]
20. Yu, M.; Zhai, T.; Lu, X.; Chen, X.; Xie, S.; Li, W.; Tong, Y. Manganese dioxide nanorod arrays on carbon fabric for flexible solid-state supercapacitors. *J. Power Sources* **2013**, *239*, 64–71. [CrossRef]
21. Wang, Q.; Xu, J.; Wang, X.; Liu, B.; Hou, X.; Yu, G.; Wang, P.; Chen, D.; Shen, G. Core–Shell CuCo$_2$O$_4$@MnO$_2$ Nanowires on Carbon Fabrics as High-Performance Materials for Flexible, All-Solid-State, Electrochemical Capacitors. *ChemElectroChem* **2014**, *1*, 559–564. [CrossRef]
22. Liao, Q.; Li, N.; Jin, S.; Yang, G.; Wang, C. A flexible all-solid-state asymmetric supercapacitors based on hierarchical carbon cloth@ CoMoO$_4$@ NiCo layered double hydroxide core-shell heterostructures. *Chem. Eng. J.* **2018**, *352*, 29–38.
23. Majumdar, D.; Ghosh, S. Recent advancements of copper oxide-based nanomaterials for supercapacitor applications. *J. Energy Storage* **2021**, *34*, 101995. [CrossRef]
24. Gao, Y.; Zheng, S.; Fu, H.; Ma, J.; Xu, X.; Guan, L.; Wu, H.; Wu, Z.-S. Three-dimensional nitrogen doped hierarchically porous carbon aerogels with ultrahigh specific surface area for high-performance supercapacitors and flexible micro-supercapacitors. *Carbon* **2020**, *168*, 701–709. [CrossRef]
25. Qian, H.; Kucernak, A.R.; Greenhalgh, E.S.; Bismarck, A.; Shaffer, M.S.P. Multifunctional structural supercapacitor composites based on carbon aerogel modified high performance carbon fiber fabric. *ACS Appl. Mater. Interfaces* **2013**, *5*, 6113–6122. [CrossRef]
26. Deka, B.K.; Hazarika, A.; Kim, J.; Kim, N.; Jeong, H.E.; Park, Y.B.; Park, H.W. Bimetallic copper cobalt selenide nanowire-anchored woven carbon fiber-based structural supercapacitors. *Chem. Eng. J.* **2019**, *355*, 551–559. [CrossRef]
27. Kong, K.; Deka, B.K.; Seo, J.W.; Park, Y.-B.; Park, H.W. Effect of CuO nanostructure morphology on the mechanical properties of CuO/woven carbon fiber/vinyl ester composites. *Compos. Part A* **2015**, *78*, 48–59. [CrossRef]
28. Artigas-Arnaudas, J.; Muñoz, B.K.; Sánchez, M.; de Prado, J.; Utrilla, M.V.; Ureña, A. Surface modifications of carbon fiber electrodes for structural supercapacitors. *Appl. Compos. Mater.* **2022**, *29*, 889–900. [CrossRef]
29. Yokoyama, S.; Motomiya, K.; Jeyadevan, B.; Tohji, K. Environmentally friendly synthesis and formation mechanism of copper nanowires with controlled aspect ratios from aqueous solution with ascorbic acid. *J. Colloid Interface Sci.* **2018**, *531*, 109–118. [CrossRef]
30. Yuan, L.; Lu, X.-H.; Xiao, X.; Zhai, T.; Dai, J.; Zhang, F.; Hu, B.; Wang, X.; Gong, L.; Chen, J.; et al. Flexible Solid-State Supercapacitors Based on Carbon Nanoparticles/MnO$_2$ Nanorods Hybrid Structure. *ACS Nano* **2012**, *6*, 656–661. [CrossRef]
31. Qi, G.; Nguyen, S.; Anthony, D.B.; Kucernak, A.R.J.; Shaffer, M.S.P.; Greenhalgh, E.S. The influence of fabrication parameters on the electrochemical performance of multifunctional structural supercapacitors. *Multifunct. Mater.* **2021**, *4*, 034001. [CrossRef]
32. Deka, B.K.; Hazarika, A.; Kim, J.; Park, Y.-B.; Park, H.W. Multifunctional CuO nanowire embodied structural supercapacitor based on woven carbon fiber/ionic liquid–polyester resin. *Compos. Part A Appl. Sci. Manuf.* **2016**, *87*, 256–262. [CrossRef]
33. Zuo, W.; Li, R.; Zhou, C.; Li, Y.; Xia, J.; Liu, J. Battery-supercapacitor hybrid devices: Recent progress and future prospects. *Adv. Sci.* **2017**, *4*, 1600539. [CrossRef] [PubMed]
34. Shirshova, N.; Qian, H.; Houllé, M.; Steinke, J.H.; Kucernak, A.R.; Fontana, Q.P.; Greenhalgh, E.S.; Bismarck, A.; Shaffer, M.S.P. Multifunctional structural energy storage composite supercapacitors. *Faraday Discuss.* **2014**, *172*, 81–103. [CrossRef] [PubMed]
35. Noh, J.; Yoon, C.-M.; Kim, Y.K.; Jang, J. High performance asymmetric supercapacitor twisted from carbon fiber/MnO$_2$ and carbon fiber/MoO$_3$. *Carbon* **2017**, *116*, 470–478. [CrossRef]
36. Shen, C.; Xie, Y.; Zhu, B.; Sanghadasa, M.; Tang, Y.; Lin, L. Wearable woven supercapacitor fabrics with high energy density and load-bearing capability. *Sci. Rep.* **2017**, *7*, 14324. [CrossRef]
37. Zhang, T.; Kim, C.H.; Cheng, Y.; Ma, Y.; Zhang, H.; Liu, J. Making a commercial carbon fiber cloth having comparable capacitances to carbon nanotubes and graphene in supercapacitors through a "top-down" approach. *Nanoscale* **2015**, *7*, 3285–3291. [CrossRef]
38. Senokos, E.; Anthony, D.B.; Rubio Carrero, N.; Crespo Ribadeneyra, M.; Greenhalgh, E.S.; Shaffer, M.S.P. Robust Single-Walled Carbon Nanotube-Infiltrated Carbon Fiber Electrodes for Structural Supercapacitors: From Reductive Dissolution to High Performance Devices. *Adv. Funct. Mater.* **2023**, *33*, 2212697. [CrossRef]
39. Greenhalgh, E.S.; Ankersen, J.; Asp, L.E.; Bismarck, A.; Fontana, Q.P.V.; Houlle, M.; Kalinka, G.; Kucernak, A.; Mistry, M.; Nguyen, S.; et al. Mechanical, electrical and microstructural characterisation of multifunctional structural power composites. *J. Compos. Mater.* **2015**, *49*, 1823. [CrossRef]

40. Pernice, M.F.; Qi, G.; Senokos, E.; Anthony, D.B.; Nguyen, S.; Valkova, M.; Greenhalgh, E.S.; Shaffer, M.S.P.; Kucernak, A.R.J. Mechanical, electrochemical and multifunctional performance of a CFRP/carbon aerogel structural supercapacitor and its corresponding monofunctional equivalents. *Multifunct. Mater.* **2022**, *5*, 025002. [CrossRef]
41. Yue, C.; Bismarck, A.; Qian, H. Carbon Aerogel Coated Carbon Fibre Fabrics as Electrode Materials for Multifunctional Supercapacitors. Master's Thesis, Imperial College London, London, UK, 2011.

Disclaimer/Publisher's Note: The statements, opinions and data contained in all publications are solely those of the individual author(s) and contributor(s) and not of MDPI and/or the editor(s). MDPI and/or the editor(s) disclaim responsibility for any injury to people or property resulting from any ideas, methods, instructions or products referred to in the content.

Article

Silver Nanoparticle–PEDOT:PSS Composites as Water-Processable Anodes: Correlation between the Synthetic Parameters and the Optical/Morphological Properties

Stefania Zappia [1,*], Marina Alloisio [2,*], Julio Cesar Valdivia [3], Eduardo Arias [3], Ivana Moggio [3,*], Guido Scavia [1,*] and Silvia Destri [1]

1. Istituto di Scienze e Tecnologie Chimiche "Giulio Natta" (SCITEC), Consiglio Nazionale delle Ricerche (CNR), Via Alfonso Corti 12, 20133 Milano, Italy; silvia.destri@scitec.cnr.it
2. Dipartimento di Chimica e Chimica Industriale (DCCI), Università di Genova, Via Dodecaneso 31, 16146 Genova, Italy
3. Centro de Investigación en Química Aplicada (CIQA), Boulevard Enrique Reyna 140, Saltillo 25294, Mexico; eduardo.arias@ciqa.edu.mx (E.A.)
* Correspondence: stefania.zappia@scitec.cnr.it (S.Z.); marina.alloisio@unige.it (M.A.); ivana.moggio@ciqa.edu.mx (I.M.); guido.scavia@scitec.cnr.it (G.S.)

Citation: Zappia, S.; Alloisio, M.; Valdivia, J.C.; Arias, E.; Moggio, I.; Scavia, G.; Destri, S. Silver Nanoparticle–PEDOT:PSS Composites as Water-Processable Anodes: Correlation between the Synthetic Parameters and the Optical/Morphological Properties. *Polymers* 2023, 15, 3675. https://doi.org/10.3390/polym15183675

Academic Editors: Vineet Kumar and Md Najib Alam

Received: 12 August 2023
Revised: 30 August 2023
Accepted: 31 August 2023
Published: 6 September 2023

Copyright: © 2023 by the authors. Licensee MDPI, Basel, Switzerland. This article is an open access article distributed under the terms and conditions of the Creative Commons Attribution (CC BY) license (https://creativecommons.org/licenses/by/4.0/).

Abstract: The morphological, spectroscopic and rheological properties of silver nanoparticles (AgNPs) synthesized in situ within commercial PEDOT:PSS formulations, labeled PP@NPs, were systematically investigated by varying different synthetic parameters ($NaBH_4/AgNO_3$ molar ratio, PEDOT:PSS formulation and silver and PEDOT:PSS concentration in the reaction medium), revealing that only the reagent ratio affected the properties of the resulting nanoparticles. Combining the results obtained from the field-emission scanning electron microscopy analysis and UV-Vis characterization, it could be assumed that PP@NPs' stabilization occurs by means of PSS chains, preferably outside of the PEDOT:PSS domains with low silver content. Conversely, with high silver content, the particles also formed in PEDOT-rich domains with the consequent perturbation of the polaron absorption features of the conjugated polymer. Atomic force microscopy was used to characterize the films deposited on glass from the particle-containing PEDOT:PSS suspensions. The film with an optimized morphology, obtained from the suspension sample characterized by the lowest silver and $NaBH_4$ content, was used to fabricate a very initial prototype of a water-processable anode in a solar cell prepared with an active layer constituted by the benchmark blend poly(3-hexylthiophene) and [6,6]-Phenyl C61 butyric acid methyl ester ($PC_{60}BM$) and a low-temperature, not-evaporated cathode (Field's metal).

Keywords: PEDOT:PSS; silver nanoparticles; anode; organic photovoltaics; nanocomposites

1. Introduction

In the last decade, polymer-based solar cells (PSCs, hereafter) have attracted enormous interest as next-generation green energy sources because of their promising potential due to the cheap cost of manufacturing, flexibility, roll-to-roll and large area processing. From this point of view, electrodes represent a bottleneck for the printing and high-speed processing of PSCs [1–3].

Poly(3,4-ethylenedioxythiophene) polystyrene sulfonate (PEDOT:PSS) is currently the most successful water-processable conductive polymer with the highest reported conductivity [2,4], which makes it essential in the fabrication of photovoltaics, optoelectronics, bioelectronics, printed electronics and so on [5]. PEDOT is a polythiophene derivative with high stability and conductivity that could be obtained from the oxidative polymerization of the corresponding 3,4-ethylenedioxythiophene (EDOT) monomer. Unfortunately, the obtained PEDOT displayed poor solubility in polar solvents. To overcome this problem, Bayer's scientists optimized the synthetic procedure by oxidizing EDOT monomers in the presence of the water-soluble sodium salt of polystyrene sulfonate (PSS) [6]. In the resulting stable aqueous dispersion, PEDOT was stabilized with PSS as a counter anion, which also acted as a dopant for the PEDOT backbone [5,7].

As is known, the PEDOT molecular weight is much lower than that of PSS, and several PEDOT oligomeric chains interact with a single PSS chain [8]. In addition, the stability of the aqueous suspension is due to the shielding of the hydrophobic PEDOT by PSS, leading to the formation of grain structures via hydrophilic, PSS-rich shell and PEDOT chains arranged in the core. The grains are linked together with hydrogen bonding between the PSS shell according to a model proposed by Dupont et al. [9]; thus, the content of PSS is strictly connected to the conductivity as it influences the charge-carrier mobility [10].

The PEDOT:PSS film has the advantage of low cost, high transmittance, stability after film formation, and adjustable conductivity [11–13]. For these reasons, it has been long-since used as both a hole-transporting layer (injection for organic light-emitting diodes and extraction for PSCs) and an anode in different devices [14–17]. Highly conductive H_2SO_4-treated PEDOT:PSS films replaced the indium tin oxide (ITO) layer as the transparent anode in PSCs, achieving a power conversion efficiency (PCE) of 3.56%, which was comparable to that of the control devices using an ITO anode [17]. Large-area and flexible PSCs need ITO-free anodes due to the limited availability of ITO, as well as its rather high cost, its stiff and brittle nature which generates cracks when the material is subjected to repeat tensile and compressive bending, and finally, its lack of adhesion to organic and polymeric materials.

Besides highly conductive PEDOT:PSS, other materials have been tested for this aim. Among them, silver nanowires (AgNWs) have been considered as a promising alternative due to their excellent electrical conductivity and mechanical flexibility. However, the junction resistance between nanowires and the remaining non-conducting area in the electrode layer may worsen the device performance. A simple method to overcome these drawbacks is the fabrication of a AgNW/PEDOT:PSS nanocomposite obtained by depositing a highly conductive PEDOT:PSS layer on a AgNW-coated substrate followed by annealing [18,19].

Another nanocomposite approach has been applied which involved directly embedding silver nanoparticles (AgNPs) in highly conductive PEDOT:PSS. The resulting device led to a 32% increase in PCE compared to the device without AgNPs [20].

In that research, the AgNPs were prepared separately and then added to PEDOT:PSS, whereas other authors used a different strategy, synthetizing AgNPs directly in PEDOT:PSS via the reduction of $AgNO_3$ salt [21] and applying the film of this material as the anode [22]. The electrical and electronic properties of the AgNPs/PEDOT:PSS composite strongly depend on the AgNP size and shape, which in turn are influenced by the $AgNO_3$/reducing agent molar ratio and the procedure protocols [22]. Under this context, in the present work, we discussed the leverage of some parameters over AgNP preparation in a PEDOT:PSS matrix, i.e., the $AgNO_3$/reducing agent molar ratio, the features of the commercial starting PEDOT:PSS suspensions, and the concentration of the conducting polymer and the silver salt in the reaction medium. In particular, we performed the spectroscopic and morphological characterization of the aqueous colloidal suspensions prepared at different $AgNO_3$/reducing agent molar ratios. Then, we performed the morphological and electrical study of the films obtained from the deposition of the dispersions varying these parameters. This empirical study allowed us to achieve a prototypical photovoltaic device characterized by AgNP/PEDOT:PSS composites as the anode and the benchmark blend phenyl-C61-butyric acid methyl ester:poly(3-hexylthiophene) (PC61BM:P3HT) as the active layer. Moreover, Field's metal was used as the cathode through easy transfer on the active layer in a molten state at 70 °C.

2. Materials and Methods

2.1. Chemicals

The poly(3,4-ethylenedioxythiophene) polystyrene sulfonate (PEDOT:PSS) aqueous suspensions used in this work were purchased from Sigma-Aldrich (Merck Life Science

S.r.l., Milan, Italy), product name Orgacon™ ICP1050 (Agfa, Motzel, Belgium) (product code #739332), and from Heraeus (Heraeus Italia, Monza-Brianza, Italy), product name Clevios™ PH1000, and are referred to hereafter as Orgacon and Clevios, respectively.

The phenyl-C61-butyric acid methyl ester (PC61BM) and poly(3-hexylthiophene) (P3HT) used for the active layers were purchased from Sigma-Aldrich (Merck Life Science S.r.l., Toluca, Mexico) and were used as received without further purifications. ITO slides were purchased from Ossila (Ossila Ltd., Sheffield, UK).

Sodium borohydride ($NaBH_4$, Sigma-Aldrich, Merck Life Science S.r.l, Milan, Italy) and silver nitrate ($AgNO_3$, Alfa Aesar, Thermo Fisher GmBH, Kandel, Germany) were commercial products used as received.

Aqueous solutions were prepared with ultra-high-purity Milli-Q water distilled twice prior to use.

Freshly prepared "piranha" solution, obtained by mixing concentrated sulfuric acid and cooled hydrogen peroxide (30% v/v) in the ratio of 2:1 v/v, was used to thoroughly clean the glassware before the silver nanoparticles' synthesis.

2.2. Synthesis of Silver Nanoparticles in Presence of PEDOT:PSS

Silver nanoparticles stabilized with PEDOT:PSS (PP@AgNPs) were synthesized "in situ" following different protocols for the wet chemical reduction of $AgNO_3$. The first series of samples was synthetized with a method adapted from a previous set-up for the preparation of $AgNO_3$ embedded in a polysaccharide matrix [23–25]. Briefly, 10 mL of Orgacon was placed in a thoroughly cleaned flask, and 0.1 g of $AgNO_3$ was added. The mixture was kept under stirring conditions at room temperature until the complete solubilization of the silver salt. Then, a proper aliquot of a freshly prepared solution of $NaBH_4$ in water (1.2 mol/L) was introduced drop by drop and the mixture was (eventually) diluted to 15 mL with water. The procedure was repeated three times to obtain different molar ratios between $NaBH_4$ and $AgNO_3$. In detail, 5 mL of $NaBH_4$ (1.2 mol/L) was added in the PP@AgNPs-1 sample ($NaBH_4/AgNO_3$ ratio = 9.9), 2.5 mL of $NaBH_4$ (1.2 mol/L) and 2.5 mL of water were added in the PP@AgNPs-2 sample ($NaBH_4/AgNO_3$ ratio = 4.9), 0.5 mL of $NaBH_4$ (1.2 mol/L) and 4.5 mL of water were added in the PP@AgNPs-3 sample ($NaBH_4/AgNO_3$ ratio = 1.0). The mixtures immediately turned from colorless to deep green and were maintained under constant stirring conditions overnight to ensure a full reaction. The end-products were aqueous suspensions with nominal concentrations of ~40 mmol/L in terms of Ag content and were stored at room temperature.

The second series of samples was synthetized according to ref. [26]. Briefly, 10 mL of PEDOT:PSS was placed in a thoroughly cleaned flask and 0.1 g of $AgNO_3$ were added. The mixture was kept under stirring conditions at room temperature for 3 h, and then, a freshly prepared aqueous solution of $NaBH_4$ (12 mmol/L, 5 mL) was added. After 30 min of stirring, the end-product was an aqueous suspension with a nominal concentration of 40 mmol/L in terms of Ag content and a 0.1 molar ratio between $NaBH_4$ and $AgNO_3$, which was stored at room temperature. The procedure was repeated twice using both formulations of PEDOT:PSS to obtain samples PP@AgNPs-4 and PP@AgNPs-5, respectively.

The third series of samples was synthetized with an adaptation of the first two methods carried out at a lower concentration of both silver salt and PEDOT:PSS. Briefly, 20 mL of an aqueous solution of $AgNO_3$ (5 mmol/L) was placed in a thoroughly cleaned flask and 1 mL of Clevios was added under constant stirring conditions. Then, a proper aliquot of a freshly prepared solution of $NaBH_4$ in water (45 mmol/L) was introduced drop by drop and the mixture was diluted to 50 mL with water. The procedure was repeated twice to obtain different molar ratios between $NaBH_4$ and $AgNO_3$. In detail, 0.25 mL of $NaBH_4$ (45 mmol/L) was added in the PP@AgNPs-6 sample ($NaBH_4/AgNO_3$ ratio = 0.1) and 25 mL of $NaBH_4$ (45 mmol/L) was added in the PP@AgNPs-7 sample ($NaBH_4/AgNO_3$ ratio = 11). The end-products were aqueous suspensions of nominal concentrations of 2 mmol/L in terms of Ag content, which were stored at room temperature.

The synthesis conditions are reported in Table 1.

Table 1. Experimental conditions set up for the synthesis of PP@AgNPs.

Sample	PEDOT:PSS Formulation	NaBH$_4$/AgNO$_3$ Molar Ratio	Ag Conc (mmol/L)
PP@AgNPs-1	Orgacon	9.9	40
PP@AgNPs-2	Orgacon	4.9	40
PP@AgNPs-3	Orgacon	1.0	40
PP@AgNPs-4	Orgacon	0.1	40
PP@AgNPs-5	Clevios	0.1	40
PP@AgNPs-6	Clevios	0.1	2
PP@AgNPs-7	Clevios	11.0	2

2.3. Characterization Techniques

UV-vis-NIR spectra of the PP@AgNPs samples were acquired at room temperature via a Shimadzu UV-1800 (Shimadzu USA Manufacturing, Inc., Canby, OR, USA) spectrophotometer, using silica cuvettes of different pathlengths.

PP@AgNPs were investigated via field-emission scanning electron microscopy (FE-SEM) by using a ZEISS SUPRA 40 VP (Carl Zeiss NST, GmbH, Oberkochen, DEU) microscope operating at 20 keV in both direct (in-Lens mode) and back-scattered (QBSD mode) configurations. Before the analysis, the nanoparticle-containing samples were thinly sputter-coated with carbon, using a Polaron E5100 sputter coater (2M Strumenti, Rome, Italy) to obtain good electrical conductivity. Size measurements of metal clusters and silver nanoparticles were carried out by means of the open-source Image J^{TM} software on no less than 100 specimens from images at different magnifications.

The rheological characterization of the PP@AgNPs suspensions was carried out using a rotational rheometer, Physica MCR 301 (Anton Paar GmbH, Graz, Austria). The experimental temperature was set at 25.0 ± 0.2 °C by means of a Peltier heating system coupled with a solvent trap kit to prevent the solvent evaporation. A cone–plate geometry (CP50) with a diameter of 50 mm, an angle of 1° and a truncation of 99 μm was used.

The morphological characterization of the PP@AgNP layers after deposition was carried out using AFM NTMDT NTEGRA (NT-MDT Spectrum Instruments Llc. Moscow, Russia) operating in contact and tapping mode (NSG10 with cantilever resonant frequency: 150–300 kHz). Local surface potential/work function was determined via AFM operating in Kelvin mode with conductive tips (NSG10/Pt). The electrical sheet resistance measurements of the films were performed by using a Jandel Multiheight Four Probe Head and a Keithley 2601 A System Source Meter with the appropriate acquisition software.

The electrical measurements of the films were realized at ambient conditions with a Quantum Design magnetometer. UV-Vis-NIR spectra were acquired on an Agilent Cary 60 instrument, with baseline in air.

2.4. Layer Deposition and Prototype Fabrication

Organic solar cells were fabricated with the following configuration: glass/Orgacon/ PP@AgNPs-6/P3HT:PCBM/Field's metal (FM), where the bilayer Orgacon/PP@AgNPs-6 works as the anode. For comparative studies, ITO slides were also used. The ITO or glass slides were cleaned in a Branson ultrasonic bath with different solvents—(1) methylene chloride (10 min), (2) isopropyl alcohol (10 min) and (3) acetone (10 min)—then dried at room temperature and exposed to UV-Ozone treatment for 10 min. All the layers were deposited via spin coating with a Laurell spin coater. For the anode preparation, Orgacon was first spun on the glass substrates, at 1000 rpm for 2 min, followed by 1 min at 4000 rpm and then annealed at 125 °C for 10 min. The following PP@AgNPs-6 layer was then deposited with the same spinning conditions but the thermal treatment at 125 °C was conducted for just 2 min. The active layer was prepared from a 1:1 wt chlorobenzene solution of P3HT:PC61BM. Both materials were dissolved in chlorobenzene with a total concentration of 20 mg/mL and kept overnight at 60 °C under magnetic stirring conditions. The mixture was then spun at 1200 rpm for 50 s with an acceleration of 2500 rpm. After deposition, the film was thermally annealed at 150 °C for 25 min. FM as the cathode was deposited at 70 °C. The I-V curves were obtained using a workstation X100 Source Measure Unit from Ossila by illuminating the devices from the ITO side with a 100 mW cm^2 white light from a Solar Light Co. (Glenside, PA, USA) Model XPS 400 solar simulator with a Xenon lamp and AM1.5 filter. All the cells were prepared and measured under ambient conditions.

3. Results and Discussion

3.1. Characterization of Silver Nanoparticles Stabilized with PEDOT:PSS (PP@AgNP) Suspensions

At first, the samples were characterized by means of UV-Vis-NIR spectroscopy. The spectra acquired from the corresponding diluted aqueous suspensions are shown in Figure 1. In detail, Figure 1a reports the spectra of samples obtained from the first synthesis, whereas Figure 1b,c report the spectra of samples obtained from the second and the third syntheses, respectively. The spectra obtained from the diluted aqueous suspensions of the starting PEDOT:PSS formulations are reported in Figure S1 in the supporting information for comparison.

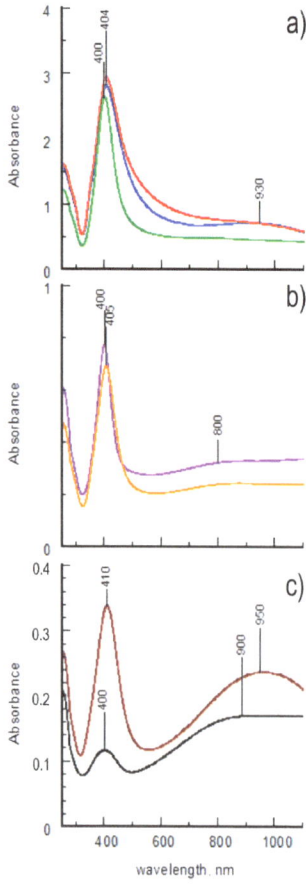

Figure 1. UV-Vis-NIR spectra of diluted suspensions. From top to bottom: (**a**) PP@AgNPs-1 (blue line), PP@AgNPs-2 (red line) and PP@AgNPs-3 (green line); (**b**) PP@AgNPs-4 (purple line) and PP@AgNPs-5 (orange line); (**c**) PP@AgNPs-6 (gray line) and PP@AgNPs-7 (brown line).

All spectral profiles are dominated by the intense, well-defined plasmonic band at 400 nm, which confirms the formation of AgNPs within the PEDOT:PSS matrix. As far as the products from the first synthesis are concerned, the narrowest spectral lineshape of PP@AgNPs-3 (Figure 1a, green line) agrees with the increased size homogeneity of the metal nanoparticles in the sample. Conversely, the polaronic band of PEDOT, usually positioned around 900 nm, is not detectable for PP@AgNPs-2 and PP@AgNPs-3 and is only slightly hinted at for PP@AgNPs-1. These results are indicative of a partial doping loss of the conductive polymer induced by the high concentrations of the reductant $NaBH_4$ [27]. Moreover, the AgNPs synthesized through "in situ" protocols are supposed to be preferentially stabilized with the PSS excess in the sample [27].

Further information can be derived from the spectral lineshapes of PP@AgNPs-4 and PP@AgNPs-5 in Figure 1b and of PP@AgNPs-6 and PP@AgNPs-7 in Figure 1c. The spectra of PP@AgNPs-4 and PP@AgNPs-5 are almost overlapped to highlight that PEDOT:PSS formulation negligibly affects the synthesis reaction of the AgNPs. Moreover, the lineshape of the plasmonic bands are very similar to that of PP@AgNPs-3, confirming that more homogeneous nanoparticles can be obtained by lowering the $NaBH_4/AgNO_3$ molar ratio under the same experimental conditions. In the spectral region corresponding to PEDOT absorptions, a weak shoulder at around 800 nm is observed in both cases, attributable to the polaron band of PEDOT being subjected to conformational changes [27]. This result indicates that the AgNPs are formed in the domains of the conductive polymer due to the high concentration of silver precursor in the reaction mixture, even if the lowest $NaBH_4/AgNO_3$ molar ratio is used. This hypothesis seems to be confirmed by the spectral profiles of PP@AgNPs-6 and PP@AgNPs-7, obtained by using a 20-fold lower $AgNO_3$ concentration. In these cases, the polaron band is positioned at around 930 nm, which is typical of unperturbed PEDOT. It is reasonable to assume that at a low metal precursor concentration, the nanoparticles are formed in smaller quantities and outside the conductive domains, even in PEDOT:PSS formulations with low PSS content such as Clevios, and independently from the reductant concentration. However, the reaction yields drastically increase when increasing the $NaBH_4/AgNO_3$ molar ratio, as shown by the higher intensity of the plasmonic band of PP@AgNPs-7.

In addition, high-resolution FE-SEM images of PP@AgNP suspensions, acquired in both direct (left) and back-scattered (right) configurations, are shown in Figure 2 for PP@AgNPs-1, PP@AgNPs-2, PP@AgNPs-3, PP@AgNPs-6 and PP@AgNPs-7. As the comparison between the UV-Vis-NIR spectra of the samples PP@AgNPs-4 and PP@AgNPs-5 allows overlapping to sample PP@AgNPs-3, only the latter sample has been further characterized with high-resolution FE-SEM. The morphological parameters extracted from the images are listed in Table 2.

The images in Figure 2a–c highlight the presence of nearly star-shaped metal clusters of dimensions greater than or equal to 200 nm. The aggregates are composed of elongated (aspect ratio around 1.4) AgNPs arranged radially around the center of the cluster. As expected, the nanoparticles grow in average dimensions and size homogeneity from PP@AgNPs-1 to PP@AgNPs-3, that is, with the $NaBH_4/AgNO_3$ molar ratio decreasing in the synthesis reaction. Moreover, it is important to notice that the observed metal clusters in Figure 2a could represent the nanoparticle aggregates formed out of the conductive PEDOT:PSS domains. This remark is consistent with the UV-Vis-NIR spectra observed for sample PP@AgNPs-1 in Figure 1a.

Different features are observed in Figure 2d,e. Elongated nanoparticles of reduced average size (about 20 nm) and axial ratio (1.3) are highlighted, with better dispersed and organized in clusters of dimensions around 70 nm only in the case of PP@AgNPs-6. This result is ascribable to the 20-fold lower quantity of the silver salt precursor used in the synthesis of this sample.

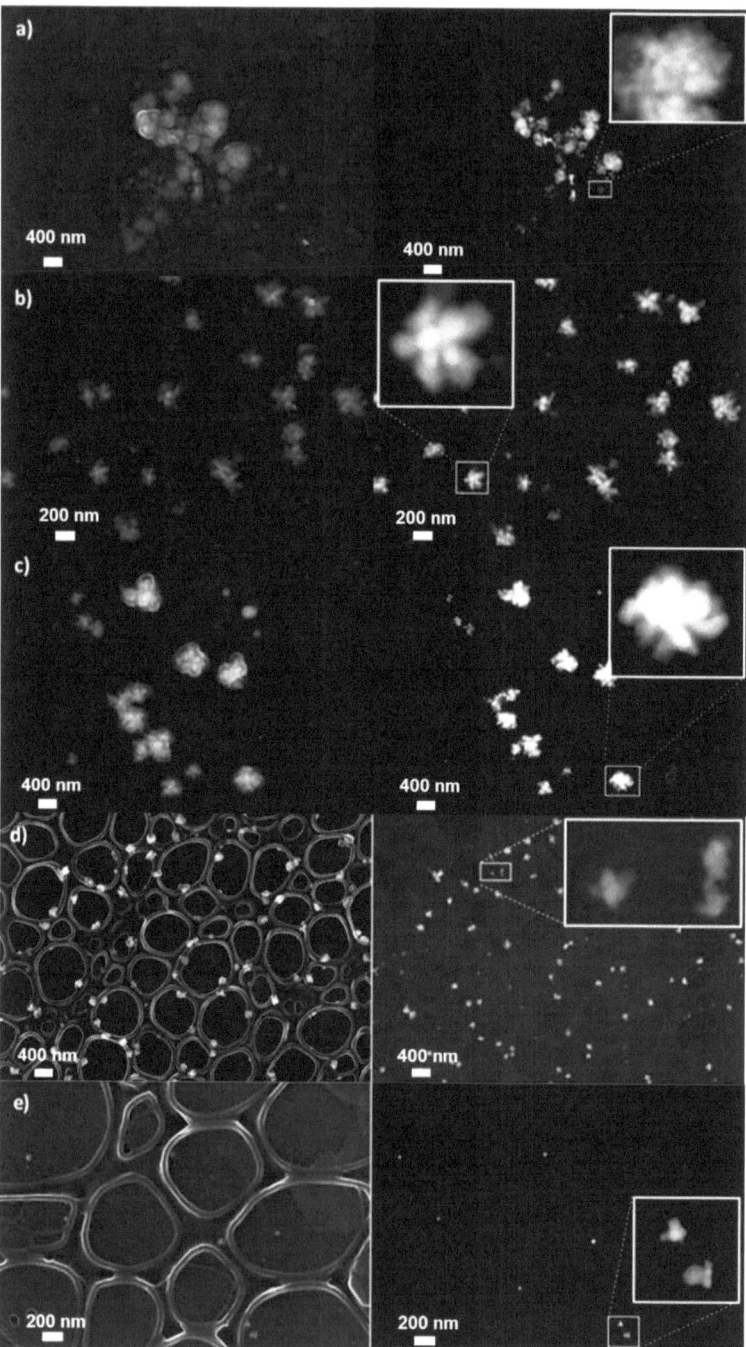

Figure 2. FE-SEM images acquired in direct (In-Lens, left) and back-scattered (QBSD, right) configurations from the samples of PP@AgNPs, for all the images (**a**–**e**). From top to bottom: (**a**) PP@AgNPs-1, (**b**) PP@AgNPs-2, (**c**) PP@AgNPs-3, (**d**) PP@AgNPs-6 and (**e**) PP@AgNPs-7. Magnifications of structural details are reported in the insets of the QBSD pictures.

Table 2. Summary of the PP@AgNP morphological parameters measured from FE-SEM images of Figure 2.

Sample	Cluster Range Size (nm)	Cluster Average Size (nm)	Nanoparticle Average Size (Standard Deviation) (nm)	Nanoparticle Aspect Ratio (Standard Deviation)
PP@AgNPs-1	90–500	-	26 ± 1 (7)	1.44 ± 0.06 (0.41)
PP@AgNPs-2	120–300	220 ± 10	32 ± 1 (9)	1.39 ± 0.04 (0.31)
PP@AgNPs-3	60–600	200 ± 5	60 ± 1 (15)	1.45 ± 0.04 (0.45)
PP@AgNPs-6	30–160	240 ± 20	21 ± 1 (5)	1.31 ± 0.04 (0.25)
PP@AgNPs-7	-	70 ± 1	18 ± 1 (9)	1.26 ± 0.03 (0.24)

Finally, the rheological behavior of the samples was also investigated to achieve information about the film-forming capability of the AgNP-containing suspensions [28]. As an example, the curves of viscosity (η) obtained from PP@AgNPs-1, PP@AgNPs-2, PP@AgNPs-3, PP@AgNPs-6 and PP@AgNPs-7 are reported in Figure 3. The main rheological parameters extracted from the curves are listed in Table 3.

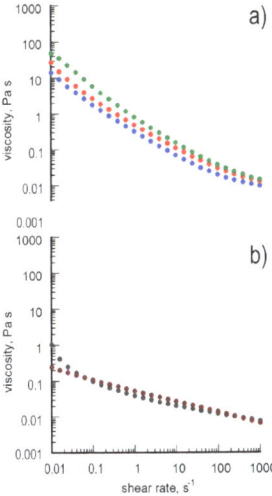

Figure 3. Curves of viscosity obtained from samples. From top to bottom: (a) PP@AgNPs-1 (blue full circles), PP@AgNPs-2 (red full circles) and PP@AgNPs-3 (green full circles); (b) PP@AgNPs-6 (gray full circles) and PP@AgNPs-7 (brown full circles).

Table 3. Summary of the rheological parameters extracted from the curves of viscosity in Figure 3.

Sample	η_0 (Pa s)
PP@AgNPs-1	14.1
PP@AgNPs-2	26.7
PP@AgNPs-3	47.2
PP@AgNPs-6	1.0
PP@AgNPs-7	0.2

The three samples show the characteristic viscosity curves of pseudoplastic Bingham fluids, which are as expected in that the suspensions contain aggregated, non-deformable nanoparticles. The absolute viscosity values at shear rate equal to zero (η_0), evaluated via extrapolation from the regression data model supplied by the instrument, are generally low.

In detail, they increase from PP@AgNPs-1 to PP@AgNPs-3. These results are consistent with the differences in the AgNP size found in the samples, since larger, non-deformable nanofillers are expected to further increase the η_0 value of the fluid in which they are dispersed and extend its Newtonian behavior. The trends with the shear rate of Figure 3b, corresponding to PP@AgNPs-6 and PP@AgNPs-7, are almost superimposed and settle for lower η values. These results were expected given that the samples are characterized by a reduced concentration of particles with comparable size. Taking into account this study, the PP@AgNP_6 and _7 are the most promising samples for making films.

3.2. Deposition of the PP@AgNP Suspensions

The first deposited layer onto glass substrates is constituted by an Orgacon layer in place of ITO, and the AFM images of this layer are shown in Figure 4a (root mean square, RMS: 1.2 nm; Table 4). The choice to use Orgacon onto glass is due not only to electronic reasons in order to guarantee high conductance in the absence of ITO, but also to improve the adhesion of the following PP@AgNP layer. At the same time, it also provides a flat and homogeneous layer for the subsequent active layer deposition, as shown in Figure 4f.

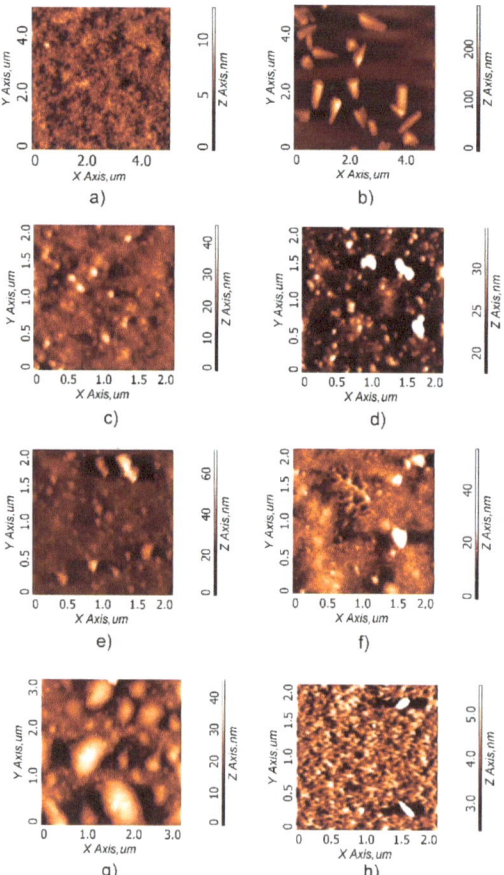

Figure 4. Atomic force microscopy (AFM) images of the PP@NP samples: (**a**) Orgacon (RMS = 1.2 nm); (**b**) PP@AgNPs-1 (RMS = 33 nm); (**c**) PP@AgNPs-2 (RMS = 4.3 nm); (**d**) PP@AgNPs-3 (RMS = 5.3 nm); (**e**) PP@AgNPs-4 (RMS = 7.7 nm); (**f**) PP@AgNPs-5 (RMS = 4.4 nm); (**g**) PP@AgNPs-07 (RMS = 8.1 nm); (**h**) PP@AgNPs-6 (RMS = 1.3 nm).

Table 4. Summary of the root mean square (RMS) related to the synthesis conditions of the samples extracted from Figure 4.

Sample Label	$NaBH_4/AgNO_3$ Molar Ratio	Ag Conc (mmol/L)	Root Mean Square, RMS (nm)
PEDOT:PSS (Orgacon)	-	-	1.2
PP@AgNPs-1	9.9	40	33.0
PP@AgNPs-2	4.9	40	4.3
PP@AgNPs-3	1.0	40	5.3
PP@AgNPs-4	0.1	40	7.7
PP@AgNPs-5	0.1	40	4.4
PP@AgNPs-6	0.1	2	1.3
PP@AgNPs-7	11.0	2	8.1

3.2.1. Effect of Reducing Agent/$AgNO_3$ Ratio

In the PP@AgNPs-1 sample displaying a higher $NaBH_4/AgNO_3$ molar ratio, elongated structures with lengths ranging from 400 nm up to 1 µm can be observed (Figure 4b), over a quite smooth layer. The presence of crystallite-like structures is responsible for the large corresponding RMS of 33 nm (Table 4). When considering the FE-SEM analysis (Figure 2a) and based on the proposed mechanism for stabilizing the nanostructures, it is likely that the elongated structures correspond to the silver nanoparticles aggregates directly covered with the excess of PSS and embedded in the PEDOT matrix that forms a smooth layer, similarly to the underlying film. Otherwise, the morphologies of PP@AgNPs-2 (Figure 4c) and PP@AgNPs-3 (Figure 4d) are granular-type, with a distribution of spherical grains with a mean diameter of 63 nm (PP@AgNPs-2) and 84 nm (PP@AgNPs-3), respectively. Moreover, the roughness of the PP@AgNP layer decreases from 33 nm for PP@AgNPs-1 to 4.3 nm and 5.3 nm for PP@AgNPs-2 and PP@AgNPs-3, respectively (Table 4). It is to be noted that a similar morphology was observed for PP@AgNPs-5 (see further) prepared with a $NaBH_4/AgNO_3$ molar ratio that is even lower, resulting in a lower RMS of 4.4 nm. This trend could indicate that the AgNPs are better dispersed in the PEDOT:PSS matrix along with the decrease in the $NaBH_4$ content during the sample preparation.

3.2.2. Effect of the Formulation of PEDOT:PSS Suspension

In the second series of samples, a comparison between the two typologies of commercial PEDOT:PSS within the nanocomposite blend (Clevios and Orgacon for PP@AgNPs-4 and PP@AgNPs-5, respectively) is carried out, while the $NaBH_4/AgNO_3$ molar ratio is kept constant (Table 1). Figure 4e,f report the AFM images of the PP@AgNPs-4 and PP@AgNPs-5 samples, respectively. It is possible to observe the formation of aggregates over a rough texture, similarly to PP@AgNPs-3, giving rise to an RMS of 7.7 nm for the Clevios-based sample and 4.4 nm using the Orgacon one (Figure 4e,f, respectively; Table 4). As already observed from the absorption spectra, the morphology study confirms that the formulation of the PEDOT:PSS aqueous suspension does not significantly affect the synthesis of AgNPs.

Once again in agreement with the spectroscopic characterization, the images of Figure 4d–f indicate that AgNPs were also formed in the domains of the conductive polymer PEDOT, notwithstanding the lowest $NaBH_4/AgNO_3$ molar ratio adopted in the cases of PP@AgNPs-4 and PP@AgNPs-5, most likely because of the high concentration of silver precursor in the reaction medium.

3.2.3. Effect of the $AgNO_3$ and PEDOT:PSS Concentration in the Reaction Medium

In the third sample series, PP@AgNPs-6 and 7, the influence of the medium concentration was tested. The ratio of Ag/PEDOT:PSS was almost doubled with respect to the previous ratio (0.1 mmol/mL PEDOT vs. 0.06 mmol/mL). For the PP@AgNPs-7 sample with a $NaBH_4/AgNO_3$ molar ratio of 11, the AgNP layer is highly rough compared to the other samples with an RMS of 8.1 nm (Table 4) and is characterized by big clusters with a diameter reaching 1 μm (Figure 4g). This is probably due to the coalescence of small AgNPs and PEDOT:PSS aggregates. On the other hand, in the case of the PP@AgNPs-6 sample prepared with a $NaBH_4/AgNO_3$ molar ratio equal to 0.1, the lack of the reducing agent (BH_4^-) with respect to the silver salt gives rise to a flat network with interconnected nanoparticles, responsible for the lowest roughness (Figure 4h and RMS = 1.3 nm Table 4). Hence, an excess of silver salt in the starting reagents induces a coalescence of reduced AgNPs instead of a homogenous distribution of smallest AgNPs.

From these remarks, it is safe to infer that the combination of a low $NaBH_4/AgNO_3$ ratio and a lower concentration of silver ions and PEDOT:PSS in the reaction medium leads to the production of the most homogeneous AgNP layer, characterized by the highest AgNP interconnection and flatness (as for PP@AgNPs-6), while an excess of the reducing agent $NaBH_4$ and/or a high concentration of silver ions produces big aggregates and a rougher layer (as for PP@AgNPs-1).

3.3. Solar Cells Prototypes: Preliminary Study

As a proof of concept to demonstrate the potentialities of PP@AgNPs, a preliminary polymer solar cell was fabricated using the PEDOT:PSS/AgNP layer as the anode, and its performance was compared with devices with the same active layer and the classic ITO anode. As discussed previously, the first PEDOT:PSS layer is required for a continuous and smooth deposit of the next nanoparticle-based layer. Among all the nanoparticles synthesized, the PP@AgNP-6 sample was chosen because of (i) its high morphological quality in that the RMS roughness is minimized compared to all the other cases; (ii) the UV-Vis-NIR features in which the polaron band of PEDOT:PSS is still observable, revealing the absence of doping losses or conformational changes that could affect the electrical conductivity; (iii) in the PP@AgNP-6 sample, the AgNPs are expected to be formed outside the conductive domains of PEDOT and then to act as electrical bridges between one domain and another, as mentioned in previous works [26,27]; (iv) from preliminary electrical sheet resistance measurements with the four-probe method, PP@AgNPs-6 results to have a sheet resistance (ohm/cm^2) sensibly lower than the other samples (i.e., <10^3 ohm/cm^2 for PP@AgNPs-6 against values above 10^4 ohm/cm^2 for all the other samples), probably due to the higher morphological homogeneity and flatness of PP@AgNPs-6 compared to the other samples.

The anode glass/Orgacon/PP@AgNPs-6 presented a σ value of 1.61×10^{-2} S/cm, an order of magnitude greater than the neat Orgacon layer, suggesting that the incorporation of the AgNPs allows for better electrical continuity among the conductive PEDOT chains.

Before preparing the device, the transmittance of the anode was evaluated. Figure 5 shows the UV-Vis-NIR spectrum of the Orgacon/PP@AgNP-6 film and, for the sake of comparison, those of the Orgacon layer, the glass substrate and of ITO slides with two different resistances (8–12 Ωsq and 30–60 Ωsq).

The glass substrate presents a quite constant and strong transmittance (87–90%) from 1100 nm up to 370 nm. At shorter wavelengths, the transmittance starts to decrease and finally falls to zero at 260 nm. The commercial ITO slides present some absorptions in the visible region. In particular, the slides with 30–60 Ωsq resistance exhibit interference fringes, likely suggesting a quite thin thickness. In general, ITO transmittance is a bit lower (≈75–80%) than that of glass, in agreement with the presence of a tin oxide layer. The transmittance of the PP@AgNP-6 anode has a value of 74–80% between 300 and 500 nm and then decreases to ≈60% because of the PEDOT polaron absorption, showing behavior similar to that of the corresponding Orgacon layer. At 446 nm, a weak valley that corresponds to the plasmon silver absorption can be observed, as better visualized in the absorption spectrum of the figure inset. The red shift from the value of 400 nm found for the corresponding aqueous suspension of Figure 2c is consistent with the cluster formation in the solid state, as previously observed for other AgNP–PEDOT composites [20]. Despite the slight decrease, the T% value in the visible range is reasonably satisfactory for an anode.

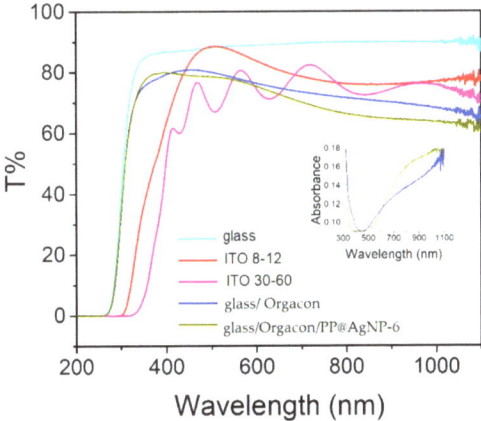

Figure 5. UV-Vis-NIR spectra in transmittance mode of glass/Orgacon/PP@AgNP-6 and, for comparison, the glass/Orgacon layer, the glass substrate and two commercial ITO slides with different resistances. Inset: corresponding absorbance spectra of the glass/Orgacon/PP@AgNP-6 and Orgacon layers on glass.

In the anode characterization, work function (WF) measurements have an important role, since WF position determines the energy needed for the hole extraction process from the active layer to the anode itself. Usually, when ITO is used as an anode, a PEDOT layer is deposited in order to increase the WF from 4.80–5.00 to 5.10 eV to better match the HOMO level of poly(3-hexylthiophene) (P3HT) in the active layer (5.1 eV) and thus favor the hole extraction from donor to anode. In our case, ITO is substituted by the layer of Orgacon/PP@AgNP-6. Since Orgacon has a WF 4.87 eV, an increase in the WF is still needed, and AgNPs partially satisfy this requirement, raising the level to 4.95 eV.

Considering all these properties, we fabricated the device by using the 1:1 blend of P3HT with PC$_{60}$BM as the active layer, which is a performing OPV system commonly used in our laboratory as a benchmark.

Figure 6 shows the J-V curves for the solar cell, the device configuration of which is represented in the inset. Reference cells were also prepared by using ITO with different resistances (Figure S2 in the supporting information). The overall photovoltaic parameters are displayed in Table 5.

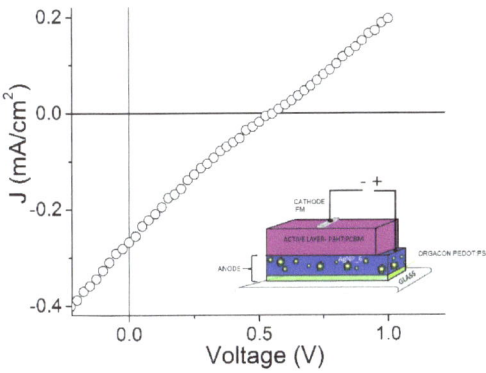

Figure 6. J-V curve for the solar cell under study, the configuration of which is sketched in the inset.

Table 5. Photovoltaic properties of organic solar cells studied in this work.

Electrode	V_{OC} (Volts)	J_{SC} (mA/cm^2)	FF	PCE (%)
ITO 8–12 Ωsq	0.60	5.13	0.55	1.70
ITO 30–60 Ωsq	0.53	3.16	0.30	0.51
Orgacon/PP@AgNP-6	0.55	0.27	0.21	0.03

It can be observed that the open circuit V_{OC} of all the devices is almost constant, corresponding to 0.5–0.6 V. Since this value is related to the difference between the HOMO level of the electron donor (P3HT) and the LUMO level of the acceptor (PC61BM), it is expected to have the same value in all the devices. Conversely, the short circuit current J_{SC} and the fill factor FF differ significantly, exhibiting a much lower value for the cell with the nanostructured anode. As FF is usually associated with the morphological quality of the active layers, the observed decrease can be explained by the fact that the active layer will be deposited on the subjacent substrate following its roughness.

On the other hand, the lower J_{SC} can be attributed to the larger resistivity of the anode. In fact, the ITO substrates, with the nominal resistance of 8–12 and 30–60 Ωsq according to the provider, give a ρ value of 0.20 Ωcm and 1.09 Ωcm when measured via magnetometry, whereas the nanostructured anode under study shows a ρ value of 62 Ωcm under the same conditions. Moreover, the fact that the performance of the cells deteriorates as the resistance of the two ITO-based reference solar cells increases seems to indicate that the electrical property of the anode is the most important parameter yet to be optimized for the composite based on AgNPs and PEDOT. It is worth mentioning that, in this preliminary assay, devices were fabricated and analyzed in ambient conditions. Moreover, a room-temperature processable cathode (Field's metal) was used to make the process cheaper, easier and environmentally competitive.

4. Conclusions

Silver nanoparticles embedded in PEDOT:PSS (PP@AgNPs) were synthesized via the chemical reduction of silver nitrate with NaBH$_4$, in the presence of commercial PEDOT:PSS, varying the following reaction conditions: NaBH$_4$/AgNO$_3$ molar ratio, PEDOT-PSS formulation (Orgacon or Clevios) and the concentration of both AgNO$_3$ and PEDOT:PSS in the medium. The best-dispersed nanoparticles corresponded to the sample obtained with the lowest silver content, lowest NaBH$_4$/AgNO$_3$ molar ratio and highest dilution of the PEDOT:PSS in the formulation and concentration, labeled as PP@AgNP_6.

The combined investigation carried out using UV-Vis-NIR spectroscopy and the FE-SEM technique revealed that the best-dispersed nanoparticles corresponded to the sample obtained with the lowest silver content, lowest NaBH$_4$/AgNO$_3$ molar ratio and highest dilution of the PEDOT:PSS in the formulation and concentration, labeled as PP@AgNP_6. In this case, AgNPs are rather formed outside the PEDOT complex domains and stabilized by the PSS in excess, which in turn guarantees higher electrical conductivity since the conjugation in the conducting polymer turns out to not be affected by the presence of the nanoparticles.

Accordingly, the rheological study of the sample hydrosol showed pseudoplastic Bingham behavior with the lowest viscosity value, which was expected to correspond to the best filmability properties. The AFM images confirmed that the best homogeneity in terms of flatness and nanoparticles interconnection was obtained with the PP@AgNP_6 specimen.

Based on all the results, the PP@AgNP_6 sample was selected to prepare ITO-free anodes for preliminary photovoltaic studies with P3HT:PCBM as the active layer. We remark that we used Field metal as the cathode, which is a low-temperature melting material, to give a further ecological benefit to the fabrication process. Under our conditions, and comparative to ITO-based reference cells, the open circuit voltage (V_{OC}) was maintained but the fill factor (FF) and short circuit current (J_{SC}) were lower, because of the higher resistivity of the nanostructured anode and morphological poorer quality of the active layer deposited on it. However, the preliminary value of 0.03% obtained for the efficiency confirms the potentiality of a nanostructured, water-processable anode in the fabrication of new-generation solar cells after the proper optimization of both AgNP morphology and device configuration.

Supplementary Materials: The following supporting information can be downloaded at: https://www.mdpi.com/article/10.3390/polym15183675/s1, Figure S1: UV-Vis-NIR spectra of diluted starting suspensions; (a) Orgacon, (b) Clevios. Figure S2: J-V curve for solar cells on ITO of different resistances.

Author Contributions: Conceptualization, S.Z., M.A., I.M., G.S. and S.D.; methodology, S.Z., M.A., I.M. and G.S.; validation, S.Z., J.C.V., M.A., I.M. and G.S.; formal analysis, S.Z., J.C.V., M.A., I.M. and G.S.; investigation, S.Z., J.C.V., M.A., I.M. and G.S.; resources, M.A., E.A., I.M., G.S. and S.D.; data curation, S.Z., J.C.V., M.A., I.M. and G.S.; writing—original draft preparation, M.A., I.M. and G.S.; writing—review and editing, S.Z., M.A., E.A., I.M., G.S. and S.D.; visualization, M.A., I.M. and G.S.; supervision, M.A., I.M., G.S. and S.D.; project administration, G.S. and S.D.; funding acquisition, E.A., G.S. and S.D. All authors have read and agreed to the published version of the manuscript.

Funding: This research received no external funding.

Institutional Review Board Statement: Not applicable.

Data Availability Statement: Not applicable.

Acknowledgments: The authors wish to acknowledge the Mobility and Great Relevance Bilateral (Italy MAECI-Mexico AMEXCID) Project-Ambient and Energy 2018 Prot. nr. MAE0044292. J.C.V., E.A. and I.M. want to acknowledge the technical assistance of Geraldina Rodriguez, Maria del Rosario Rangel and Gilberto Hurtado.

Conflicts of Interest: The authors declare no conflict of interest. The funders had no role in the design of the study; in the collection, analyses, or interpretation of data; in the writing of the manuscript; or in the decision to publish the results.

References

1. Lu, X.; Zhang, Y.; Zheng, Z. Metal-Based Flexible Transparent Electrodes: Challenges and Recent Advances. *Adv. Electron. Mater.* **2021**, *7*, 2001121. [CrossRef]
2. Hu, L.; Song, J.; Yin, X.; Su, Z.; Li, Z. Research Progress on Polymer Solar Cells Based on PEDOT:PSS Electrodes. *Polymers* **2020**, *12*, 145. [CrossRef] [PubMed]
3. Li, Q.; Zhang, J.; Li, Q.; Li, G.; Tian, X.; Luo, Z.; Qiao, F.; Wu, X.; Zhang, J. Review of Printed Electrodes for Flexible Devices. *Front. Mater.* **2019**, *5*, 77. [CrossRef]
4. Shi, H.; Liu, C.; Jiang, Q.; Xu, J. Effective Approaches to Improve the Electrical Conductivity of PEDOT:PSS: A Review. *Adv. Electron. Mater.* **2015**, *1*, 1500017. [CrossRef]
5. Gueye, M.N.; Carella, A.; Faure-Vincent, J.; Demadrille, R.; Simonato, J.P. Progress in Understanding Structure and Transport Properties of PEDOT-Based Materials: A Critical Review. *Prog. Mater. Sci.* **2020**, *108*, 100616. [CrossRef]
6. EP0339340A2; Neue Polythiophene, Verfahren Zu Ihrer Herstellung und Ihre Verwendung. European Patent Office: Munich, Germany, 1988.
7. Jonas, F.; Morrison, J.T. 3,4-Polyethylenedioxythiophene (PEDT): Conductive Coatings Technical Applications and Properties. *Synth. Met.* **1997**, *85*, 1397–1398. [CrossRef]
8. Lang, U.; Muller, E.; Naujoks, N.; Dual, J. Microscopical Investigations of PEDOT:PSS Thin Films. *Adv. Funct. Mater.* **2009**, *19*, 1215–1220. [CrossRef]
9. Dupont, S.R.; Novoa, F.; Voroshazi, E.; Dauskardt, R.H. Decohesion Kinetics of PEDOT:PSS Conducting Polymer Films. *Adv. Funct. Mater.* **2014**, *24*, 1325–1332. [CrossRef]
10. Stocker, T.; Kohler, A.; Moos, R. Why Does the Electrical Conductivity in PEDOT:PSS Decrease with PSS Content? A Study Combining Thermoelectric Measurements with Impedance Spectroscopy. *J. Polym. Sci. Part B Polym. Phys.* **2012**, *50*, 976–983. [CrossRef]
11. Jeong, W.; Gwon, G.; Ha, J.H.; Kim, D.; Eom, K.J.; Park, J.H.; Kang, S.J.; Kwak, B.; Hong, J.-I.; Lee, S.; et al. Enhancing the Conductivity of PEDOT:PSS Films for Biomedical Applications via Hydrothermal Treatment. *Biosens. Bioelectron.* **2021**, *171*, 112717. [CrossRef]
12. Ouyang, L.; Musumeci, C.; Jafari, M.J.; Ederth, T.; Inganäs, O. Imaging the Phase Separation between PEDOT and Polyelectrolytes during Processing of Highly Conductive PEDOT:PSS Films. *ACS Appl. Mater. Interfaces* **2015**, *7*, 19764–19773. [CrossRef] [PubMed]
13. Kim, Y.; Ballantyne, A.M.; Nelson, J.; Bradley, D.D.C. Effects of Thickness and Thermal Annealing of the PEDOT:PSS Layer on the Performance of Polymer Solar Cells. *Org. Electron.* **2009**, *10*, 205–209. [CrossRef]

14. Fan, X.; Xu, B.; Liu, S.; Cui, C.; Wang, J.; Yan, F. Transfer-Printed PEDOT:PSS Electrodes Using Mild Acids for High Conductivity and Improved Stability with Application to Flexible Organic Solar Cells. *ACS Appl. Mater. Interfaces* **2016**, *8*, 14029–14036. [CrossRef] [PubMed]
15. Kim, N.; Kang, H.; Lee, J.H.; Kee, S.; Lee, S.H.; Lee, K. Highly Conductive All-Plastic Electrodes Fabricated Using a Novel Chemically Controlled Transfer-Printing Method. *Adv. Mater.* **2015**, *27*, 2317–2323. [CrossRef]
16. Kim, N.; Kee, S.; Lee, S.H.; Lee, B.H.; Kahng, Y.H.; Jo, Y.R.; Kim, B.J.; Lee, K. Highly Conductive PEDOT:PSS Nanofibrils Induced by Solution-Processed Crystallization. *Adv. Mater.* **2014**, *26*, 2268–2272. [CrossRef]
17. Xia, Y.; Sun, K.; Ouyang, J. Solution-Processed Metallic Conducting Polymer Films as Transparent Electrode of Optoelectronic Devices. *Adv. Mater.* **2012**, *24*, 2436–2440. [CrossRef]
18. Kim, Y.U.; Park, S.H.; Nhan, N.T.; Hoang, M.H.; Cho, M.J.; Choi, D.H. Optimal Design of PEDOT:PSS Polymer-Based Silver Nanowire Electrodes for Realization of Flexible Polymer Solar Cells. *Macromol. Res.* **2021**, *29*, 75–81. [CrossRef]
19. Kim, H.G.; Kim, M.; Kim, S.S.; Paek, S.H.; Kim, Y.C. Silver Nanowire/PEDOT:PSS Hybrid Electrode for Flexible Organic Light-Emitting Diodes. *J. Sci. Adv. Mater. Devices* **2021**, *6*, 372–378. [CrossRef]
20. Ko, S.-J.; Choi, H.; Lee, W.; Kim, T.; Lee, B.R.; Jung, J.-W.; Jeong, J.-R.; Song, M.H.; Lee, J.C.; Woo, H.Y.; et al. Highly Efficient Plasmonic Organic Optoelectronic Devices Based on a Conducting Polymer Electrode Incorporated with Silver Nanoparticles. *Energy Environ. Sci.* **2013**, *6*, 1949–1955. [CrossRef]
21. Ghazy, O.A.; Ibrahim, M.M.; Abou Elfadl, F.I.; Hosni, H.M.; Shehata, E.M.; Deghiedy, N.M.; Balboul, M.R. PEDOT:PSS Incorporated Silver Nanoparticles Prepared by Gamma Radiation for the Application in Organic Solar Cells. *J. Radiat. Res. Appl. Sci.* **2015**, *8*, 166–172. [CrossRef]
22. Morvillo, P.; De Girolamo Del Mauro, A.; Nenna, G.; Diana, R.; Ricciardi, R.; Minarini, C. ITO-Free Anode with Plasmonic Silver Nanoparticles for High Efficient Polymer Solar Cells. *Energy Procedia* **2014**, *60*, 13–22. [CrossRef]
23. Ottonelli, M.; Zappia, S.; Demartini, A.; Alloisio, M. Chitosan-stabilized Noble Metal Nanoparticles: Study of Their Shape Evolution and Post- Functionalization Properties. *Nanomaterials* **2020**, *10*, 224. [CrossRef] [PubMed]
24. Castellano, M.; Alloisio, M.; Darawish, R.; Dodero, A.; Vicini, S. Electrospun Composite Mats of Alginate with Embedded Silver Nanoparticles: Synthesis and Characterization. *J. Therm. Anal. Calorim.* **2019**, *137*, 767–778. [CrossRef]
25. Alloisio, M.; Zappia, S.; Demartini, A.; Petrillo, G.; Ottonelli, M.; Thea, S.; Dellepiane, G.; Muniz-Miranda, M. Enhanced and Reproducible Photogeneration of Blue Poly(Pentacosadiacetylene) Chemisorbed onto Silver Nanoparticles: An Optimized Synthetic Protocol. *Mater. Chem. Phys.* **2014**, *147*, 293–303. [CrossRef]
26. Moreno, K.J.; Moggio, I.; Arias, E.; Llarena, I.; Moya, S.E.; Ziolo, R.F.; Barrientos, H. Silver Nanoparticles Functionalized in Situ with the Conjugated Polymer (PEDOT:PSS). *J. Nanosci. Nanotechnol.* **2009**, *9*, 3987–3992. [CrossRef]
27. Valdivia, J.C.; Pérez, A.; Rodríguez, G.; Hurtado, G.; Moggio, I.; Arias, E.; Zappia, S.; Destri, S.; Scavia, G.; Alloisio, M.; et al. Estudio Espectroscópico de Nanopartículas de Plata Pasivadas Con El Polímero Conjugado PEDOT:PSS Spectroscopic Study of Silver Nanoparticles Passivated with the Conjugated Polymer PEDOT:PSS. *Superf. Vacío* **2021**, *34*, 211101. [CrossRef]
28. Glasser, A.; Cloutet, É.; Hadziioannou, G.; Kellay, H. Tuning the Rheology of Conducting Polymer Inks for Various Deposition Processes. *Chem. Mater.* **2019**, *31*, 6936–6944. [CrossRef]

Disclaimer/Publisher's Note: The statements, opinions and data contained in all publications are solely those of the individual author(s) and contributor(s) and not of MDPI and/or the editor(s). MDPI and/or the editor(s) disclaim responsibility for any injury to people or property resulting from any ideas, methods, instructions or products referred to in the content.

Article

Fabrication of High-Performance Natural Rubber Composites with Enhanced Filler–Rubber Interactions by Stearic Acid-Modified Diatomaceous Earth and Carbon Nanotubes for Mechanical and Energy Harvesting Applications

Md Najib Alam, Vineet Kumar, Han-Saem Jung and Sang-Shin Park *

School of Mechanical Engineering, Yeungnam University, 280, Daehak-ro, Gyeongsan 38541, Republic of Korea; mdnajib.alam3@gmail.com (M.N.A.); vineetfri@gmail.com (V.K.); depictme@naver.com (H.-S.J.)
* Correspondence: pss@ynu.ac.kr

Citation: Alam, M.N.; Kumar, V.; Jung, H.-S.; Park, S.-S. Fabrication of High-Performance Natural Rubber Composites with Enhanced Filler–Rubber Interactions by Stearic Acid-Modified Diatomaceous Earth and Carbon Nanotubes for Mechanical and Energy Harvesting Applications. *Polymers* **2023**, *15*, 3612. https://doi.org/10.3390/polym15173612

Academic Editor: Roman A. Surmenev

Received: 7 August 2023
Revised: 29 August 2023
Accepted: 30 August 2023
Published: 31 August 2023

Copyright: © 2023 by the authors. Licensee MDPI, Basel, Switzerland. This article is an open access article distributed under the terms and conditions of the Creative Commons Attribution (CC BY) license (https:// creativecommons.org/licenses/by/ 4.0/).

Abstract: Mechanical robustness and high energy efficiency of composite materials are immensely important in modern stretchable, self-powered electronic devices. However, the availability of these materials and their toxicities are challenging factors. This paper presents the mechanical and energy-harvesting performances of low-cost natural rubber composites made of stearic acid-modified diatomaceous earth (mDE) and carbon nanotubes (CNTs). The obtained mechanical properties were significantly better than those of unfilled rubber. Compared to pristine diatomaceous earth, mDE has higher reinforcing efficiencies in terms of mechanical properties because of the effective chemical surface modification by stearic acid and enhanced filler–rubber interactions. The addition of a small amount of CNT as a component in the hybrid filler systems not only improves the mechanical properties but also improves the electrical properties of the rubber composites and has electromechanical sensitivity. For example, the fracture toughness of unfilled rubber (9.74 MJ/m^3) can be enhanced by approximately 484% in a composite (56.86 MJ/m^3) with 40 phr (per hundred grams of rubber) hybrid filler, whereas the composite showed electrical conductivity. At a similar mechanical load, the energy-harvesting efficiency of the composite containing 57 phr mDE and 3 phr CNT hybrid filler was nearly double that of the only 3 phr CNT-containing composite. The higher energy-harvesting efficiency of the mDE-filled conductive composites may be due to their increased dielectric behaviour. Because of their bio-based materials, rubber composites made by mDE can be considered eco-friendly composites for mechanical and energy harvesting applications and suitable electronic health monitoring devices.

Keywords: rubber nanocomposites; diatomaceous earth; carbon nanotubes; mechanical properties; electrical properties; energy harvesting

1. Introduction

Rubber composites have attracted considerable attention in recent technological applications [1]. Rubber composites have advantages over other polymer composites, such as high stretchability, low glass transition temperatures, excellent resiliency, and good abrasion resistance. These properties make rubber composites suitable for applications in tyres, tubes, conveyor belts, and shoes. Owing to their stretchability, rubber composites are used in many advanced applications, such as electromagnetic interference shielding [2–4], strain sensors [5–7], nanogenerators [8,9], and other stretchable devices [10–12].

Fillers play a vital role in determining the functionality of rubber composites. Depending on their reinforcement capability, fillers can be classified as reinforcing, semi-reinforcing, or non-reinforcing. Non-reinforcing fillers are used to reduce the cost of rubber composites. However, proper modification of these fillers has changed them to semi-reinforcing or even reinforcing fillers, which can improve certain mechanical properties of rubber composites [13]. Reinforcing fillers are generally on the nanometer scale and have a large interacting surface area that can strongly bind rubber chains to their surface. Silica and carbon black are the two most abundant fillers used in the tyre industry for mechanical reinforcement. Carbon nanomaterials interact very well with rubber, followed by van der Waals and π–π stacking interactions [14]. Among different carbon nanomaterials, carbon nanotubes (CNTs) are highly useful for improving the mechanical, thermal, and electrical properties of rubber composites [15]. Because of their one-dimensional rod-like morphology, high electrical conductivity (>10^2–10^6 S/cm), and high thermal conductivity (>2000 $Wm^{-1} K^{-1}$), a small amount of CNT can significantly enhance the electrical and thermal properties of rubber compounds.

Rubbers are dielectric materials with negligible or very low electrical conductivity. Conducting materials must be incorporated to make rubber a conducting material. A wide range of conducting rubber composites can be fabricated depending on the structure and conductivity of the filler. In addition to their mechanical applications, conducting rubber composites have gained wide attention in modern electronic devices [16,17]. Owing to their mechanical robustness, conducting rubber composites can replace metal-based conductors, in which stretchability is a vital factor. Piezoresistive and piezoelectric behaviours of rubber composites have been observed in various applications such as sensors, actuators, and electromechanical transducers [18,19]. Electrical power generation using dielectric elastomers is highly renewable and has gained popularity in recent years [20–22]. Dielectric materials play a significant role in capacitance-based energy harvesting devices. Rubber has a very low dielectric loss and is suitable for energy harvesting at low frequencies. To further improve the dielectric constant, the addition of ceramic materials with high dielectric constants is useful [23,24]. However, these materials should be used in higher proportions to achieve the percolation of polymer composites [24]. Salaeh et al. [25] used barium titanate and lead titanate to improve the dielectric behaviour of natural rubber. Although these ceramic materials provide improved dielectric rubber composites at higher proportions, they significantly reduce the mechanical properties of the composite and enhance the hysteresis loss. Moreover, ceramic materials are expensive, and heavy-metal-based ceramics may be toxic. Hence, energy-harvesting composites with cheap and nontoxic filler materials may be of great interest for energy-harvesting devices, especially health-monitoring self-sensing devices.

Recently, biofillers have become popular in rubber compounds [26]. They are attractive because of their low cost, natural availability, non-toxicity, and hierarchical structure. Among the different natural fillers, natural fibres, crystalline cellulose, bone dust, biochar, and bio-calcium carbonate have been successfully applied in rubber compounding, either in pristine or modified forms [27]. From the viewpoint of toxicity and the limitations of petroleum resources, scientists have tried to establish silica as an alternative filler to carbon black in the tyre industry, with some special advantages [28–30]. Silica fillers mainly reduce the rolling resistance of tyres, which reduces the fuel cost of vehicles [30]. Although silica filler partially or fully replaces carbon black, depending on the rubber products with reduced toxicity, synthetic silica filler remains costly in rubber compounding, as it requires some critical steps [30].

In addition to synthetic silica, biogenic silica can serve as an alternative to silica resources. Beidaghy et al. [31] discussed different procedures for producing silica from rice husk and straw. Choophun et al. [32] successfully utilised rice husk-derived silica for rubber compounding, which may have applications in tyre tread formulations. Recently, Barrios and colleagues [33] have discussed various sustainable fillers, such as chitin, chitosan, lignin, and cellulose, for elastomeric compounds. Sustainable fillers might possess a high content of polar groups that are not compatible with non-polar rubber. Additionally, they contain a high level of absorbed moisture that inhibits the interfacial interactions between the filler and rubber [34]. Rubber–filler interactions can be enhanced with these filler systems, followed by drying the filler at an elevated temperature and surface modification. Occasionally, the methods are not straightforward or inexpensive, which has hindered the industrial applications of these fillers in rubber compounding. Diatomaceous earth is another major source of biogenic silica [35]. The diatomaceous earth comprises the focal remains of diatoms (hard-shelled microalgae). It mainly consists of 80–90% silica, 2–4% alumina, and 0.5–2% iron oxide. In a review paper, Reka et al. [36] discussed diatomaceous earth in detail, along with its different applications. They discovered that diatomaceous earth had a low level of moisture content that could be easily removed by heating and exhibited good thermal stability. Diatomaceous earth naturally contains hierarchical structures because of the hard-shelled structure of diatoms. Filler morphology plays an important role in rubber reinforcement [37]. Higher filler structures may improve the mechanical strength of rubber compounds, owing to the contribution of filler–filler mechanical interactions [38]. Diatomaceous earth can also be used to improve the dielectric properties of rubber [39]. It also contains certain metal oxides that can participate as cure activators in sulphur vulcanisation [40]. However, despite possessing several advantageous properties for use as a filler in rubber compounding, diatomaceous earth has drawbacks associated with polar functional groups on its surface. Because diatomaceous earth is a naturally abundant, cheap material with a highly porous structure, it is used as a filler in different rubber composites [41–45].

Among different types of rubber, natural rubber is the most widely used in industrial products. Some benefits, such as abundant natural availability, non-toxicity, low cost, excellent resilience, high stretchability, and good scratch resistance, can only be observed in natural rubber [46]. Liliane Bokobza [47], Sethulekshmi et al. [48], and Sethulekshmi et al. [48] have reviewed natural rubber-based composites with numerous filler systems. Natural rubber shows very good mixing behaviour with all types of fillers and enhances the desired properties [47,48]. Inorganic nanofillers or surface-modified fillers have a greater effect on the reinforcement of rubber properties. Similar to other clay minerals, diatomaceous earth can also be used as a reinforcing filler in rubber composites. However, owing to the larger particle size and polarity, it remains challenging to improve the tensile properties other than the stiffness of rubber composites. Although there are many possibilities for enhancing the properties of rubber composites with pure or modified diatomaceous earth, only a few reports have been published [41–45,49]. To the best of our knowledge, this is the first study on diatomaceous earth and carbon nanotubes as hybrid reinforcing fillers in natural rubber composites for suitable applications in mechanical and self-powered electromechanical sensing.

In this study, we aim to fabricate novel natural rubber composites comprising stearic acid-modified diatomaceous earth and carbon nanotubes. Motivated by the excellent oil absorption properties of diatomaceous earth, a simple and efficient method was developed to alter its polarity using stearic acid. Thus, the conversion of the polar surface to the non-polar surface of diatomaceous earth could significantly improve the mechanical properties of natural rubber by improving the interfacial interactions. Because the mechanical stability of rubber composites is the most important factor for all types of stretchable applications, we mainly discuss the mechanical properties of rubber composites, along with some basic electrical properties. After the successful fabrication of rubber composites using modified diatomaceous earth and CNT-based hybrid filler systems, the energy-harvesting performances of the composites were evaluated and discussed.

2. Materials and Methods

2.1. Materials

An ultra-soft masterbatch rubber compound was prepared by mixing the natural rubber (STR 5L), zinc oxide, stearic acid, accelerator tetramethyl thiuram disulfide, accelerator N-tert-butyl-2-benzothiazolimesulfonamide, and sulphur in 100, 5, 2, 1, 1.75, and 1.5 amounts in gram, respectively, in a two-roll mill for about 30 min. The rubber and curing ingredients were sourced from local rubber goods manufacturing companies, Gyeongsang, Republic of Korea. Multi-walled carbon nanotubes (specific surface area: ~250 m^2/g) were obtained from Hanwha Nanotech Corporation, Seoul, Republic of Korea. Good quality filtration-grade diatomaceous earth (DE) powder was purchased from Sigma–Aldrich (St. Louis, MO, USA). Stearic acid was purchased from Sigma–Aldrich (USA). Toluene was purchased from Daejung Chemicals & Metals Ltd., Siheung, Republic of Korea.

2.2. Preparation of Masterbatch Rubber

The raw rubber was initially masticated on a two-roll mill for approximately 20 min. Subsequently, zinc oxide and stearic acid were incorporated and milled for about 5 min. Lastly, accelerators and sulphur were added simultaneously and milled for another 5 min before being removed from the mill. The parameters and fabrication method for the masterbatch rubber can be located elsewhere [40].

2.3. Stearic Acid Treatment of Diatomaceous Earth

First, 2 g of stearic acid was dissolved in 100 mL of toluene via ultrasonication for a few minutes. After that, 20 g of pristine DE was added to the solution and kept for 30 min in an ultra-sonication bath at 25 °C. After sonication, the colloidal-like mixture was dried at 80 °C for complete drying. The dried compounds (mDE) were stored in a desiccator. The ratio of diatomaceous earth to stearic acid for filler modification was fixed at 10:1 (w/w). Typically, coupling agents below 10 wt% were employed in relation to the silica filler based on the quantity of functional groups in terms of molar content [28].

2.4. Characterizations of Fillers

The morphologies of the filler and tensile fractured rubber sections were characterised by field-emission scanning electron microscopy (FE-SEM, S-4800, Hitachi, Tokyo, Japan). Before the SEM analysis, the samples were sputter-coated with platinum. The diatomaceous earth was characterised before and after treatment with stearic acid using Fourier transform infrared spectroscopy (FT-IR) to investigate the chemical changes in the functional groups at 4 cm^{-1} resolution and two scan numbers.

2.5. Fabrication of Rubber Composites

The solvent-bending method was applied to fabricate rubber composites by maintaining filler structures similar to their pristine forms. First, 25 g of masterbatch rubber was soaked in 150 mL of toluene in a glass jar for one day. A smooth rubber slurry was obtained via mechanical stirring. In another vessel, the required amount of filler(s) was added to 100 mL of toluene and sonicated for approximately 30 min. The two slurries were mixed and vigorously stirred for 10 min. The final slurry was then transferred to a flat tray and dried in an oven at 80 °C. It should be noted that appropriate handling and reutilisation of the solvent (toluene) can have a negligible environmental impact. Furthermore, achieving a nano-level filler distribution is only feasible through solvent blending methods. Nevertheless, an excessive use of solvent can lead to the sedimentation of the filler particles. The conventional dry mixing on the two-roll mill was avoided because it could disrupt the anisotropy of the filler particles and potentially result in the loss of electrical conductivity in the rubber composite. The advantages of employing the solvent-blending technique can be found elsewhere [50,51]. The dried compounds were vulcanised in a hot press at 150 °C for 15 min as sheets, cylindrical samples, and electrodes, as previously described [50,51]. Details of the mixing compositions are listed in Table 1.

2.6. Mechanical Properties

Compressive mechanical properties were evaluated using cylindrical samples (d = 20 mm × h = 10 mm), and tensile mechanical properties were evaluated using dumbbell-shaped (ISO-37, Type-2) [52] test pieces in a universal testing machine (Lloyd, Westminster, UK) using a 1 kN load cell. The deformation rates for the compressive and tensile tests were 2 mm/min and 300 mm/min, respectively. The load-carrying capacities of the rubber composites after multiple cycles were determined using dimensions identical to those of the cylindrical samples. Different mechanical properties, such as Young's modulus,

modulus at 10% elongation, tensile strength, elongation at break, and fracture toughness, were obtained from the stress–strain data, and their average values are presented. Four specimens were scrutinised for each composition, and the most representative stress–strain data were presented.

Table 1. Mixing composition for rubber composites per hundred grams of rubber (phr).

Formulations	Masterbatch	Amounts of DE/mDE (phr)	Amounts of CNT (phr)
NR-unfilled	100	-	-
NR/20-DE	100	20	-
NR/20-mDE	100	20	-
NR/40-mDE	100	40	-
NR/60-mDE	100	60	-
NR/3-CNT	100	-	3
NR/17-mDE/3-CNT	100	17	3
NR/37-mDE/3-CNT	100	37	3
NR/57-mDE/3-CNT	100	57	3

2.7. Swelling Properties

To determine the solvent swelling index and cross-link density in the rubber composites, cylindrical samples of the above dimensions were kept in toluene for 7 days to reach equilibrium. After 7 days, the surface toluene was immediately removed using blotting paper, and the swollen weight was measured. The swelling index was calculated as follows from Equation (1):

$$\text{Swelling Index} = \frac{(\text{Swelled weight} - \text{Initial weight})}{\text{Initial weight}} \quad (1)$$

The chemical cross-link densities of the vulcanised compounds were calculated from the equilibrium swelling data according to the Flory–Rehner equation [53] as presented in Equation (2):

$$V_c = -\frac{\left\{\ln(1-V_r) + V_r + \chi V_r^2\right\}}{\left\{V_s d_r \left(V_r^{\frac{1}{3}} - \frac{V_r}{2}\right)\right\}} \quad (2)$$

where V_c is the cross-link density of the rubber vulcanisate, V_r is the volume fraction of rubber in the swollen compound, $\chi = 0.3795$ is the interaction parameter of the natural rubber and toluene system, $V_s = 106.2$ cm^3/mole is the molar volume of toluene (solvent), and d_r is the density of the rubber vulcanisate.

The volume fractions of rubber compounds were calculated by this formula in Equation (3),

$$V_r = \frac{\left(\frac{w_r}{d_r}\right)}{\left\{\left(\frac{w_r}{d_r}\right) + \left(\frac{w_s}{d_s}\right)\right\}} \quad (3)$$

where w_r is the weight of the rubber taken, d_r is the density of the rubber vulcanisate, which was obtained by the formula density = mass/volume considering the cylindrical sample (d = 20 mm × h = 10 mm), w_s is the weight of the swollen solvent, and $d_s = 0.87$ g/cm^3 is the density of the swollen toluene (solvent).

2.8. Electrical and Electromechanical Sensing Properties

The electrical resistivity was calculated by this formula in Equation (4),

$$\rho = \frac{RA}{L} \quad (4)$$

where ρ is the electrical resistivity, R is the electrical resistance of the rubber composite, A is the area of the electrode, and L is the distance between the electrodes. The resistance (R) was measured using a source meter with copper electrodes placed on opposite sides of the cylindrical sample (d = 20 mm × h = 10 mm).

The electrical conductivity of the rubber composites was found by this formula in Equation (5),

$$\text{Conductivity} = \frac{1}{\rho} \quad (5)$$

Electromechanical energy-harvesting devices were prepared by placing the composites as 1 mm thick electrodes on opposite sides of a 5 mm thick unfilled rubber sheet. The electromechanical activity of the rubber composites in energy-harvesting systems was measured using the output voltages. The electrodes were connected to a source metre (Agilent, Model: 34401A, Santa Clara, CA, USA), and a dynamic load of 50 kPa was applied to the top of the DC electrode using a loading tip. During cyclic loading–unloading, the energy-harvesting device exhibited changes in the output voltages. At similar mechanical loads, the voltage output can be considered the electromechanical energy-harvesting efficiency of these composites.

3. Results and Discussion

3.1. Characteristics of the Filler Materials

The morphology of the CNTs and pristine DE can be seen in Figure 1a,b. From Figure 1a, it can be seen that the diameters of the CNTs are less than 50 nm, and the lengths vary from a few hundred nanometers to micrometres. The SEM images suggest that the CNTs have a higher aspect ratio. Because of the high aspect ratio of the CNT, a small amount was sufficient to achieve electrical percolation in the rubber composite. Figure 1b suggests that the DE particles are micrometres in size and have hierarchical porous structures. Owing to the porous structures and cavities inside the DE, rubber molecules can easily enter the particles, and effective reinforcement is possible because of physical interactions. However, owing to the higher polarity of DE, chemical interactions are expected to be poor between unmodified DE and non-polar rubber molecules. Figure 1c shows the morphology of the stearic-modified DE. It is difficult to visualise the surface modification of DE with a morphology similar to that of untreated DE. Hence, stearic acid may have better chemical interactions than film formation on the DE surface. The detailed chemical interactions between stearic acid and DE were ascertained from the FT-IR spectra shown in Figure 1d.

The black line in Figure 1d represents the untreated DE. The strong band at 3675 cm^{-1} may arise from the in-phase symmetric vibration of OH groups on either the outer or inner surface of the octahedral sheets that are weakly hydrogen-bonded with the next tetrahedral sheet, as found in kaolinite [54,55]. Because DE contains kaolinite structures, it is reasonable to determine this band. The peak at 2989 cm^{-1} may be due to unknown organic impurities in the DE. The 1063 cm^{-1} band could be attributed to a Si-O-Si stretching vibration [54]. The band at ~790 cm^{-1} was assigned to the OH translational vibration [54,56].

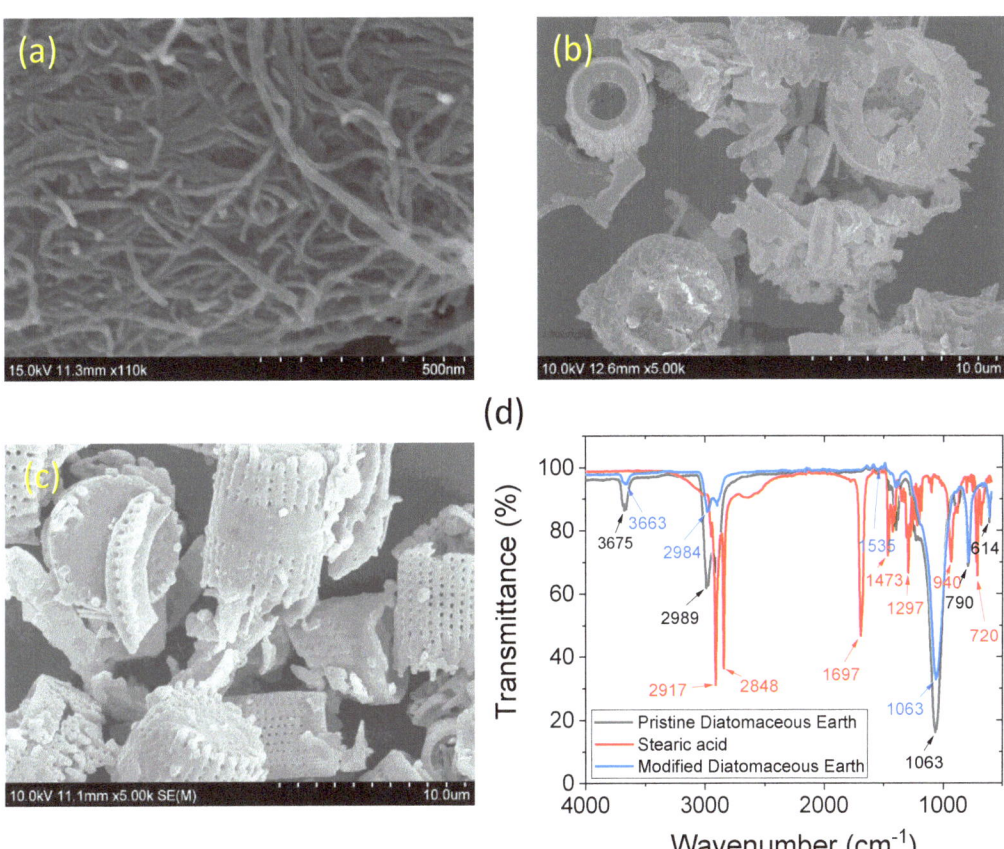

Figure 1. SEM images of (**a**) multi-walled carbon nanotube, (**b**) pristine diatomaceous earth, and (**c**) stearic acid-modified diatomaceous earth; (**d**) FT-IR spectroscopies of pristine DE (black line), stearic acid (red line), and modified DE (mDE) (blue line).

The red line in Figure 1d represents the FTIR spectrum of stearic acid. The peaks at approximately 2917 and 2848 cm^{-1} are representative of the stretching vibrations of the CH$_2$ groups [57]. The peak at ~1697 cm^{-1} represents the stretching of the stearic acid carbonyl group. From the observed spectra of the modified DE (blue line), it can be seen that the peak of the carbonyl group was completely absent, although the modified DE was not washed to remove the unreacted stearic acid. The absence of this peak strongly indicates the conversion of carbonyl to carboxyl anions through chemical bonding with DE, which results in a new peak at 1535 cm^{-1} for the antisymmetric stretching vibration of the C=O bond in the carboxyl group [57,58]. It is believed that, in addition to silica and kaolinite in DE, there are many basic materials [36,54] that react with stearic acid and form strong bonds. Because of these strong interactions, the intensities of the other functional groups in DE were either suppressed or tended to be lower when we compared the black and blue lines of the pristine and modified DE, respectively, as shown in Figure 1d.

The mechanism of action of stearic acid medication is illustrated in Figure 2. It is believed that stearic acid molecules dissolved in toluene can attach to the DE surface via H bonding and form a metastable compound [59]. Upon ultrasonication, the metastable compound finally formed a stable modified compound, followed by a condensation-type chemical reaction [58]. It is believed that, upon vibration, the gallery between the octahedral and tetrahedral sheets may expand and become feasible for exchanging the hydroxyl group from the octahedral site with the carboxyl group from stearic acid. At a 10:1 ratio of DE to stearic acid, there may be a complete condensation reaction between DE and stearic acid because the stearic acid peaks for the carbonyl group are absent in the modified DE. Thus, stearic acid modification can considerably lower the surface polarity of DE and can interact strongly with non-polar natural rubber, as evidenced by the various properties in later sections.

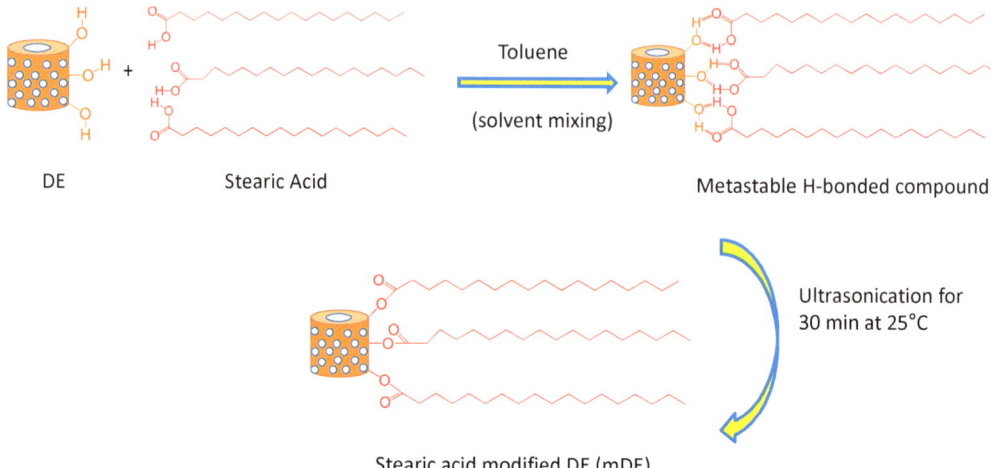

Figure 2. Schematics for the modification of diatomaceous earth by stearic acid.

3.2. Mechanical and Physical Properties of the Rubber Composites

Various compressive mechanical properties are shown in Figure 3a–d. From the compressive stress–strain curves in Figure 3a, it is clear that the mDE-filled compound (NR/20-mDE) has a better compressive modulus than the pristine DE-filled (NR/20-DE) compound. This could be due to additional interactions between the rubber and modified DE (mDE). In addition, the compressive modulus increased with increasing mDE content (Figure 3a). Interestingly, after 40 phr of mDE, the stress–strain slope lost linearity at a higher compressive strain. This could be due to filler percolation above 40 phr DE. The clear slope change of approximately 5% of the compressive strain in the 60 phr mDE-containing composite (NR/60-mDE) may be due to strain-dependent filler percolation. However, with up to 5% compression, the stress–strain slope remained linear. Hence, mDE can be useful as a filler for up to 60 phr when low deformation and high compressive strength are required. The addition of CNTs to the mDE-filled compounds further increased the compressive modulus (Figure 3b). This could be due to the improved filler dispersion in the hybrid filler systems and the higher reinforcing power of the CNT fillers. The highest compressive modulus is obtained for the NR/57-mDE/3-CNT system. The variation in Young's modulus with the amount of filler is plotted in Figure 3c,d. From these figures, it can be observed that the increase in Young's modulus with the filler amount is more exponential for the hybrid filler systems than for the mDE-only filler systems. This suggests that overall interactions combining physical and chemical hybrid filler systems have a greater value than mDE-filled compounds at similar filler amounts. This may be due to the higher aspect ratio and higher interfacial interactions between the rubber and CNT [60].

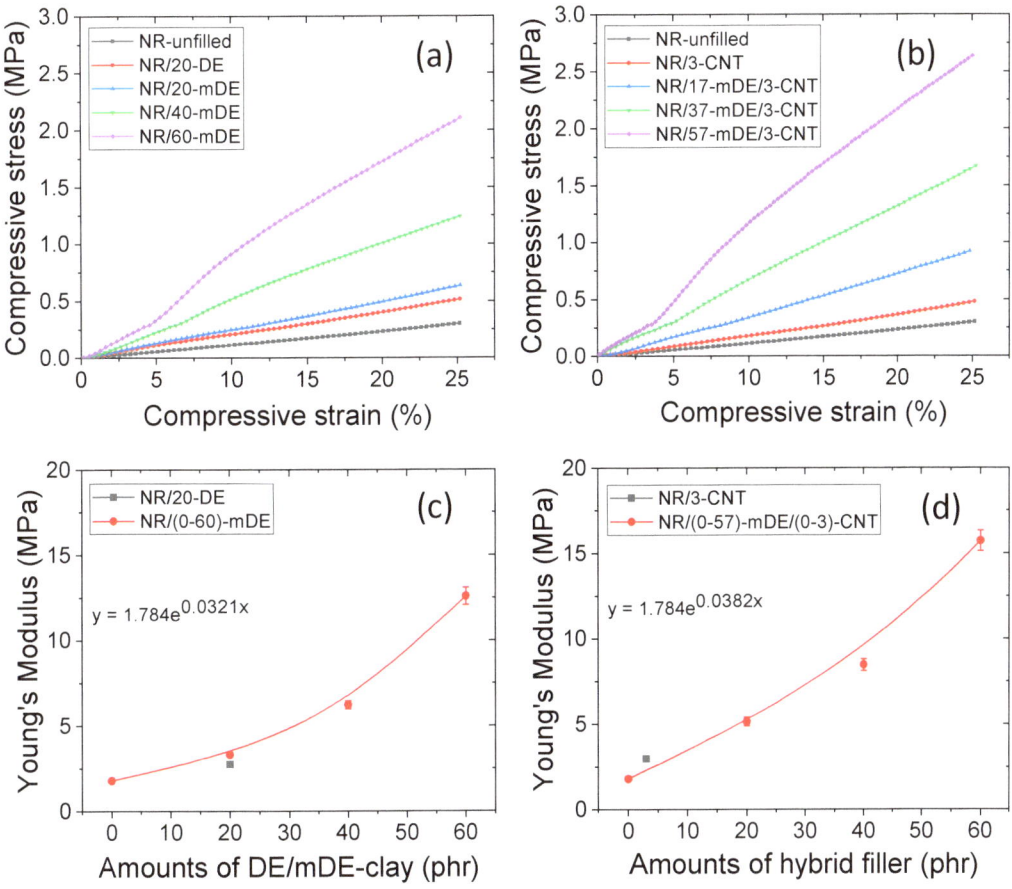

Figure 3. Compressive mechanical properties, (**a**,**b**) compressive stress–strain curves, and (**c**,**d**) Young's modulus as a function of filler amounts.

Cyclic compressive loading–unloading tests were performed to study actual dynamic compressive mechanical applications. The results are shown in Figure 4a–f. From these figures, it can be concluded that the hybrid filler systems have a higher load-carrying capacity than the mDE filler systems with similar filler amounts. Interestingly, for the mDE-only filler systems, the load value stabilised after approximately 5–10 cycles. However, the load stabilised at higher cycles for the hybrid filler systems. It can be seen that the load-carrying capacity is significantly higher up to 40 phr of filler amounts in the hybrid filler systems compared to mDE filler systems only at similar filler amounts. From Figure 4e,f, it can be observed that the load value decreases with increasing cycles. However, stable load values were obtained after 50 cycles at higher filler loadings. The decreasing load-carrying capacity was due to hysteresis loss, which might be due to the permanent breakdown of some filler networks and stress softening [61]. The better load-carrying capacity, even after many cycles, indicates that mDE-based rubber composites can be useful for compressive applications, even at higher filler loadings.

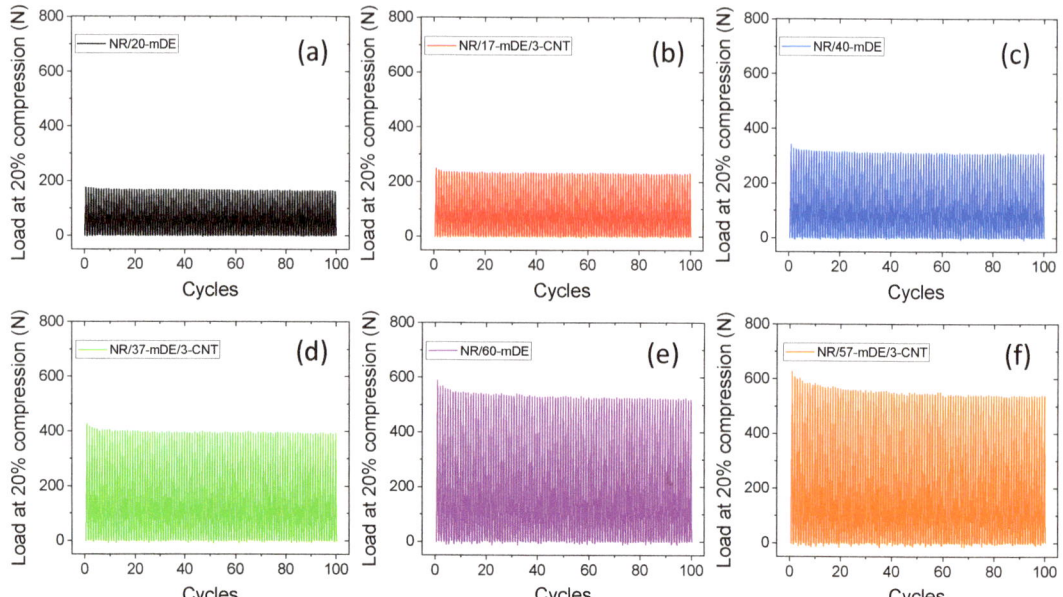

Figure 4. Effect on compressive load at 20% deformation with loading–unloading cycles in different rubber composites: (**a**) NR/20-mDE, (**b**) NR/17-mDE/3-CNT, (**c**) NR/40-mDE, (**d**) NR/37-mDE/3-CNT, (**e**) NR/60-mDE, and (**f**) NR/57-mDE/3-CNT.

Figure 5a–f shows the various tensile mechanical properties. From the curves in Figure 5a,b, it is evident that the stress–strain relationships are hyperelastic. With increasing filler content, the stress–strain curve became more nonlinear at lower strains. At the beginning of the curve, a higher slope value indicates the existence of filler–filler interactions that inhibit extension at lower deformations [38]. With increasing filler amounts, the filler–filler interactions become more prominent. After a critical distance of nearly 10% elongation, the filler–filler interactions were minimised. Above this deformation, the stress–strain slopes better represented the filler–polymer and polymer–polymer interactions [38]. According to the lower stress–strain slope of the unfilled rubber, it is difficult to say that stress-induced crystallisation occurs in these vulcanised systems. The addition of filler greatly improved the stress–strain slope value, which can be attributed to the improved filler–filler and filler–polymer interactions in the rubber composites.

Because the modulus at a small deformation better indicates filler–filler interactions [38], the modulus value at this deformation can provide information regarding the filler percolation threshold. From Figure 5c, the greater change in the modulus from 40 to 60 phr could be due to filler percolation above 40 phr. Hybrid fillers show a higher change in modulus with filler amounts, which could be due to the improved filler distribution and anisotropic structure of the CNT, which produces hybrid filler networks [62,63].

Figure 5d shows the variation in tensile strength with the filler amount. From this figure, it can be understood that the tensile strength was significantly improved at 20–40 phr filler amounts for the unfilled rubber. This may be due to the formation of a sufficient amount of the rubber–filler network at this filler level. The highest tensile strength was obtained in the hybrid filler system (NR/17-mDE/3-CNT) with 20 phr of filler. This further suggests that the hybrid filler showed improved filler distribution and reached filler percolation at lower filler amounts. Up to a 40 phr filler amount, both filler systems showed better tensile strength, but beyond that amount, the tensile strength decreased more rapidly with increasing filler content. It was observed that with a hybrid filler system at 40 phr, the tensile strength was approximately 429% higher compared to the unfilled rubber vulcanisate. A similar observation was noted in the case of a silica–kaolinite mixed mineral filler, similar to diatomaceous earth, where after 40 phr of filler, there were no significant changes in the tensile strength values [13]. At higher filler amounts, the filler–filler interactions were enhanced at the expense of the filler–polymer interactions, owing to the lowering of the rubber fraction. It was found that similar mDE filler amounts provided better tensile strengths than the pristine ones. This suggests that the modified DE improved the filler–polymer interactions. Significant improvements in the tensile strength of the hybrid filler systems were also possible because of the higher surface area of the CNT.

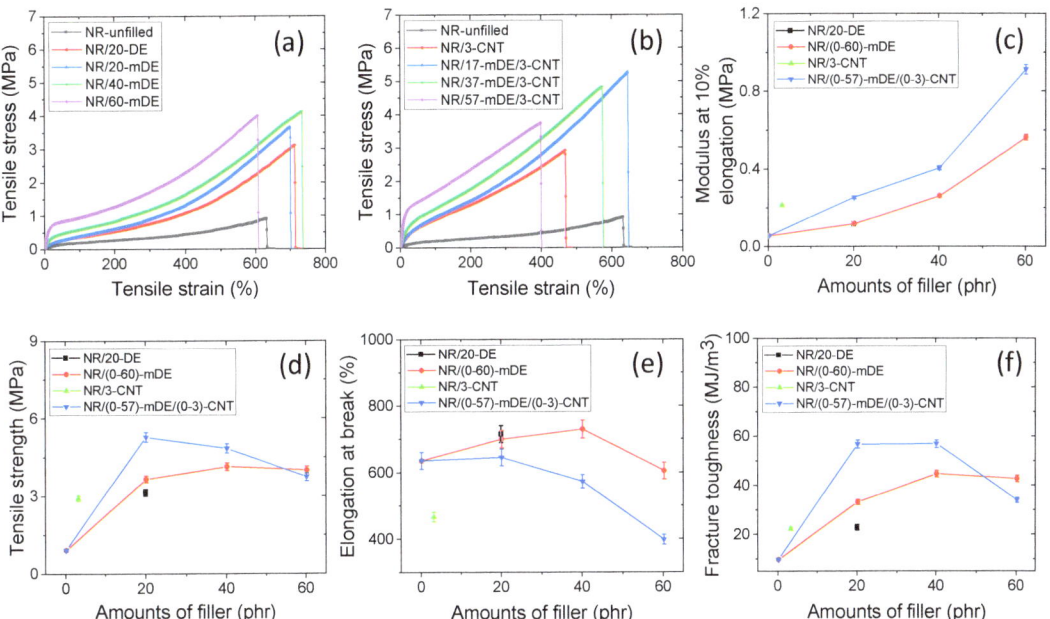

Figure 5. Tensile mechanical properties: (**a**,**b**) stress–strain curves, (**c**) modulus at 10% deformation, (**d**) tensile strength, (**e**) elongation at break, and (**f**) fracture toughness.

Figure 5e shows the variation in the elongation at break with respect to the filler amount. From this figure, it can be seen that the elongation at the break can be enhanced by the addition of up to a critical amount of filler. This strongly supports the formation of filler–polymer networks at the filler percolation level that are more flexible and reversible than chemical bonds and can be stretched more than unfilled rubber [11]. The highest elongation at break was obtained for the 40 phr mDE filler. This could be due to the strong interconnecting filler–polymer networks, with fewer filler–filler networks below the filler percolation level [11]. Hybrid filler systems may have improved filler dispersion, and hybrid filler networks cause more effectively bonded rubber [62,63]. Strong rubber–filler interaction inhibits the extension of rubber chains and has a similar or reduced elongation at break values as unfilled rubber.

Figure 5f shows the variation in fracture toughness with the filler amount. From this figure, it is clear that the toughness of the composites can be enhanced by up to 40 phr of filler content. After 40 phr, the hybrid filler showed reduced fracture toughness, whereas the mDE showed only an insignificant change in toughness. Although the 60 phr hybrid filler system has a lower toughness value compared to 20 or 40 phr, it is still higher than the 3 phr MWCNT and unfilled rubber compounds. Toughness generally depends on the overall bonding strength, considering both the physical and chemical properties [64].

In a later section, it is observed that the cross-link densities of the rubber composites are optimised at an optimum filler concentration. In addition, optimum rubber–filler networks can only be achieved with an optimum amount of both filler and rubber. At higher filler amounts, the rubber fraction was reduced to obtain fewer rubber–filler networks, thereby reducing the fracture toughness. Owing to the better filler distribution and rubber–filler interactions, the fracture toughness of NR/37-mDE/3-CNT was 484% higher than that of the unfilled natural rubber.

The swelling and cross-linking characteristics of the rubber compounds are shown in Figure 6a–c. As shown in Figure 6a, the swelling index gradually decreases with increasing filler content. This suggests that an increase in the amount of filler enhanced the rubber–filler interactions. The hybrid filler showed a higher rate of decrease in the swelling index than the mDE-only filler system. Hybrid filler systems are expected to have higher surface areas than mDE filler systems at similar filler amounts and more interactions with rubber. From Figure 6b, it is evident that the composites containing the hybrid fillers showed higher cross-linking densities than the mDE filler systems with similar filler amounts. The highest cross-linking density was obtained for the 40 phr hybrid filler containing the NR/37-mDE/3-CNT composite. When the filler amount was greater than 40 phr, the cross-link density was slightly reduced, which may have been due to a significant reduction in the rubber fraction. Although the 40 phr hybrid filler shows the highest cross-link density, it does not merely indicate the highest toughness value but also roughly indicates a better modulus value compared to 20 phr. It is believed that because of the decrease in the rubber fraction at higher filler contents, particularly over filler percolation, the extensibility of rubber chains may be reduced, as is evident from the reduced elongation at break values.

Figure 6c shows the filler efficiencies of the hybrid filler systems used to investigate the synergism in the cross-linking density. The cross-linking efficiencies of the hybrid filler systems were determined by dividing the cross-linking densities by the filler amount. It can be seen that the experimentally obtained cross-link densities in the hybrid filler systems are much better than the theoretically calculated cross-link densities from the contributions of the individual fillers. This synergism in the cross-link densities of the hybrid filler systems may be due to the improved mDE filler distribution in the composites, which may have a greater effect on the curing activity [13].

The improved mechanical and physical characteristics of the rubber composites can be further understood from a microscopic view of the filler distribution, as shown in Figure 7a–h. From Figure 7a,b, it is clear that the stearic acid modification of DE significantly improved the adhesion of rubber to the filler surface. From Figure 7b–d, it is clear that the rubber–filler compatibility can be reduced at higher filler amounts when mDE is used as a single filler system. Further improvement in the filler–rubber compatibility can be achieved with the addition of CNT in the hybrid filler systems (Figure 7e–g) compared to the mDE-filled composites (Figure 7b–d) at similar filler amounts. This may be due to the more homogeneous filler distribution in the hybrid filler systems during solvent-aided mixing [51]. Also, from Figure 7h, it can be seen that rubber molecules can enter the DE particles through the pores, and effective rubber–filler binding is possible. However, at higher filler amounts, the separation of the contact areas between the rubber and filler increased with strain (Figure 7h) and hence significantly reduced the elongation at break.

From the study of filler modification and the observed mechanical properties, a simple mechanism can be drawn for the reinforcement of rubber using stearic-acid-modified DE, as shown in Figure 8. Stearic acid primarily forms strong bonds with DE through its polar groups. However, the non-polar tail may have interacted with the non-polar rubber chains through van der Waals-type interactions. Because stearic acid has long-chain hydrocarbons, the interactions become strong, and significant rubber reinforcement is possible.

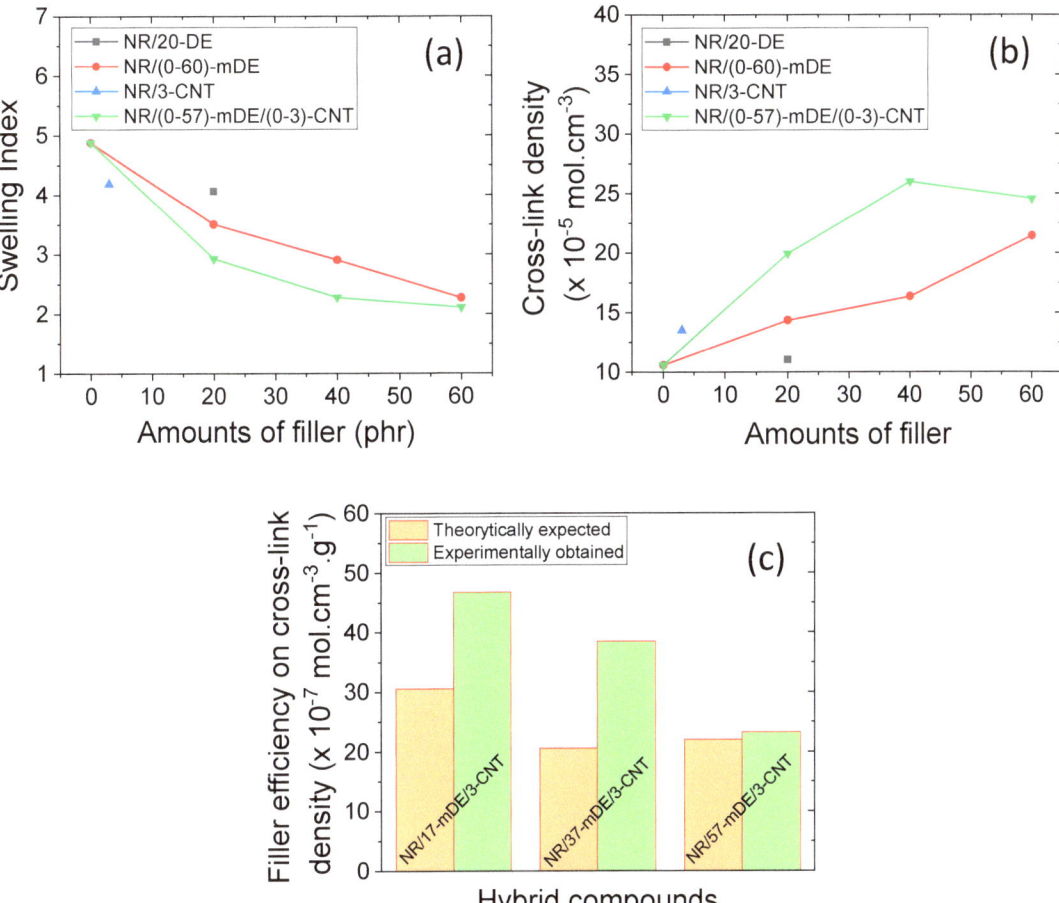

Figure 6. Swelling properties: (**a**) variation in swelling indexes with filler amounts, (**b**) variation in cross-link densities with filler amounts, and (**c**) filler efficiencies on cross-link densities to the hybrid filler systems.

3.3. Electrical and Energy Harvesting Performances of the Rubber Composites

It is well known that only electrically conducting materials can improve the electrical properties of rubber composites. However, the choice of filler is an important factor in obtaining better electrical and mechanical properties. CNTs have a high surface area and aspect ratio, exhibit very good interactions with rubber, and can enhance their mechanical properties. The basic electrical properties, such as resistivity and conductivity, of the rubber composites, are shown in Figure 9a,b. From Figure 9a, it is evident that the resistivity increases almost linearly with an increase in the mDE content with a fixed amount of CNT in the rubber composites up to 40 phr of filler. Subsequently, it increased exponentially with an increase in the filler amount. This could be because of the percolation of more dielectric mDE fillers in the rubber composites [39]. Higher amounts of dielectric materials can enhance the dielectric properties of rubber and, hence, lower its conductivity, as shown in Figure 9b. Nevertheless, all rubber composites showed very good electrical conductivity in the semiconducting region.

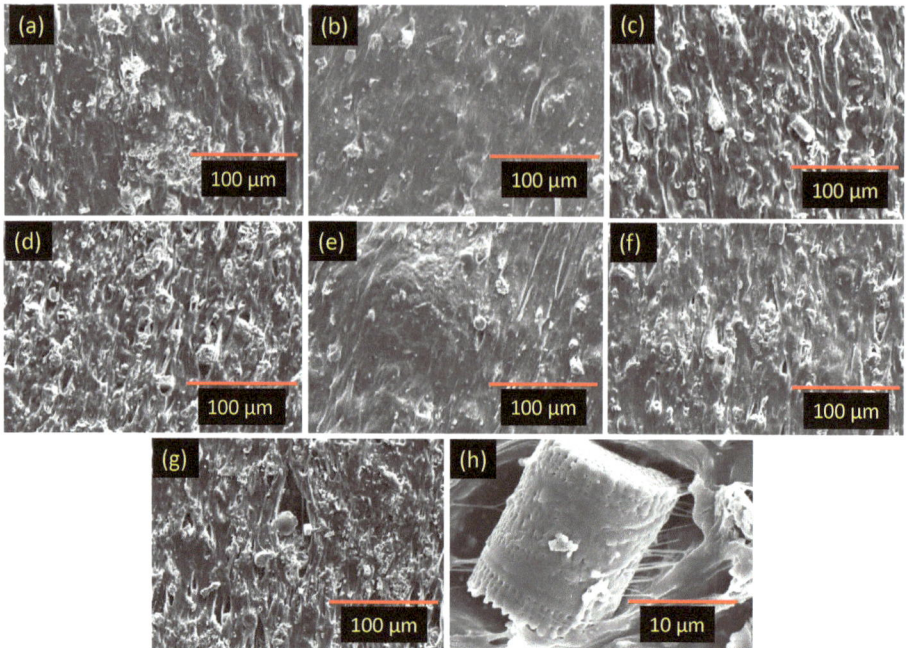

Figure 7. SEM images of rubber composites; (**a**) NR/20-DE, (**b**) NR/20-mDE, (**c**) NR/40-mDE, (**d**) NR/60-mDE, (**e**) NR/17-mDE/3-CNT, (**f**) NR/37-mDE/3-CNT, (**g**) NR/57-mDE/3-CNT, and (**h**) NR/57-mDE/3-CNT at higher resolution.

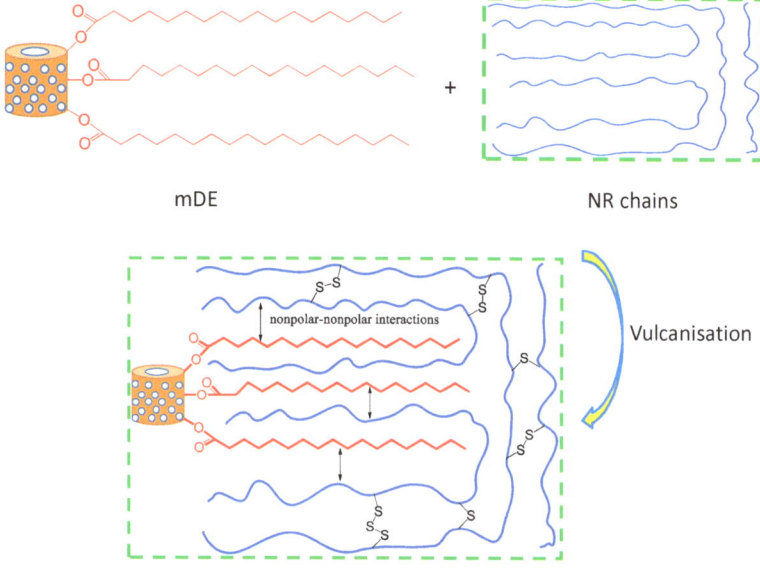

Figure 8. Possible reinforcing mechanism of rubber by stearic acid-modified diatomaceous earth.

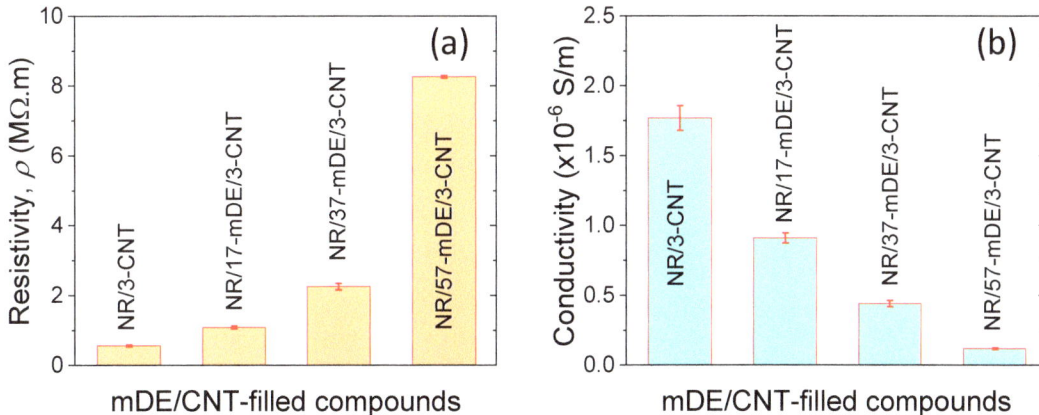

Figure 9. Electrical properties of mDE/CNT–filled rubber composites: (**a**) resistivity and (**b**) conductivity.

The specimens used for the energy-harvesting study are shown in Figure 10a, and the energy-harvesting efficiencies are shown in Figure 10b–h. According to the capacitor network model [24,65], a large number of microcapacitors can be formed when a conducting material incorporates a dielectric material. It is also seen that these capacitors are sensitive to mechanical force and can act as transducers of mechanical force to electrical energy during the charging and discharging processes during the change in the capacitance value upon mechanical deformation [66,67]. The change in the capacitance of the composites with the loading–unloading cycles is evident from the output voltages. From Figure 10b–d, it is evident that the output voltages are sometimes much higher than the regular voltages. It is believed that upon the addition of conducting materials (CNT), microcapacitors have different capacitance values depending on the distance between the conducting and non-conducting paths, assuming a parallel-plate-type capacitor. Upon the application of force, a small change in the distance can change the large voltage gradient for capacitors with higher capacitance values. The capacitor exhibits a higher capacitance at the minimum distance between the conducting and non-conducting paths. Below this critical distance, the two electrodes may collapse and lose their capacitance and sensitivity drift in the next cycle. However, these incidents are irregular because the conducting paths are unstable in viscoelastic materials [68,69]. Some new conducting paths were generated, and some were destroyed by the applied deformation. As shown in Figure 10b–e, increasing the mDE content enhanced the intensities of the regular peaks, and the NR/57-mDE/3-CNT composites showed very regular peaks with higher intensities after 250 loading–unloading cycles. This could be because of the generation of the highest number of microcapacitors at the electrical percolation threshold and uniform capacitance values through uniform dispersion of CNT in the rubber matrix. Figure 10f shows regular peaks between −10 and +10 mV, which is sufficient for sensing very low mechanical deformations because modern electronic devices can measure as low as microvolts. It is evident that the energy output of the NR/57-mDE/3-CNT composite is nearly double that of the CNT-filled composite, considering the representative areas per cycle in Figure 10g,h. Because of their robust mechanical and energy-conversion efficiencies, mDE and CNT hybrid filler-based composites may find suitable applications in electromechanical energy harvesting and other mechanical sensing devices.

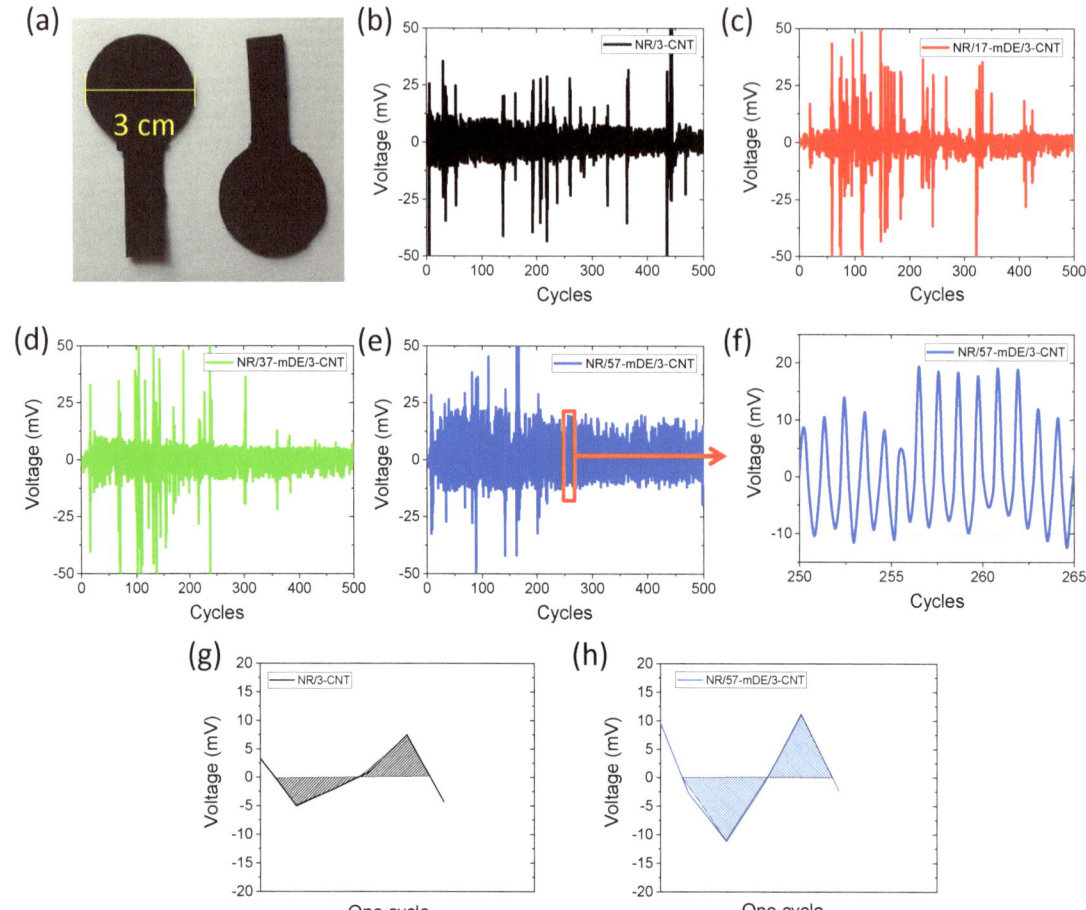

Figure 10. (**a**) Energy harvesting specimens, (**b–f**) voltage output with dynamic loading–unloading cycles for different rubber composites, and (**g,h**) areas in one cyclic loading–unloading.

4. Conclusions

This study presents the successful development of rubber composites based on naturally available diatomaceous earth with successful filler modification by stearic acid using a convenient method. The FT-IR study and the mechanical properties of the rubber composites suggest that stearic acid effectively modifies the diatomaceous earth. Compared to pristine diatomaceous earth, stearic acid-modified diatomaceous earth exhibited better reinforcing properties. Modified diatomaceous earth can be useful up to 60 phr, with improved mechanical properties compared with unfilled rubber. Hybrid fillers comprising modified diatomaceous earth and multi-walled carbon nanotubes in suitable amounts can further improve the mechanical and electrical properties of rubber composites. For example, a composite based on a hybrid filler of 37 phr mDE and 3 phr CNT has higher tensile strength and fracture toughness values of around 429% and 484%, respectively, in comparison to the unfilled rubber vulcanisate. Hybrid fillers comprising modified diatomaceous earth and CNT improved the energy-harvesting efficiencies of rubber composites at higher mDE filler amounts. Thus, the fabricated rubber composites can be useful for harvesting electrical power from renewable mechanical energy sources and for electromechanical sensing applications. Currently, the hybrid composite can be commercialised for low mechanical energy purposes due to its low fracture toughness value at higher mDE content. However, the fracture toughness and mechanical stability can be further improved by employing a certain amount of reinforcing piezoelectric nanomaterial, a step that will be undertaken in future extended work.

Author Contributions: Conceptualization, M.N.A. and V.K.; methodology, M.N.A., V.K. and H.-S.J.; validation, M.N.A., V.K., H.-S.J. and S.-S.P.; formal analysis, M.N.A., V.K. and H.-S.J.; investigation, M.N.A.; data curation, M.N.A. and V.K.; writing—original draft preparation, M.N.A.; writing—review and editing, M.N.A., V.K., H.-S.J. and S.-S.P.; visualization, M.N.A. and S.-S.P.; supervision, S.-S.P.; project administration, S.-S.P.; funding acquisition, S.-S.P. All authors have read and agreed to the published version of the manuscript.

Funding: This research is funded by the Korean Government (MOTIE, 2022) (grant no. P0002092, the Competency Development Program for Industry Specialist).

Institutional Review Board Statement: Not applicable.

Data Availability Statement: Data will be available upon request to the corresponding author.

Acknowledgments: This research was supported by the Korea Institute for Advancement of Technology (KIAT).

Conflicts of Interest: The authors declare that they have no conflicts of interest.

References

1. Thomas, S.; Maria, H.J. (Eds.) *Progress in Rubber Nanocomposites*; Woodhead Publishing: Cambridge, UK, 2017.
2. Shahapurkar, K.; Gelaw, M.; Tirth, V.; Soudagar, M.E.M.; Shahapurkar, P.; Mujtaba, M.A.; MC, K.; Ahmed, G.M.S. Comprehensive Review on Polymer Composites as Electromagnetic Interference Shielding Materials. *Polym. Polym. Compos.* **2022**, *30*, 09673911221102127. [CrossRef]
3. Zhan, Y.; Oliviero, M.; Wang, J.; Sorrentino, A.; Buonocore, G.G.; Sorrentino, L.; Lavorgna, M.; Xia, H.; Iannace, S. Enhancing the EMI Shielding of Natural Rubber-Based Supercritical CO_2 Foams by Exploiting Their Porous Morphology and CNT Segregated Networks. *Nanoscale* **2019**, *11*, 1011–1020. [CrossRef] [PubMed]
4. Zhang, X.; Fan, C.; Ma, Y.; Zhao, H.; Sui, J.; Liu, J.; Sun, C. Elastic Composites Fabricating for Electromagnetic Interference Shielding Based on MWCNTs and Fe_3O_4 Unique Distribution in Immiscible NR/NBR Blends. *Polym. Eng. Sci.* **2022**, *62*, 2019–2030. [CrossRef]
5. Kim, H.J.; Thukral, A.; Yu, C. Highly Sensitive and Very Stretchable Strain Sensor Based on a Rubbery Semiconductor. *ACS Appl. Mater. Interfaces* **2018**, *10*, 5000–5006. [CrossRef] [PubMed]
6. Souri, H.; Banerjee, H.; Jusufi, A.; Radacsi, N.; Stokes, A.A.; Park, I.; Sitti, M.; Amjadi, M. Wearable and Stretchable Strain Sensors: Materials, Sensing Mechanisms, and Applications. *Adv. Intell. Syst.* **2020**, *2*, 2000039. [CrossRef]
7. Alam, M.N.; Kumar, V.; Lee, D.J.; Choi, J. Synergistically Toughened Silicone Rubber Nanocomposites Using Carbon Nanotubes and Molybdenum Disulfide for Stretchable Strain Sensors. *Compos. Part B Eng.* **2023**, *259*, 110759.
8. Zhang, H.; Yao, L.; Quan, L.; Zheng, X. Theories for Triboelectric Nanogenerators: A Comprehensive Review. *Nanotechnol. Rev.* **2020**, *9*, 610–625. [CrossRef]
9. Yi, F.; Lin, L.; Niu, S.; Yang, P.K.; Wang, Z.; Chen, J.; Zhou, Y.; Zi, Y.; Wang, J.; Liao, Q.; et al. Stretchable Rubber-Based Triboelectric Nanogenerator and Its Application as Self-Powered Body Motion Sensors. *Adv. Funct. Mater.* **2015**, *25*, 3688–3696. [CrossRef]
10. Li, Y.; Li, J.; Li, W.; Du, H. A State-of-the-Art Review on Magnetorheological Elastomer Devices. *Smart Mater. Struct.* **2014**, *23*, 123001. [CrossRef]
11. Alam, M.N.; Kumar, V.; Jo, C.R.; Ryu, S.R.; Lee, D.J.; Park, S.S. Mechanical and Magneto-Mechanical Properties of Styrene-Butadiene-Rubber-Based Magnetorheological Elastomers Conferred by Novel Filler-Polymer Interactions. *Compos. Sci. Technol.* **2022**, *229*, 109669.
12. Liu, Y.; Pharr, M.; Salvatore, G.A. Lab-on-Skin: A Review of Flexible and Stretchable Electronics for Wearable Health Monitoring. *ACS Nano* **2017**, *11*, 9614–9635. [CrossRef] [PubMed]
13. Alam, M.N.; Kumar, V.; Debnath, S.C.; Jeong, T.; Park, S.S. Naturally Abundant Silica-Kaolinite Mixed Minerals as an Outstanding Reinforcing Filler for the Advancement of Natural Rubber Composites. *J. Polym. Res.* **2023**, *30*, 59. [CrossRef]
14. Ma, P.C.; Siddiqui, N.A.; Marom, G.; Kim, J.K. Dispersion and Functionalization of Carbon Nanotubes for Polymer-Based Nano-composites: A Review. *Compos. Part A Appl. Sci. Manuf.* **2010**, *41*, 1345–1367. [CrossRef]
15. Mensah, B.; Kim, H.G.; Lee, J.H.; Arepalli, S.; Nah, C. Carbon Nanotube-Reinforced Elastomeric Nanocomposites: A Review. *Int. J. Smart Nano Mater.* **2015**, *6*, 211–238.
16. Geetha, S.; Satheesh Kumar, K.K.; Rao, C.R.; Vijayan, M.; Trivedi, D.C. EMI Shielding: Methods and Materials-A Review. *J. Appl. Polym. Sci.* **2009**, *112*, 2073–2086.

17. Park, M.; Park, J.; Jeong, U. Design of Conductive Composite Elastomers for Stretchable Electronics. *Nano Today* **2014**, *9*, 244–260. [CrossRef]
18. Zhao, Y.; Yin, L.J.; Zhong, S.L.; Zha, J.W.; Dang, Z.M. Review of Dielectric Elastomers for Actuators, Generators, and Sensors. *IET Nanodielectrics* **2020**, *3*, 99–106. [CrossRef]
19. Park, K.I.; Jeong, C.K.; Kim, N.K.; Lee, K.J. Stretchable Piezoelectric Nanocomposite Generator. *Nano Converg.* **2016**, *3*, 12. [CrossRef]
20. Koh, S.J.A.; Keplinger, C.; Li, T.; Bauer, S.; Suo, Z. Dielectric Elastomer Generators: How Much Energy Can Be Converted? *IEEE/ASME Trans. Mechatron.* **2010**, *16*, 33–41. [CrossRef]
21. Chiba, S.; Waki, M.; Jiang, C.; Takeshita, M.; Uejima, M.; Arakawa, K.; Ohyama, K. The Possibility of a High-Efficiency Wave Power Generation System Using Dielectric Elastomers. *Energies* **2021**, *14*, 3414. [CrossRef]
22. Di, K.; Bao, K.; Chen, H.; Xie, X.; Tan, J.; Shao, Y.; Li, Y.; Xia, W.; Xu, Z.; E, S. Dielectric Elastomer Generator for Electromechanical Energy Conversion: A Mini Review. *Sustainability* **2021**, *13*, 9881. [CrossRef]
23. Uyor, U.O.; Popoola, A.P.I.; Popoola, O.M.; Aigbodion, V.S. Thermal, Mechanical, and Dielectric Properties of Functionalized Sandwich BN-BaTiO$_3$-BN/Polypropylene Nanocomposites. *J. Alloys Compd.* **2022**, *894*, 162405. [CrossRef]
24. Zeng, Y.; Xiong, C.; Li, W.; Rao, S.; Du, G.; Fan, Z.; Chen, N. Significantly Improved Dielectric and Mechanical Performance of Ti$_3$C$_2$T$_x$ MXene/Silicone Rubber Nanocomposites. *J. Alloys Compd.* **2022**, *905*, 164172. [CrossRef]
25. Salaeh, S.; Muensit, N.; Bomlai, P.; Nakason, C. Ceramic/Natural Rubber Composites: Influence Types of Rubber and Ceramic Materials on Curing, Mechanical, Morphological, and Dielectric Properties. *J. Mater. Sci.* **2011**, *46*, 1723–1731. [CrossRef]
26. Chang, B.P.; Gupta, A.; Muthuraj, R.; Mekonnen, T.H. Bioresourced Fillers for Rubber Composite Sustainability: Current Development and Future Opportunities. *Green Chem.* **2021**, *23*, 5337–5378. [CrossRef]
27. Bhagavatheswaran, E.S.; Das, A.; Rastin, H.; Saeidi, H.; Jafari, S.H.; Vahabi, H.; Najafi, F.; Khonakdar, H.A.; Formela, K.; Jouyandeh, M.; et al. The Taste of Waste: The Edge of Eggshell over Calcium Carbonate in Acrylonitrile Butadiene Rubber. *J. Polym. Environ.* **2019**, *27*, 2478–2489. [CrossRef]
28. Ten Brinke, J.W.; Debnath, S.C.; Reuvekamp, L.A.; Noordermeer, J.W. Mechanistic Aspects of the Role of Coupling Agents in Silica-Rubber Composites. *Compos. Sci. Technol.* **2003**, *63*, 1165–1174. [CrossRef]
29. Mora-Barrantes, I.; Rodríguez, A.; Ibarra, L.; González, L.; Valentín, J.L. Overcoming the Disadvantages of Fumed Silica as Filler in Elastomer Composites. *J. Mater. Chem.* **2011**, *21*, 7381–7392. [CrossRef]
30. Sarkawi, S.S.; Kaewsakul, W.; Sahakaro, K.; Dierkes, W.K.; Noordermeer, J.W. A Review on Reinforcement of Natural Rubber by Silica Fillers for Use in Low-Rolling Resistance Tires. *J. Rubber Res.* **2015**, *18*, 203–233.
31. Beidaghy Dizaji, H.; Zeng, T.; Hartmann, I.; Enke, D.; Schliermann, T.; Lenz, V.; Bidabadi, M. Generation of High-Quality Biogenic Silica by Combustion of Rice Husk and Rice Straw Combined with Pre- and Post-Treatment Strategies-A Review. *Appl. Sci.* **2019**, *9*, 1083. [CrossRef]
32. Choophun, N.; Chaiammart, N.; Sukthavon, K.; Veranitisagul, C.; Laobuthee, A.; Watthanaphanit, A.; Panomsuwan, G. Natural Rubber Composites Reinforced with Green Silica from Rice Husk: Effect of Filler Loading on Mechanical Properties. *J. Compos. Sci.* **2022**, *6*, 369. [CrossRef]
33. Utrera-Barrios, S.; Bascuñán, A.; Verdejo, R.; Lopez-Manchado, M.A.; Aguilar-Bolados, H.; Marianella Hernández, M. Sustainable Fillers for Elastomeric Compounds. In *Green-Based Nanocomposite Materials and Applications*; Engineering Materials; Springer: Cham, Switzerland, 2023.
34. Masłowski, M.; Miedzianowska, J.; Strzelec, K. Natural Rubber Composites Filled with Crop Residues as an Alternative to Vulcanizates with Common Fillers. *Polymers* **2019**, *11*, 972. [CrossRef] [PubMed]
35. Sardo, A.; Orefice, I.; Balzano, S.; Barra, L.; Romano, G. Mini-Review: Potential of Diatom-Derived Silica for Biomedical Applications. *Appl. Sci.* **2021**, *11*, 4533. [CrossRef]
36. Reka, A.A.; Smirnov, P.V.; Belousov, P.; Durmishi, B.; Abbdesettar, L.; Aggrey, P.; Kabra Malpani, S.; Idrizi, H. Diatomaceous Earth: A Literature Review. *J. Nat. Sci. Math.* **2022**, *7*, 256–268.
37. Scotti, R.; Conzatti, L.; D'Arienzo, M.; Di Credico, B.; Giannini, L.; Hanel, T.; Stagnaro, P.; Susanna, A.; Tadiello, L.; Morazzoni, F. Shape Controlled Spherical (0D) and Rod-like (1D) Silica Nanoparticles in Silica/Styrene Butadiene Rubber Nanocomposites: Role of the Particle Morphology on the Filler Reinforcing Effect. *Polymers* **2014**, *55*, 1497–1506. [CrossRef]
38. Fröhlich, J.; Niedermeier, W.; Luginsland, H.D. The Effect of Filler-Filler and Filler-Elastomer Interaction on Rubber Reinforcement. *Compos. Part A Appl. Sci. Manuf.* **2005**, *36*, 449–460. [CrossRef]
39. Olewnik-Kruszkowska, E.; Brzozowska, W.; Adamczyk, A.; Gierszewska, M.; Wojtczak, I.; Sprynskyy, M. Effect of Diatomaceous Biosilica and Talc on the Properties of Dielectric Elastomer-Based Composites. *Energies* **2020**, *13*, 5828. [CrossRef]
40. Alam, M.N.; Kumar, V.; Park, S.-S. Advances in Rubber Compounds Using ZnO and MgO as Co-Cure Activators. *Polymers* **2022**, *14*, 5289. [CrossRef]
41. Kucuk, F.; Sismanoglu, S.; Kanbur, Y.; Tayfun, U. Optimization of Mechanical, Thermo-Mechanical, Melt-Flow, and Thermal Performance of TPU Green Composites by Diatomaceous Earth Content. *Clean Eng. Technol.* **2021**, *4*, 100251. [CrossRef]
42. Marković, G.; Marinović-Cincović, M.; Jovanović, V.; Samaržija-Jovanović, S.; Budinski-Simendić, J. NR/CSM/Biogenic Silica Rubber Blend Composites. *Compos. Part B Eng.* **2013**, *55*, 368–373. [CrossRef]
43. Porobić, S.J.; Marković, G.; Ristić, I.; Samaržija-Jovanović, S.; Jovanović, V.; Budinski-Simendić, J.; Marinović-Ćinčović, M. Hybrid Materials Based on Rubber Blend Nanocomposites. *Polym. Compos.* **2019**, *40*, 3056–3064. [CrossRef]

44. Kucuk, F.; Sismanoglu, S.; Kanbur, Y.; Tayfun, U. Effect of Silane-Modification of Diatomite on its Composites with Thermoplastic Polyurethane. *Mater. Chem. Phys.* **2020**, *256*, 123683. [CrossRef]
45. Wu, W.L.; Chen, Z. Modified-Diatomite Reinforced Rubbers. *Mater. Lett.* **2017**, *209*, 159–162. [CrossRef]
46. Arias, M.; Van Dijk, P. What is Natural Rubber and Why Are We Searching for New Sources. *Front. Young Minds* **2019**, *7*, 100. [CrossRef]
47. Bokobza, L. Natural Rubber Nanocomposites: A Review. *Nanomaterials* **2018**, *9*, 12. [CrossRef]
48. Sethulekshmi, A.S.; Saritha, A.; Joseph, K. A Comprehensive Review on the Recent Advancements in Natural Rubber Nanocomposites. *Int. J. Biol. Macromol.* **2022**, *194*, 819–842. [CrossRef]
49. Ambegoda, V.T.; Egodage, S.M.; Blum, F.D.; Maddumaarachchi, M. Enhancement of Hydrophobicity of Natural Rubber Latex Films Using Diatomaceous Earth. *J. Appl. Polym. Sci.* **2021**, *138*, 50047. [CrossRef]
50. Alam, M.N.; Kumar, V.; Ryu, S.R.; Ko, T.J.; Lee, D.J.; Choi, J. Synergistic Magnetorheological NR-NBR Elastomer Blend with Electrolytic Iron Particles. *Rubber Chem. Technol.* **2021**, *94*, 642–656. [CrossRef]
51. Alam, M.N.; Choi, J. Highly Reinforced Magneto-Sensitive Natural-Rubber Nanocomposite Using Iron Oxide/Multilayer Graphene as Hybrid Filler. *Compos. Commun.* **2022**, *32*, 101169. [CrossRef]
52. ISO-37; Rubber, Vulcanized or Thermoplastic—Determination of Tensile Stress-Strain Properties. ISO: Geneva, Switzerland, 2017.
53. Flory, P.J.; Rehner, J., Jr. Statistical Mechanics of Cross-Linked Polymer Networks I. Rubberlike Elasticity. *J. Chem. Phys.* **1943**, *11*, 512–520. [CrossRef]
54. Ilia, I.; Stamatakis, M.; Perraki, T. Mineralogy and Technical Properties of Clayey Diatomites from North and Central Greece. *Open Geosci.* **2009**, *1*, 393–403. [CrossRef]
55. Balan, E.; Saitta, A.M.; Mauri, F.; Calas, G. First-Principles Modeling of the Infrared Spectrum of Kaolinite. *Am. Mineral.* **2001**, *86*, 1321–1330. [CrossRef]
56. Madejová, J.; Janek, M.; Komadel, P.; Herbert, H.J.; Moog, H.C. FTIR Analyses of Water in MX-80 Bentonite Compacted from High Salinity Salt Solution Systems. *Appl. Clay Sci.* **2002**, *20*, 255–271. [CrossRef]
57. Zeng, Y.X.; Zhong, X.W.; Liu, Z.Q.; Chen, S.; Li, N. Preparation and Enhancement of Thermal Conductivity of Heat Transfer Oil-Based MoS_2 Nanofluids. *J. Nanomater.* **2013**, *2013*, 270490. [CrossRef]
58. Li, Z.; Shen, S.Y.; Peng, J.R.; Yang, C.R. Mechanochemical Modification of Wollastonite and Its Application to Polypropylene. *Key Eng. Mater.* **2003**, *249*, 409–412. [CrossRef]
59. Sari, A.; Işildak, Ş. Adsorption Properties of Stearic Acid onto Untreated Kaolinite. *Bull. Chem. Soc. Ethiop.* **2006**, *20*, 259–267. [CrossRef]
60. Kumar, R.M.; Sharma, S.K.; Kumar, B.M.; Lahiri, D. Effects of Carbon Nanotube Aspect Ratio on Strengthening and Tribological Behavior of Ultra High Molecular Weight Polyethylene Composite. *Compos. Part A Appl. Sci. Manuf.* **2015**, *76*, 62–72. [CrossRef]
61. Kucherskii, A.M. Hysteresis Losses in Carbon-Black-Filled Rubbers Under Small and Large Elongations. *Polym. Test.* **2005**, *24*, 733–738. [CrossRef]
62. Alam, M.N.; Kumar, V.; Potiyaraj, P.; Lee, D.J.; Choi, J. Mutual Dispersion of Graphite-Silica Binary Fillers and Its Effects on Curing, Mechanical, and Aging Properties of Natural Rubber Composites. *Polym. Bull.* **2022**, *79*, 2707–2724. [CrossRef]
63. Ivanoska-Dacikj, A.; Bogoeva-Gaceva, G.; Valić, S.; Wießner, S.; Heinrich, G. Benefits of Hybrid Nano-Filler Networking Between Organically Modified Montmorillonite and Carbon Nanotubes in Natural Rubber: Experiments and Theoretical Interpretations. *Appl. Clay Sci.* **2017**, *136*, 192–198. [CrossRef]
64. Mazzotta, M.G.; Putnam, A.A.; North, M.A.; Wilker, J.J. Weak Bonds in a Biomimetic Adhesive Enhance Toughness and Performance. *J. Am. Chem. Soc.* **2020**, *142*, 4762–4768. [CrossRef]
65. Yuan, J.K.; Li, W.L.; Yao, S.H.; Lin, Y.Q.; Sylvestre, A.; Bai, J. High Dielectric Permittivity and Low Percolation Threshold in Polymer Composites Based on SiC-Carbon Nanotubes Micro/Nano Hybrid. *Appl. Phys. Lett.* **2011**, *98*, 032901. [CrossRef]
66. Zang, Y.; Zhang, F.; Di, C.A.; Zhu, D. Advances of Flexible Pressure Sensors toward Artificial Intelligence and Health Care Applications. *Mater. Horiz.* **2015**, *2*, 140–156. [CrossRef]
67. Alam, M.N.; Kumar, V.; Jeong, T.; Park, S.S. Nanocarbon Black and Molybdenum Disulfide Hybrid Filler System for the Enhancement of Fracture Toughness and Electromechanical Sensing Properties in the Silicone Rubber-Based Energy Harvester. *Polymers* **2023**, *15*, 2189. [CrossRef]
68. Wang, L.; Choi, J. Highly Stretchable Strain Sensors with Improved Sensitivity Enabled by a Hybrid of Carbon Nanotube and Graphene. *Micro Nano Syst. Lett.* **2022**, *10*, 17. [CrossRef]
69. Persons, A.K.; Ball, J.E.; Freeman, C.; Macias, D.M.; Simpson, C.L.; Smith, B.K.; Burch V, R.F. Fatigue Testing of Wearable Sensing Technologies: Issues and Opportunities. *Materials* **2021**, *14*, 4070. [CrossRef]

Disclaimer/Publisher's Note: The statements, opinions and data contained in all publications are solely those of the individual author(s) and contributor(s) and not of MDPI and/or the editor(s). MDPI and/or the editor(s) disclaim responsibility for any injury to people or property resulting from any ideas, methods, instructions or products referred to in the content.

Article

Isinglass as an Alternative Biopolymer Membrane for Green Electrochemical Devices: Initial Studies of Application in Electric Double-Layer Capacitors and Future Perspectives

Paweł Jeżowski [1,*] and Przemysław Łukasz Kowalczewski [2]

1 Institute of Chemistry and Technical Electrochemistry, Poznan University of Technology, 4 Berdychowo St., 60-965 Poznań, Poland
2 Department of Food Technology of Plant Origin, Poznań University of Life Sciences, 31 Wojska Polskiego St., 60-624 Poznań, Poland
* Correspondence: pawel.jezowski@put.poznan.pl

Citation: Jeżowski, P.; Kowalczewski, P.Ł. Isinglass as an Alternative Biopolymer Membrane for Green Electrochemical Devices: Initial Studies of Application in Electric Double-Layer Capacitors and Future Perspectives. *Polymers* 2023, *15*, 3557. https://doi.org/10.3390/polym15173557

Academic Editors: Vineet Kumar and Md Najib Alam

Received: 31 July 2023
Revised: 22 August 2023
Accepted: 24 August 2023
Published: 26 August 2023

Copyright: © 2023 by the authors. Licensee MDPI, Basel, Switzerland. This article is an open access article distributed under the terms and conditions of the Creative Commons Attribution (CC BY) license (https://creativecommons.org/licenses/by/4.0/).

Abstract: The presented work discusses in detail the preparation of a cheap and environmentally friendly biopolymer membrane from isinglass and its physicochemical characterisation. One of the possible uses of the obtained membrane can be as a separator between electrodes in novel green electrochemical devices as in, for example, electric double-layer capacitors (EDLCs). The functionality of the mentioned membrane was investigated and demonstrated by classical electrochemical techniques such as cyclic voltammetry (CV), galvanostatic cycling with potential limitation (GCPL), and electrochemical impedance spectroscopy (EIS). The obtained values of capacitance (approximately 30 F g^{-1}) and resistance (approximately. 3 Ohms), as well as the longevity of the EDLC during electrochemical floating at a voltage of 1.6 V (more than 200 h), show that the proposed biopolymer membrane could be an interesting alternative among the more environmentally friendly energy storage devices, while additionally it could be more economically justified.

Keywords: biopolymer; membrane; energy storage; green chemistry; EDLC

1. Introduction

Recent years have been dominated by synthetic polymers; from everyday-use simple appliances to highly sophisticated equipment, you can be certain to find it in your nearest surroundings. Since 2004, more than 4.6 billion tons of plastics have been produced [1]. The increasing use of synthetic polymers causes a worldwide dilemma due to pollution [2]. Biopolymers could substitute for synthetic ones in any possible field; for example, in electrochemistry, all devices are composed of two electrodes that are separated by a porous membrane that allows for the flow of ions. Usually, the membrane is made of materials such as glass fibre, which has potential cancerogenic properties [3], nafion, which is classified as a hazardous material, or poly(propylene), which is responsible for the release of microplastic due to degradation [4]. Biopolymers are currently widely investigated as a possible way to exchange problematic and hazardous materials [5]. The use of cellulose, chitin [6,7], chitosan [8,9], dextrin [10,11], dextran [12,13], agar [14,15], lignin [16,17], and other biopolymers [18–22] have been reported in the literature, with their promising use in energy storage devices such as lithium-ion batteries (LIBs) [23–25] or electric double-layer capacitors (EDLCs) [26–28]. Biopolymers can be used for almost any aspect of electrode construction, from an electrode material binder [8], to conductor glue for improved adhesion and electrical conductivity [29], to an electrolyte [30], and even for preparation of active material responsible for energy storage mechanisms [31,32]. Most of the studies on the use of biopolymer membranes in EDLCs are focused on the use of aqueous electrolytes and can be summarised in Table 1.

Table 1. Short literature review of the biopolymer membranes and their use in the aqueous EDLCs.

Ref.	Membrane	Electrolyte	Electrode Active Material	U [V]	ESR [Ω]	EDR [Ω]	C [F g^{-1}]	E [Wh kg^{-1}]	Longevity
[6]	Chitin/Cellulose	1.0 mol L^{-1} lithium sulphate	AC Kynol ACC-507-20 AC Norti DLC Supra 30	0.8	1	1.7	90 *	2	20 k cycles, 5% capacitance drop
[28]	Cellulose	2.0 mol L^{-1} lithium acetate	AC Kynol ACC-507-20	0.8	5	15	20–25	2	10 k cycles, no observable drop
[33]	Cellulose nanofibrils	1.0 mol L^{-1} sodium sulphate	AC (not specified)	1.2	1–3	12–22	80–100 *	23–27 *	5 k cycles, up to 5% capacitance drop
[34]	Chitosan/Sodium alginate	2.0 mol L^{-1} lithium sulphate	AC Kynol ACC-507-20	1.6	1	3–5	125 *	8–10	1 k cycles, 10% capacitance drop
[35]	Chitosan/NaOH /glutaraldehyde	2.0 mol L^{-1} lithium acetate	AC Kynol ACC-507-20	0.8	1–3	5–10	100 *	2	10 k cycles, no observable drop
[36]	Chitin	2.0 mol L^{-1} lithium acetate	AC Kynol ACC-507-20	0.8	1	5	100 *	2	10 k cycles, no observable drop
[37]	Carboxylated chitosan	1.0 mol L^{-1} hydrochloric acid	AC Shenyang Kejing AC (as prepared)	0.9	1	8	40	3	(not specified)
[38]	Starch	1.0 mol L^{-1} sulphuric acid	AC Kynol ACC-507-20	0.8	0.5	9–63	100–250 *	10–20 *	2 k cycles, 3% capacitance drop
[39]	Cellulose/Agarose	1.0 mol L^{-1} sulphuric acid 1.0 mol L^{-1} sodium sulphate	AC Kuraray YP-80F	0.8	1–14	1–30	100–120 *	2	10 k cycles, 10% capacitance drop
[30]	Agar	0.5 mol L^{-1} potassium sulphate	AC Kynol ACC-507-20	1.6	1	2	80–110 *	7	10 k cycles, up to 8% capacitance drop
This work	Isinglass	1.0 mol L^{-1} sodium sulphate	AC Kynol ACC-507-20	1.6	2	7	25	8–10	10 k cycles, up to 5% capacitance drop
	Glass fibre	1.0 mol L^{-1} sodium sulphate	AC Kynol ACC-507-20	1.6	1	3	28	8–10	10 k cycles, up to 4% capacitance drop

AC—activated carbon; *—values presented per mass of one electrode.

However, the preparation of biopolymer membranes presented in the literature can be time-consuming and can even take up to 6 h at an elevated temperature of 50 °C [40], can be complicated in execution [41], or can simply be costly [36] due to the use of additional chemicals that cannot be considered environmentally friendly or green. In the mentioned studies, the use of potassium iodide or glutaraldehyde, which are considered toxic to the environment as well as human beings, is rather in opposition to the main goal of green biopolymers, which is their environmental friendliness or at least their benign character. Furthermore, price is often omitted in scientific studies, but from an economical point of view, the use of 1-butyl-3-methylimidazolium acetate (close to 1500 USD per 1 kg) could be hard to justify for a bigger-scale preparation of a sustainable biopolymer membrane for any application. Furthermore, the use of biopolymer membranes allows for the mitigation of a possible spill of liquid electrolyte from the electrochemical cell in case of cell body damage, as most of the electrolyte is stored in the membrane volume [42]. However, the use of liquid electrolytes is one of the most-used approaches because of the simplified construction and automatization of cell assembly on a larger scale. Today, the reuse of wastes from fish production, including isinglass, presents a promising avenue for addressing both environmental and technological challenges. Isinglass, a protein-rich substance obtained from fish bladders, is commonly used in the beer and wine industry as a fining agent [43–45]. However, its potential extends far beyond this traditional application. In the realm of electrochemistry, isinglass can serve as a valuable precursor for the development of advanced alternative biopolymer membranes for green electrochemical devices and can be employed as electrodes in various energy storage and conversion devices, such as supercapacitors and fuel cells. Utilising fish waste-derived isinglass as a membrane material not only reduces environmental burdens by reusing otherwise discarded by-products but also contributes to the advancement of eco-friendly and sustainable electrochemical technologies. This interdisciplinary approach holds great promise in creating a more circular economy while fostering advances in renewable energy and green technology sectors.

For the mentioned reasons, in this research, we decided to prepare a cost-effective and green biopolymer membrane that could be successfully incorporated into energy storage devices like, for example, EDLCs. An isinglass membrane can be successfully prepared by heating it with a small amount of deionized water. Furthermore, isinglass possesses flexible properties that can be used in the construction of novel wearable energy storage systems [46]; for example, different types of smartwatches, which are becoming more popular each day, not only among young generations but also for the elderly, as they can be used for heart monitoring. These electronics could implement energy storage cells inside the band and the flexible solar cell outside the band to maximize their longevity.

2. Materials and Methods

2.1. Preparation of Biopolymer Membrane

Isinglass (1 g, protein content 75–80%, humidity 10–12%, ashes 8–12%, Kremer Pigmente GmbH + Co. KG, Aichstetten, Germany) was placed in a stainless-steel beaker with 10 mL of distilled water. The heterogeneous mixture was heated (100 °C) and stirred (200 RPM) until excess solvent evaporated and a homogenous viscous yellowish mixture was obtained. It was then casted on a silicone board and spread across its surface to obtain a biopolymer membrane as thin as possible and then was placed in a universal laboratory convection oven heated up to 80 °C for approximately 1 h (UN30, Memmert GmbH + Co. KG, Schwabach, Germany). Once the membrane was dry, its thickness was measured (approximately 150 µm) and it was introduced into the aqueous electrolyte solution of 1 mol L^{-1} Li$_2$SO$_4$ and stored in such conditions until the separators were cut with hollow punchers (diameter 12 mm). The preparation of the isinglass membrane is summarised in Figure 1.

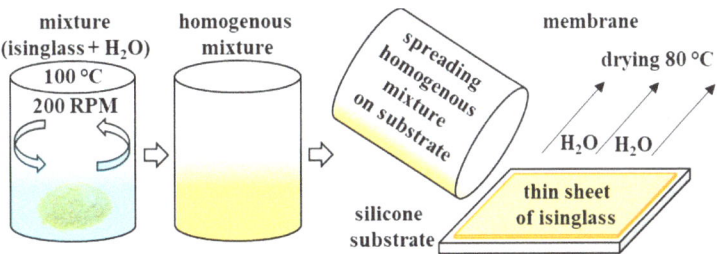

Figure 1. Schematic illustration of the isinglass membrane preparation process.

2.2. Electrochemical Cell Preparation

The electrodes for EDLCs were cut with hollow punchers (diameter 10 mm) of Kynol ACC-507-20 carbon cloth with an average mass of each electrode of approximately 8 mg ± 1 mg. The electrodes were introduced to the Swagelok laboratory electrochemical cell with stainless steel (316L) current collectors coated with conductive glue [29] and separated by a prepared biopolymer membrane (described in Section 2.1) or by a glass microfiber disk (diameter 12 mm and thickness 260 μm, Whatman® GF/A, Sigma-Aldrich, Darmstadt, Germany). Finally, each electrode was soaked with approximately 100 μL electrolyte. Additionally, to present the flexible properties of the isinglass membrane, porotype pouch cells were assembled to highlight excellent performance when an external force is applied to the cell body during electrochemical testing.

2.3. Electrochemical Testing Schedule

Cyclic voltammetry (CV) was performed at various voltage windows from 1.0 to 2.0 V and at a constant scanning rate of 5 mV s^{-1} to determine the maximum operational voltage for the rest of the electrochemical investigations, according to the calculation of the S-value presented by Weingarth et al. [47]. Later, several scanning rates were used from 1.0 to 100 mV s^{-1} at 1.6 V voltage to observe qualitative changes in the charge accumulation. Galvanostatic cycling with potential limitation (GCPL) was carried out at a voltage of 1.6 V with different current densities from 100 mA g^{-1} to 5000 mA g^{-1} to quantitatively establish the capacitance, energy, and power of the electrochemical cells. Furthermore, the constant power (CP) technique was implemented to schedule electrochemical investigations to adequately calculate the energy and power of electrochemical cells [48]. Potentiostatic electrochemical impedance spectroscopy (PEIS) was performed to measure the internal resistance (IR) of electrochemical cells (composed of equivalent series resistance (ESR) and equivalent distributed resistance (EDR)) in the range of frequencies from 1 mHz to 100 kHz at the sinusoidal amplitude of the input signal 5 mV. Change in the resistance upon floating was measured by the Current Interrupted (CI) technique, where small pulses of current separated by rest periods allows one to estimate the internal resistance of the device. The Electrochemical Floating Test (EFT) allowed for accelerated ageing of the electrochemical cell by holding the maximum voltage of the electrochemical cell for an estimated time period of 2 h several times and establishing its stability during the overall time of the test. All valueswere calculated per mass of both electrodes. The equipment used for electrochemical investigations was VMP3 (BioLogic, Seyssinet-Pariset, France) and EC-Lab software version 11.32 (BioLogic, Seyssinet-Pariset, France) to process all acquired data.

2.4. Raman Spectroscopy

To observe the structure and texture of the sample, a DXR3xi Raman Imaging Microscope (Thermo Fisher Scientific Inc., Waltham, MA, USA) was used. Photographic images were taken at the different magnifications ×10 and ×50, and the Raman spectra were taken using a laser at 532 nm wavelength with power of 10 mW. The spectrum was acquired in the range of 200 to 3500 cm^{-1} wavenumber at magnification ×50.

3. Results and Discussion

The as-received isinglass particles (Figure 2a,b) were observed under a microscope to see the changes that occurred after the preparation of the biopolymer membrane (Figure 2c,d). As seen, the texture of the substrate and the product are completely different from each other; substrate particles are rough and uneven. The dried biopolymer membrane has an intriguing drapery-like texture and the material itself is flexible but brittle. The pictures of the membrane in Figure 2e,f were taken after the drop of electrolyte (1 mol L^{-1} Li$_2$SO$_4$ in H$_2$O) was in direct contact with the biopolymer membrane. The texture of the biopolymer membrane flattened and evened out. Additionally, the porous structure of the membrane began to be visible. Moreover, the membrane started to be elastic, flexible, and easily adhered to the surface of the electrode.

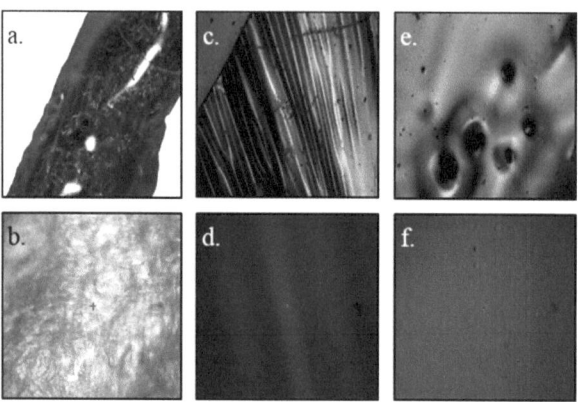

Figure 2. Microscopic images at magnification ×10 (**a,c,e**) and ×50 (**b,d,f**) for as-received isinglass particles (**a,b**), dry biopolymer isinglass membrane (**c,d**), and membrane soaked with electrolyte (**e,f**). Red plus sign is a central point of take image.

Raman spectra proved that the textural changes in the material had no impact on the structure, as isinglass particles (upper red curve) and the dry biopolymer isinglass membrane (lower green curve) have almost identical visible vibrations in Figure 3a, which are coming from amide I and amide III, which are sensitive to the changes in the structure. These vibrations are visible near the spectral range from 1200–1750 cm^{-1}. Bands visible near 1200–1400 cm^{-1} can be assigned to the following C-C stretching and N-H in-plane bending vibrations. Bands in the range 1600–1700 cm^{-1} are usually related to the C = O stretching vibrations of the peptide bond and N-H bending vibrations. Furthermore, the most visible bands in the wavenumber range 2800–3200 cm^{-1} are -CH$_2$ stretching vibrations [49]. Finally, the noticeable bump above 3200 cm^{-1} comes from the stretching vibrations of the -OH group. This shows that the proposed isinglass is mainly composed of I-type collagen [50], presented in Figure 3b.

The isinglass membrane was introduced into the electrochemical cell with two carbon electrodes and 1 mol L^{-1} Li$_2$SO$_4$. Cyclic voltammetry studies at 5 mV s^{-1} and with an increase in operational voltage from 1.0 to 2.0 V are presented in Figure 4a,b. The shape of both of the cyclic voltammograms is nearly rectangular, which implies that the energy storage mechanism is purely capacitive. In the case of isinglass, the rectangular shape of cyclic voltammograms was almost unaffected by the increased voltage; on the other hand, the electrochemical cell with the glass fibre separator presented a noticeable rise in anodic and cathodic current, especially above the value ca. 1.6 V. To correctly estimate the highest possible operational voltage for each cell, the S-value was calculated [47] at every investigated voltage. The S-value is calculated from the CV data. Firstly, the surface area is integrated above (anodic) and below (cathodic) current of 0 mA g^{-1} (y axis) at each voltage. Then, the ratio of integral values for anodic and cathodic current is calculated, which is called the S-value (Figure 4c,d). Finally, the difference between adjacent S-values is calculated, and if the value Δ > 0.005, then it is the maximum operational voltage. For the electrochemical cell with a glass fibre separator, the maximum voltage was 1.6 V since the difference in the S-values between 1.6 and 1.7 V was greater than 0.005 (0.013), while in the case of the isinglass membrane, the difference between the S-values was 0.012. For comparative purposes, both cells were limited to an operating voltage of 1.6 V, similar to other reports [30].

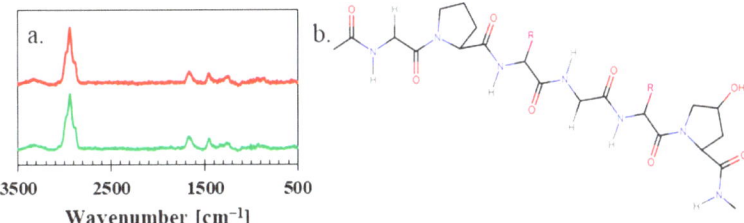

Figure 3. (a) Raman spectra of isinglass particles (upper red curve) and dry biopolymer isinglass membrane (lower green curve) and (b) chemical structure of the main chemical component of isinglass.

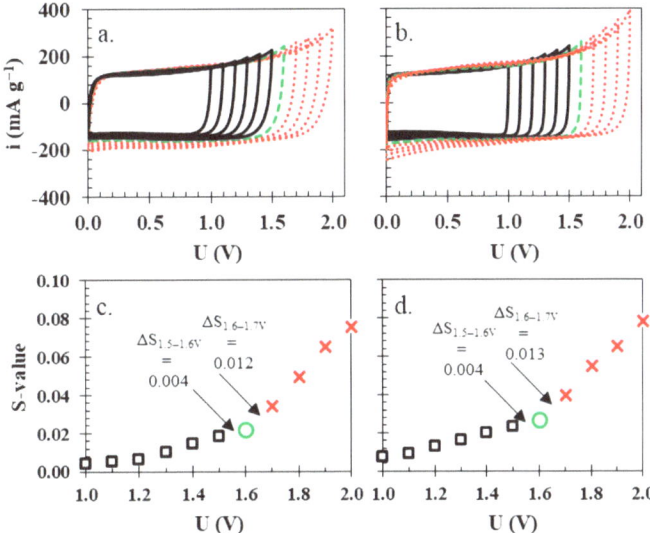

Figure 4. Cyclic voltammograms of electrochemical cells with (a) isinglass membrane and (b) glass fibre separator. The S-values calculated for cells with (c) biopolymer membrane and (d) glass fibre separator.

At lower scanning rates of 1 and 5 mV s^{-1} (Figure 5a,b, respectively), both electrochemical cells with different separators (solid green line—isinglass biopolymer membrane, red dashed line—glass fibre) presented a similar electrochemical behaviour. Still, the shape of voltammograms for the cell with a biopolymer membrane presents a more rectangular shape without any noticeable increases in the anodic nor the cathodic current, which is in accordance with observations from previous experiments. A noticeable difference between the two separators was observed when higher scanning rates were applied. At 10 mV s^{-1} (Figure 5c), the rectangular shape of the biopolymer membrane starts to be less rectangular, and at 20 mV s^{-1} (Figure 5d), it shows an even more resistive characteristic—due to worse charge propagation—than in the electrochemical cell with the glass fibre separator.

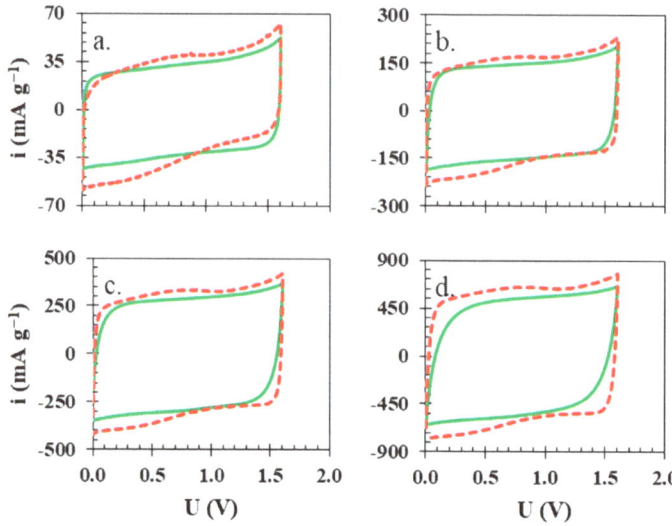

Figure 5. Cyclic voltammograms for electrochemical cells with isinglass membrane (solid green line) and glass fibre separator (red dashed line) at (**a**) 1, (**b**) 5, (**c**) 10, and (**d**) 20 mV s^{-1}, voltage 1.6 V.

Similar to the CV results, the charge/discharge profiles indicate that the attainable capacitance for both systems is similar and near 30 F g^{-1}, where both cells—the one with the glass fibre separator (red dashed lines) and the one with the isinglass (green solid lines)—reached the same value of 28 F g^{-1} at current density 100 mA g^{-1} (Figure 6a). While at the higher current densities of 200 and 500 mA g^{-1} (Figure 6b,c, respectively), the discrepancy between the two cells starts to be present due to the difference in their internal resistance; at the highest current density presented, 1000 mA g^{-1} (Figure 6d), the capacitance of the electrochemical cell with the glass fibre separator reaches ca. 27 F g^{-1}, and the isinglass separator achieves only 20 F g^{-1}. Increasing the current density highlights resistance differences in the case of the electrochemical cell with the biopolymer membrane and the one with the glass fibre separator.

The visible resistance discrepancy between the electrochemical cell with the biopolymer membrane and the glass fibre separator was quantitatively examined by potentiostatic electrochemical impedance spectroscopy (PEIS) (Figure 7a–c). Values of equivalent series resistance (ESR) and equivalent distributed resistance (EDR) for the electrochemical cell with the glass fibre separator (Figure 7a,b, red dashed line) are 1 and 3 Ohms, respectively. However, in the case of the electrochemical cell with the biopolymer membrane (Figure 7a,b, green solid line), these values were higher; ESR was 3 Ohms and EDR was 7 Ohms. Furthermore, the calculated values of capacitance vs. frequency proved what was also observed in the cyclic voltammetry and galvanostatic charging/discharging, that the biopolymer membrane can be successfully applied when milder conditions are applied to the electrochemical cell with a biopolymer membrane as the separator.

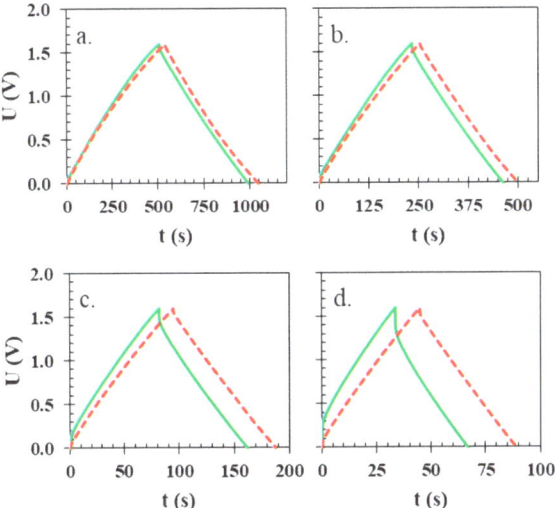

Figure 6. Galvanostatic charge/discharge profiles of electrochemical cells with different types of separators the biopolymer membrane (solid green line) and glass fibre disk (red dashed line) at different current densities (**a**) 100, (**b**) 200, (**c**) 500 and (**d**) 1000 mA g^{-1}.

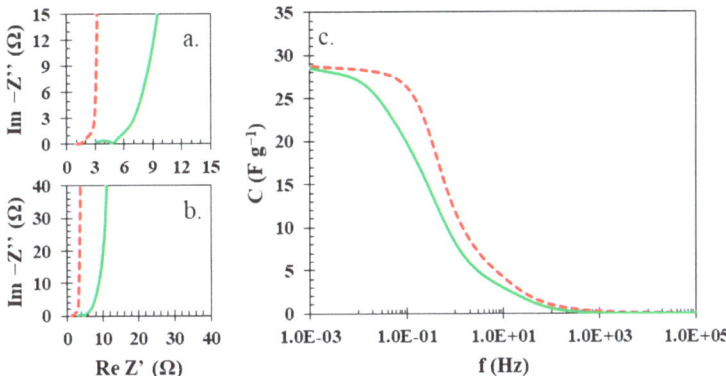

Figure 7. (**a**) Nyquist plots of electrochemical cells with biopolymer membrane (solid green line) and glass fibre separator (red dashed line) and (**b**) the magnification of the ESR and EDR region and (**c**) the characteristic of capacitance versus frequency.

Once all of the basic electrochemical experiments were finished, it was necessary to establish the end-of-life criterion (80% of initial capacitance or a two-fold increase in resistance). Figure 8a,c present the leakage current profile for the holding time at the maximum voltage of 2 h separated by ca. periods of 2 h, during which additional CVs were performed to present qualitative changes in CV curves (Figure 8b,d), as well as GCPL and CI experiments to quantitatively estimate capacitance and resistance, presented in Figure 8e. This 4 h time period can be considered one full cycle, which means that during the experiment presented in Figure 8a,c, the total time of holding at maximum voltage was 100 h. The data from the initial 200 h of the electrochemical technique were insufficient to establish the failure of the electrochemical cells, and it was repeated until the end-of-life criterion (80% capacitance), which was reached after 362 h of holding time at the maximum voltage of 1.6 V (the electrochemical cell with the glass fibre separator reached almost identical end-of-life criterion at 378 h). After 200 h of technique time, it was already noticeable that the shape of CVs (Figure 8b,d) had deteriorated from the typical rectangular shape, which indicates partial oxidation and reduction of electrodes, which are identical for both electrochemical cells and are similar to observations reported elsewhere [51]. During the overall time of the experiment, the capacitance of the electrochemical cell with the isinglass membrane was 27 to 24 F g^{-1}, while the resistance increased from 2.9 to 4.0 Ohms; in comparison, the electrochemical cell with the capacitance of the glass fibre separator fell from 29 to 27 F g^{-1} and the resistance increased from 1.0 to 1.3 Ohm, which means the overall drop in capacitance and increase in resistance are similar in both of the cases. The leakage current data indicated that the amount of current necessary to sustain the maximum voltage during the hold period was 10 mA g^{-1} in the case of the biopolymer membrane, and for the glass fibre separator, it was 23 mA g^{-1}, which can be attributed to the dielectric properties of the isinglass membrane.

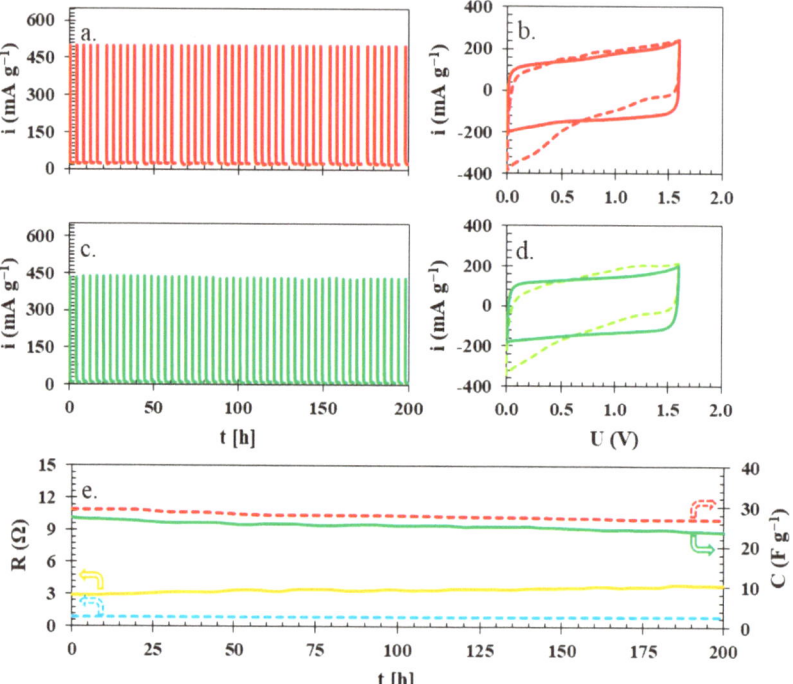

Figure 8. Leakage current for an electrochemical cell with (**a**) glass fibre separator and (**c**) biopolymer membrane. CV plots for electrochemical cells with (**b**) glass fibre and (**d**) isinglass separator before (solid lines) and after (dashed lines) 200 h of experiment time. (**e**) Change in capacitance (upper lines) and resistance (lower lines) during the initial 200 h of electrochemical testing of the cells with isinglass membrane (solid lines) and glass fibre separator (dashed lines).

In addition, both electrochemical devices were subjected to electrochemical tests to establish their cyclic life (Figure 9) by continuous charge/discharge at a current density of 0.1 A g^{-1} within the voltage range of 0.0 to 1.6 V. Changes in capacitance and specific energy values during 10,000 cycles are very similar for both electrochemical cells with the isinglass membrane (solid lines) and the glass fibre separator (dashed lines). In the case of the glass fibre separator, there was a 4% decrease in the capacitance (dashed middle red line) value after 10,000 cycles, and the Coulombic efficiency was nearly 99.6% (dashed yellow upper line). For the electrochemical cell with the isinglass membrane, there was a noticeable 5% drop in the capacitance compared to the initial value (26 F g^{-1}), while the Coulombic efficiency was ca. 99.7%. It is interesting that the initial 3% of the drop in the capacitance value was observed during the first 1000 cycles for the electrochemical cell with the isinglass membrane and the glass fibre separator. Afterwards, between cycle numbers 1000 and 10,000, the drop in capacitance value is smaller, ca. 2%. The observable capacitance in the case of electrochemical cells with a glass fibre separator and a biopolymer membrane appears to stabilise afterward. This observation is probably related to the initial penetration of the porous structure of the electrodes with the electrolyte solution and the ion entrapment inside of the carbon material.

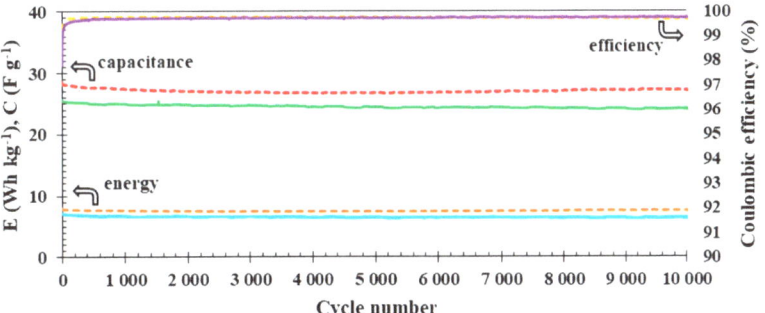

Figure 9. Cyclic life of the electrochemical cell with glass fibre separator (dashed lines) and biopolymer membrane (solid lines). The upper lines represent the Coulombic efficiencies of electrochemical cells, the middle lines represent the capacitance values, and, finally, the lower lines represent the specific energy data.

Finally, specific values of energy and power values were calculated based on galvanostatic charge–discharge profiles, as well as the constant power discharge technique (Figure 10a), because galvanostatic experiments can be prone to overestimation of data [48]. The specific energy and specific power values for both electrochemical cells are almost identical up to the specific power value of 2 kW kg^{-1}, when the higher resistance of the electrochemical cell with the isinglass membrane starts to hinder the energy of said cell (3 Wh kg^{-1}). The energy of the electrochemical cell with the glass fibre separator (5 Wh kg^{-1}) at higher values of specific power is even more noticeable. The presented data clearly show the better overall electrochemical performance of the electrochemical cell with the glass fibre separator at higher current loads. The isinglass possesses one intrinsic property that glass fibre separators do not have, which is flexibility. To compare both dielectrics, pouch cell prototypes were assembled and tested using cyclic voltammetry at a scanning rate of 5 mV s^{-1} up to a voltage of 1.6 V (Figure 10b) when an external force was applied to the pouch cell. Electrochemical cells were permanently bent during electrochemical testing (Figure 10c) to observe the current response when the external force was applied to the cell. As seen, the pouch cell with the biopolymer membrane (green solid line) still retains the rectangular shape characteristic for uninterrupted charge propagation, while the electrochemical characteristic of the cell with glass fibre is noticeably impacted by the applied external force. The difference between the glass fibre and the isinglass biopolymer membrane is in their structure. Although the glass fibre is made of tiny rod-shaped structures that are excellent for static applications, their flexibility is limited, and they are prone to separate from each other and increase overall resistance.

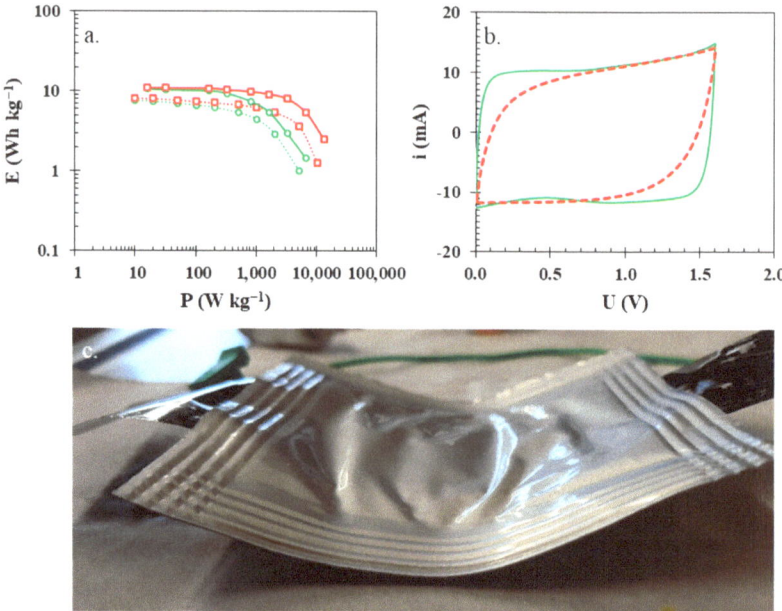

Figure 10. (**a**) Ragone plot of calculated energy and power based on GCPL (solid lines) and CP (dotted lines) for an electrochemical cell with a glass fibre separator (red lines with square markers) and an isinglass membrane (green lines with circle markers). (**b**) CVs of the electrochemical cells with an isinglass membrane (solid green line) and a glass fibre separator (solid red line) examined in (**c**) a prototype pouch cell.

4. Conclusions

The obtained isinglass membranes were successfully implemented in green energy storage devices such as EDLCs. In our opinion, the isinglass membrane is an interesting alternative to the data reported so far, due to its relatively high operating voltage of 1.6 V, good longevity of 10,000 cycles (or at least 200 h of floating), and last but not least, the flexibility proven by the prototype pouch cell that we were able to prepare. Moreover, the preparation of the biopolymer membrane does not require any hazardous chemicals and is easy and fast. The capacitance, energy, and power of the assembled EDLC with a biopolymer membrane and a glass fibre separator are close to each other under milder conditions, and they are similar. The capacitance near 28 F g^{-1}, a resistance of 3 Ohms, and an energy of ca. 9 Wh kg^{-1} at power 1 kW kg^{-1} are close to other results that can be found in the literature. However, the overall performance of isinglass still needs additional improvement to reach lower values of internal resistance, as this is the limiting factor and one of the biggest drawbacks of the presented idea. To mitigate the mentioned issues, we are already investigating several possible modifications to the biopolymer membrane construction and additives, which could give a promising improvement in ion transfer throughout the membrane and lower the overall internal resistance of the device, even at stationary conditions (without applied external force, which changes the shape of the electrochemical cell). Moreover, as the mentioned membrane can work successfully, especially at lower currents, it is possible to apply it to other energy storage devices like aqueous batteries or pseudocapacitors, where lower currents than those of EDLCs are applicable. Furthermore, compared to typical glass fibre separators, this type of biopolymer membrane has excellent flexible properties and can be easily transferred to more industrial-type cells, like pouch cells or cylindrical cells.

Author Contributions: Conceptualization, P.J.; methodology, P.J.; validation, P.J. and P.Ł.K.; formal analysis, P.J.; investigation, P.J.; resources, P.J.; data curation, P.J.; writing—original draft preparation, P.J.; writing—review and editing, P.J. and P.Ł.K.; visualization, P.J. and P.Ł.K.; supervision, P.J. All authors have read and agreed to the published version of the manuscript.

Funding: This research received no external funding.

Institutional Review Board Statement: Not applicable.

Data Availability Statement: The datasets generated for this study are available on request to the corresponding author.

Conflicts of Interest: The authors declare no conflict of interest.

References

1. United Nations Environment. *Drowning in Plastics: Marine Litter and Plastic Waste Vital Graphics*; United Nations Environment: Nairobi, Kenya, 2021.
2. Ritchie, H.; Roser, M. Plastic Pollution. Available online: https://ourworldindata.org/plastic-pollution (accessed on 24 July 2023).
3. Hesterberg, T.W.; Chase, G.; Axten, C.; Miller, W.C.; Musselman, R.P.; Kamstrup, O.; Hadley, J.; Morscheidt, C.; Bernstein, D.M.; Thevenaz, P. Biopersistence of Synthetic Vitreous Fibers and Amosite Asbestos in the Rat Lung Following Inhalation. *Toxicol. Appl. Pharmacol.* **1998**, *151*, 262–275. [CrossRef] [PubMed]
4. Li, D.; Shi, Y.; Yang, L.; Xiao, L.; Kehoe, D.K.; Gun'ko, Y.K.; Boland, J.J.; Wang, J.J. Microplastic release from the degradation of polypropylene feeding bottles during infant formula preparation. *Nat. Food* **2020**, *1*, 746–754. [CrossRef] [PubMed]
5. Leja, K.; Lewandowicz, G. Polymer Biodegradation and Biodegradable Polymers—A Review. *Polish J. Environ. Stud.* **2010**, *19*, 255–266.
6. Kasprzak, D.; Galiński, M. Chitin and chitin-cellulose composite hydrogels prepared by ionic liquid-based process as the novel electrolytes for electrochemical capacitors. *J. Solid State Electrochem.* **2021**, *25*, 2549–2563. [CrossRef]
7. Kasprzak, D.; Galiński, M. Chitin as a universal and sustainable electrode binder for electrochemical capacitors. *J. Power Sources* **2023**, *553*, 232300. [CrossRef]
8. Salleh, N.A.; Kheawhom, S.; Mohamad, A.A. Chitosan as biopolymer binder for graphene in supercapacitor electrode. *Results Phys.* **2021**, *25*, 104244. [CrossRef]
9. Cruz, J.; Kawasaki, M.; Gorski, W. Electrode Coatings Based on Chitosan Scaffolds. *Anal. Chem.* **2000**, *72*, 680–686. [CrossRef]
10. Hoti, G.; Matencio, A.; Rubin Pedrazzo, A.; Cecone, C.; Appleton, S.L.; Khazaei Monfared, Y.; Caldera, F.; Trotta, F. Nutraceutical Concepts and Dextrin-Based Delivery Systems. *Int. J. Mol. Sci.* **2022**, *23*, 4102. [CrossRef]
11. Abdulwahid, R.T.; Aziz, S.B.; Kadir, M.F.Z. Replacing synthetic polymer electrolytes in energy storage with flexible biodegradable alternatives: Sustainable green biopolymer blend electrolyte for supercapacitor device. *Mater. Today Sustain.* **2023**, *23*, 100472. [CrossRef]
12. Aziz, S.; Hamsan, M.H.; Nofal, M.M.; Karim, W.O.; Brevik, I.; Brza, M.A.; Abdulwahid, R.T.; Al-Zangana, S.; Kadir, M.F.Z. Structural, Impedance and Electrochemical Characteristics of Electrical Double Layer Capacitor Devices Based on Chitosan: Dextran Biopolymer Blend Electrolytes. *Polymers* **2020**, *12*, 1411. [CrossRef]
13. Aziz, S.B.; Ali, F.; Anuar, H.; Ahamad, T.; Kareem, W.O.; Brza, M.A.; Kadir, M.F.Z.; Abu Ali, O.A.; Saleh, D.I.; Asnawi, A.S.F.M.; et al. Structural and electrochemical studies of proton conducting biopolymer blend electrolytes based on MC: Dextran for EDLC device application with high energy density. *Alex. Eng. J.* **2022**, *61*, 3985–3997. [CrossRef]
14. Raphael, E.; Avellaneda, C.O.; Manzolli, B.; Pawlicka, A. Agar-based films for application as polymer electrolytes. *Electrochim. Acta* **2010**, *55*, 1455–1459. [CrossRef]
15. Zhang, F.; Liu, T.; Zhang, J.; Cui, E.; Yue, L.; Jiang, R.; Hou, G. The potassium hydroxide-urea synergy in improving the capacitive energy-storage performance of agar-derived carbon aerogels. *Carbon* **2019**, *147*, 451–459. [CrossRef]
16. Espinoza-Acosta, J.L.; Torres-Chávez, P.I.; Olmedo-Martínez, J.L.; Vega-Rios, A.; Flores-Gallardo, S.; Zaragoza-Contreras, E.A. Lignin in storage and renewable energy applications: A review. *J. Energy Chem.* **2018**, *27*, 1422–1438. [CrossRef]
17. Ajjan, F.N.; Casado, N.; Rębiś, T.; Elfwing, A.; Solin, N.; Mecerreyes, D.; Inganäs, O. High performance PEDOT/lignin biopolymer composites for electrochemical supercapacitors. *J. Mater. Chem. A* **2016**, *4*, 1838–1847. [CrossRef]
18. Aziz, S.B.; Brza, M.A.; Hamsan, M.H.; Kadir, M.F.Z.; Muzakir, S.K.; Abdulwahid, R.T. Effect of ohmic-drop on electrochemical performance of EDLC fabricated from PVA:dextran:NH4I based polymer blend electrolytes. *J. Mater. Res. Technol.* **2020**, *9*, 3734–3745. [CrossRef]
19. Zainuddin, N.K.; Rasali, N.M.J.; Mazuki, N.F.; Saadiah, M.A.; Samsudin, A.S. Investigation on favourable ionic conduction based on CMC-K carrageenan proton conducting hybrid solid bio-polymer electrolytes for applications in EDLC. *Int. J. Hydrogen Energy* **2020**, *45*, 8727–8741. [CrossRef]
20. Aziz, S.B.; Hamsan, M.H.; Nofal, M.M.; San, S.; Abdulwahid, R.T.; Raza Saeed, S.R.; Brza, M.A.; Kadir, M.F.Z.; Mohammed, S.J.; Al-Zangana, S. From Cellulose, Shrimp and Crab Shells to Energy Storage EDLC Cells: The Study of Structural and Electrochemical Properties of Proton Conducting Chitosan-Based Biopolymer Blend Electrolytes. *Polymers* **2020**, *12*, 1526. [CrossRef]

21. Aziz, S.B.; Hamsan, M.H.; Abdullah, R.M.; Abdulwahid, R.T.; Brza, M.A.; Marif, A.S.; Kadir, M.F.Z. Protonic EDLC cell based on chitosan (CS): Methylcellulose (MC) solid polymer blend electrolytes. *Ionics* **2020**, *26*, 1829–1840. [CrossRef]
22. Teoh, K.H.; Lim, C.-S.; Liew, C.-W.; Ramesh, S.; Ramesh, S. Electric double-layer capacitors with corn starch-based biopolymer electrolytes incorporating silica as filler. *Ionics* **2015**, *21*, 2061–2068. [CrossRef]
23. Lizundia, E.; Kundu, D. Advances in Natural Biopolymer-Based Electrolytes and Separators for Battery Applications. *Adv. Funct. Mater.* **2021**, *31*, 2005646. [CrossRef]
24. Zhang, T.-W.; Tian, T.; Shen, B.; Song, Y.-H.; Yao, H.-B. Recent advances on biopolymer fiber based membranes for lithium-ion battery separators. *Compos. Commun.* **2019**, *14*, 7–14. [CrossRef]
25. Uddin, M.-J.; Alaboina, P.K.; Zhang, L.; Cho, S.-J. A low-cost, environment-friendly lignin-polyvinyl alcohol nanofiber separator using a water-based method for safer and faster lithium-ion batteries. *Mater. Sci. Eng. B* **2017**, *223*, 84–90. [CrossRef]
26. Aziz, S.B.; Hadi, J.M.; Dannoun, E.M.A.; Abdulwahid, R.T.; Saeed, S.R.; Shahab Marf, A.; Karim, W.O.; Kadir, M.F.Z. The Study of Plasticized Amorphous Biopolymer Blend Electrolytes Based on Polyvinyl Alcohol (PVA): Chitosan with High Ion Conductivity for Energy Storage Electrical Double-Layer Capacitors (EDLC) Device Application. *Polymers* **2020**, *12*, 1938. [CrossRef] [PubMed]
27. Spina, G.E.; Poli, F.; Brilloni, A.; Marchese, D.; Soavi, F. Natural Polymers for Green Supercapacitors. *Energies* **2020**, *13*, 3115. [CrossRef]
28. Kasprzak, D.; Stępniak, I.; Galiński, M. Electrodes and hydrogel electrolytes based on cellulose: Fabrication and characterization as EDLC components. *J. Solid State Electrochem.* **2018**, *22*, 3035–3047. [CrossRef]
29. Jeżowski, P.; Kowalczewski, P.Ł. Starch as a Green Binder for the Formulation of Conducting Glue in Supercapacitors. *Polymers* **2019**, *11*, 1648. [CrossRef]
30. Menzel, J.; Frąckowiak, E.; Fic, K. Agar-based aqueous electrolytes for electrochemical capacitors with reduced self-discharge. *Electrochim. Acta* **2020**, *332*, 135435. [CrossRef]
31. Wang, Q.; Cao, Q.; Wang, X.; Jing, B.; Kuang, H.; Zhou, L. A high-capacity carbon prepared from renewable chicken feather biopolymer for supercapacitors. *J. Power Sources* **2013**, *225*, 101–107. [CrossRef]
32. Suzanowicz, A.M.; Lee, Y.; Schultz, A.; Marques, O.J.J.; Lin, H.; Segre, C.U.; Mandal, B.K. Synthesis and Electrochemical Properties of Lignin-Derived High Surface Area Carbons. *Surfaces* **2022**, *5*, 265–279. [CrossRef]
33. Islam, M.A.; Ong, H.L.; Villagracia, A.R.; Halim, K.A.A.; Ganganboina, A.B.; Doong, R.-A. Biomass–derived cellulose nanofibrils membrane from rice straw as sustainable separator for high performance supercapacitor. *Ind. Crops Prod.* **2021**, *170*, 113694. [CrossRef]
34. Nowacki, K.; Galiński, M.; Stępniak, I. Synthesis and Characterization of Chitosan/Sodium Alginate Blend Membrane for Application in an Electrochemical Capacitor. *Prog. Chem. Appl. Chitin Deriv.* **2020**, *XXV*, 174–191. [CrossRef]
35. Nowacki, K.; Galiński, M.; Stępniak, I. Synthesis and characterization of modified chitosan membranes for applications in electrochemical capacitor. *Electrochim. Acta* **2019**, *320*, 134632. [CrossRef]
36. Wysokowski, M.; Nowacki, K.; Jaworski, F.; Niemczak, M.; Bartczak, P.; Sandomierski, M.; Piasecki, A.; Galiński, M.; Jesionowski, T. Ionic liquid-assisted synthesis of chitin–ethylene glycol hydrogels as electrolyte membranes for sustainable electrochemical capacitors. *Sci. Rep.* **2022**, *12*, 8861. [CrossRef]
37. Yang, H.; Liu, Y.; Kong, L.; Kang, L.; Ran, F. Biopolymer-based carboxylated chitosan hydrogel film crosslinked by HCl as gel polymer electrolyte for all-solid-sate supercapacitors. *J. Power Sources* **2019**, *426*, 47–54. [CrossRef]
38. Kasturi, P.R.; Ramasamy, H.; Meyrick, D.; Sung Lee, Y.; Kalai Selvan, R. Preparation of starch-based porous carbon electrode and biopolymer electrolyte for all solid-state electric double layer capacitor. *J. Colloid Interface Sci.* **2019**, *554*, 142–156. [CrossRef]
39. Skunik-Nuckowska, M.; Wisińska, N.H.; Garbacz, P.; Dyjak, S.; Wieczorek, W.; Kulesza, P.J. Polysaccharide-Based Hydrogel Electrolytes Enriched with Poly(Norepinephrine) for Sustainable Aqueous Electrochemical Capacitors. *SSRN Electron. J.* **2022**, *11*, 109346. [CrossRef]
40. Konwar, S.; Singh, D.; Strzałkowski, K.; Masri, M.N.B.; Yahya, M.Z.A.; Diantoro, M.; Savilov, S.V.; Singh, P.K. Stable and Efficient Dye-Sensitized Solar Cells and Supercapacitors Developed Using Ionic-Liquid-Doped Biopolymer Electrolytes. *Molecules* **2023**, *28*, 5099. [CrossRef]
41. Asnawi, A.S.F.M.; Hamsan, M.H.; Aziz, S.B.; Kadir, M.F.Z.; Matmin, J.; Yusof, Y.M. Impregnation of [Emim]Br ionic liquid as plasticizer in biopolymer electrolytes for EDLC application. *Electrochim. Acta* **2021**, *375*, 137923. [CrossRef]
42. Sun, P.; Chen, J.; Huang, Y.; Tian, J.-H.; Li, S.; Wang, G.; Zhang, Q.; Tian, Z.; Zhang, L. High-Strength agarose gel electrolyte enables long-endurance wearable Al-air batteries with greatly suppressed self-corrosion. *Energy Storage Mater.* **2021**, *34*, 427–435. [CrossRef]
43. Scardecchia, S.; Vita, A.; Santulli, C.; Forcellese, A. A material proposed for re-use of hemp shives as a waste from fiber production. *Mater. Today Proc.* **2020**, *31*, S213–S216. [CrossRef]
44. Nazeer, R.A.; Joshi, I.; Mahankali, S.; Mazumdar, S.; Sridharan, B.; Sankar, S.J. Is Marine Waste a Boon or Bane? An Insight on Its Source, Production, Disposal Consequences, and Utilization. In *Applied Biotechnology for Emerging Pollutants Remediation and Energy Conversion*; Springer Nature: Singapore, 2023; pp. 231–250.
45. Schellmann, N.C. Animal glues: A review of their key properties relevant to conservation. *Stud. Conserv.* **2007**, *52*, 55–66. [CrossRef]
46. Li, L.; Lu, F.; Wang, C.; Zhang, F.; Liang, W.; Kuga, S.; Dong, Z.; Zhao, Y.; Huang, Y.; Wu, M. Flexible double-cross-linked cellulose-based hydrogel and aerogel membrane for supercapacitor separator. *J. Mater. Chem. A* **2018**, *6*, 24468–24478. [CrossRef]

47. Weingarth, D.; Noh, H.; Foelske-Schmitz, A.; Wokaun, A.; Kötz, R. A reliable determination method of stability limits for electrochemical double layer capacitors. *Electrochim. Acta* **2013**, *103*, 119–124. [CrossRef]
48. Zhao, J.; Burke, A.F. Electrochemical Capacitors: Performance Metrics and Evaluation by Testing and Analysis. *Adv. Energy Mater.* **2021**, *11*, 2002192. [CrossRef]
49. Połomska, M.; Kubisz, L.; Wolak, J.; Hojan-Jezierska, D. Effects of Temperature on the FT NIR Raman Spectra of Fish Skin Collagen. *Appl. Sci.* **2021**, *11*, 8358. [CrossRef]
50. Senadheera, T.R.L.; Dave, D.; Shahidi, F. Sea Cucumber Derived Type I Collagen: A Comprehensive Review. *Mar. Drugs* **2020**, *18*, 471. [CrossRef]
51. Wu, T.-H.; Hsu, C.-T.; Hu, C.-C.; Hardwick, L.J. Important parameters affecting the cell voltage of aqueous electrical double-layer capacitors. *J. Power Sources* **2013**, *242*, 289–298. [CrossRef]

Disclaimer/Publisher's Note: The statements, opinions and data contained in all publications are solely those of the individual author(s) and contributor(s) and not of MDPI and/or the editor(s). MDPI and/or the editor(s) disclaim responsibility for any injury to people or property resulting from any ideas, methods, instructions or products referred to in the content.

Communication

Redox-Active Ferrocene Polymer for Electrode-Active Materials: Step-by-Step Synthesis on Gold Electrode Using Automatic Sequential Polymerization Equipment

Hao-Xuan Guo [1,*], Yuriko Takemura [1], Daisuke Tange [1], Junichi Kurata [2] and Hiroyuki Aota [1,*]

[1] Department of Chemistry and Materials Engineering, Kansai University, Suita 564-8680, Osaka, Japan; toatiatua@gmail.com (Y.T.); k170289@kansai-u.ac.jp (D.T.)
[2] Department of Mechanical Engineering, Kansai University, Suita 564-8680, Osaka, Japan; kurata@oram.co.jp
* Correspondence: hx-guo@kansai-u.ac.jp (H.-X.G.); aota@kansai-u.ac.jp (H.A.); Tel.: +81-06-6368-0831 (H.-X.G.)

Abstract: Redox-active polymers have garnered significant attention as promising materials for redox capacitors, which are energy-storage devices that rely on reversible redox reactions to store and deliver electrical energy. Our focus was on optimizing the electrochemical performance in the design and synthesis of redox-active polymer electrodes. In this study, a redox-active polymer was prepared through step-by-step synthesis on a gold electrode. To achieve this, we designed an automatic sequential polymerization equipment that minimizes human intervention and enables a stepwise polymerization reaction. The electrochemical properties of the polymer gold electrodes were investigated. The degree of polymerization of the polymer grown on the gold electrode can be controlled by adjusting the cycle of the sequential operation. As the number of cycles increases, the amount of accumulated charge increases proportionally, indicating the potential for enhanced electrochemical performance.

Keywords: redox capacitor; redox-active polymer; gold electrode

Citation: Guo, H.-X.; Takemura, Y.; Tange, D.; Kurata, J.; Aota, H. Redox-Active Ferrocene Polymer for Electrode-Active Materials: Step-by-Step Synthesis on Gold Electrode Using Automatic Sequential Polymerization Equipment. *Polymers* **2023**, *15*, 3517. https://doi.org/10.3390/polym15173517

Academic Editors: Vineet Kumar and Md Najib Alam

Received: 2 August 2023
Revised: 20 August 2023
Accepted: 22 August 2023
Published: 23 August 2023

Copyright: © 2023 by the authors. Licensee MDPI, Basel, Switzerland. This article is an open access article distributed under the terms and conditions of the Creative Commons Attribution (CC BY) license (https://creativecommons.org/licenses/by/4.0/).

1. Introduction

Redox-active polymers have emerged as promising materials for redox capacitors, which are energy-storage devices that rely on reversible redox reactions to store and deliver electrical energy [1–13]. Redox-active polymers, which offer high energy density, serve as ideal electrode materials because of their high charge storage capacity, good cycling stability, and ability to undergo reversible redox reactions. These polymers typically comprise conjugated [3,4] or non-conjugated [5,6] main or side chains that incorporate redox-active moieties, such as transition metal complexes [7,8], organic radicals [9–11], or quinones [12,13], which are capable of undergoing reversible oxidation and reduction processes.

The design and synthesis of redox-active polymer electrodes are focused on optimizing their electrochemical performance. Several parameters, including redox-active units, polymer structure, molecular weight, and design of the molecular-level electrode surface, were carefully optimized to enhance the charge storage capacity, stability, and conductivity of the polymer electrode [14–18]. Emphasizing the design of the molecular-level electrode surface is crucial because it plays a vital role in optimizing polymer morphology and electrode configuration, thereby facilitating efficient charge transport and minimizing energy loss within the capacitor system. Although dip- and spin-coating methods are commonly employed for fabricating polymer electrodes, concerns have emerged regarding polymer orientation and electrode stability. However, these concerns can be effectively addressed by utilizing the Self-Assembled Monolayer (SAM) [19], which involves bonding gold and thiophene to prepare electrodes. SAM provides a solution for achieving a desirable polymer orientation and enhancing the bonding between the polymer and electrode [20–23].

Our primary focus was the electrode design. Our approach aims to achieve the sequential reaction of monomers on the electrode surface instead of simply attaching the polymer to the electrode. This approach offers several advantages, including the suppression of polydispersity and ease of controlling the degree of polymerization. By adjusting the degree of polymerization, precise control over the maximum charge storage capacity can be achieved. Additionally, this method allows for the incorporation of various functionalities into the molecule by modifying the aldehyde group.

Previously, we reported a novel pseudo-living polymerization technique involving 1-methylpyrrole (MePyr) and aldehydes. This technique allows for a step-by-step reaction starting from the terminal monomer and facilitates the synthesis of polymers with a precisely controlled structure [24,25]. In this study, our objective was to polymerize a redox-active polymer on a gold electrode using MePyr and ferrocenecarboxaldehyde (FcA). To achieve this, we designed an automatic sequential polymerization equipment that minimizes human intervention and enables a stepwise polymerization reaction. Herein, we demonstrate the synthesis of a redox-active polymer on a gold electrode using an automatic sequential polymerization approach. Adjusting the cycle of the sequential operation allows for precision control over the degree of polymerization of the polymer grown on the gold electrode. Furthermore, as the number of cycles increases, the amount of accumulated charge increases proportionally, indicating the potential for enhanced electrochemical performance.

2. Materials and Methods

2.1. Materials

1-Methylpyrrole (MePyr), and 3-thiophenemethanol were purchased from Tokyo Chemical Industry (Tokyo, Japan). Ferrocenecarboxaldehyde (FcA), *p*-toluenesulfonic acid monohydrate (*p*-TS), tetrabutylammonium hexafluorophosphate (TBuAPF$_6$), ferrocene (Fc), tetrahydrofuran (THF), dimethyl sulfoxide (DMSO), and propylene carbonate were purchased from FU-JIFILM Wako Pure Chemical Corporation (Osaka, Japan). The MePyr monomer was purified via distillation prior to its use.

2.2. Measurements

Electrochemical measurements were conducted using a three-electrode system. The ALS660B electrochemical analyzer was employed for the measurements. ALS660B electrochemical analyzer, Pt working electrode, and Ag/AgCl reference electrode were purchased from BAS corporation (Tokyo, Japan).

2.3. Synthesis

2.3.1. Automatic Sequential Polymerization Equipment

The automatic sequential polymerization equipment (Figure 1) used in this study was specifically designed by J. Kurata. This equipment was developed to enable precise and efficient step-by-step polymerization reactions on an electrode surface. By minimizing human intervention and integrating automated control, our equipment allows for the systematic synthesis of redox-active polymers with enhanced properties. In addition, the immersion and washing times in the experimental procedure could be adjusted according to the specific requirements of the synthesis. By modifying these parameters, we optimized the reaction conditions and controlled the growth and attachment of the redox-active polymer to the gold electrode.

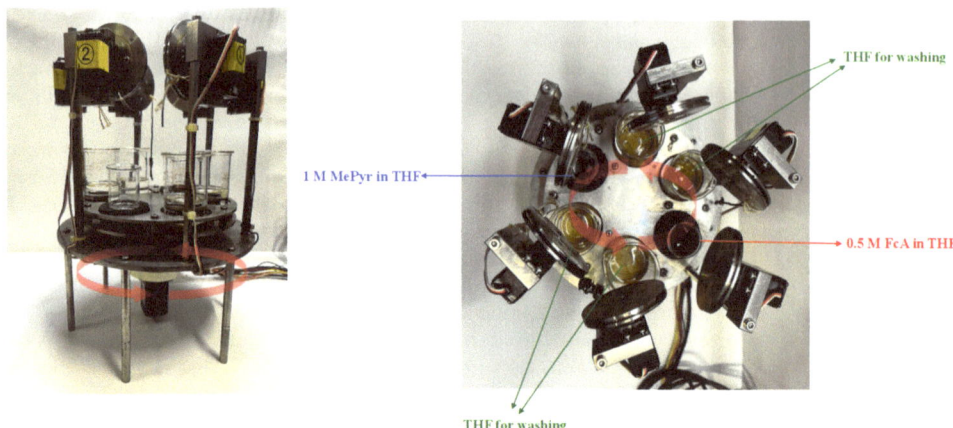

Figure 1. Image of automatic serialization equipment.

2.3.2. Preparation of Polymer Electrodes

The preparation of the gold electrode involved several steps. Firstly, a 1.0 cm × 2.0 cm Indium Tin Oxide (ITO) glass substrate was thoroughly cleaned using an ultrasonic cleaning process. The substrates were sequentially immersed in water, methanol, and acetone for 10 min each. Following the cleaning process, a thin film of gold (100 nm) was deposited onto the substrate using vacuum vapor deposition. Subsequently, the electrode was annealed to enhance the crystallinity of the gold surface and ensure flatness.

Preparation of polymer electrodes using automatic sequential polymerization equipment (Figure 1): To prepare Solution 0, 0.0086 g of 3-Thiophenemethanol, 0.122 g of MePyr, and 0.014 g of p-TS were dissolved in 0.3 g of THF. The resulting solution was allowed to react at 60 °C for 24 h. After the reaction, the solution was diluted with DMSO to obtain a 1 mM solution. The gold electrodes were then immersed in the solution for 24 h. After immersion, the gold electrodes were washed with DMSO and THF. To prepare Solution 1, FcA (0.750 mmol, 0.161 g), p-TS (0.610 mmol, 0.116 g), and THF (1.5 mL) were mixed. The gold electrode was placed in Solution 1 and allowed to react for 30 s, followed by washing with THF. To obtain Solution 2, MePyr (1.5 mmol, 0.122 g), p-TS (0.075 mmol, 0.014 g), and THF (1.5 mL) were mixed together. The gold electrode was placed in Solution 2 and allowed to react for 20 s. Alternating between Solution 1 and Solution 2, this process was repeated for 20, 40, and 80 cycles.

Important note: It is important to use freshly prepared MePyr, FcA and THF washing solutions for each set of 20 cycles when preparing modified gold electrodes.

3. Results and Discussion

Redox-active polymers were synthesized via a step-by-step approach, involving the sequential synthesis of MePyr and FcA on gold electrodes using an automatic sequential polymerization apparatus. The redox behavior of ferrocene, which is characterized by the reversible transfer of electrons between its iron centers, makes it an excellent building block for the construction of electrochemically active materials [26]. Our goal is to systematically synthesize and characterize a redox-active ferrocene polymer on a gold electrode to explore its potential as an electrode material for various applications. To facilitate unidirectional growth of the polymer from the gold electrodes, 3-thiophenemethanol was chosen. The structures of the polymers are shown in Figure 2a. The step-by-step synthesis followed a specific order: Step (0) involved synthesizing the terminal groups of 3-thiophenemethanol and methylpyrrole, followed by binding these terminal groups to the gold electrodes using SAM. Step (1) entailed the reaction with FcA. Step (2) and Step (3) involved washing with THF. Step (4) entailed the reaction with MePyr. Step (5) and Step (6) included washing with THF. The sequence from Step (1) to Step (6) constituted one cycle (further details are provided in Section 2.3.2; a video demonstrating one cycle of the step-by-step synthesis process is included in Supplementary Materials). We performed three sets of syntheses with 20, 40, and 80 cycles. The preparation process for each set is illustrated in Figure 2b.

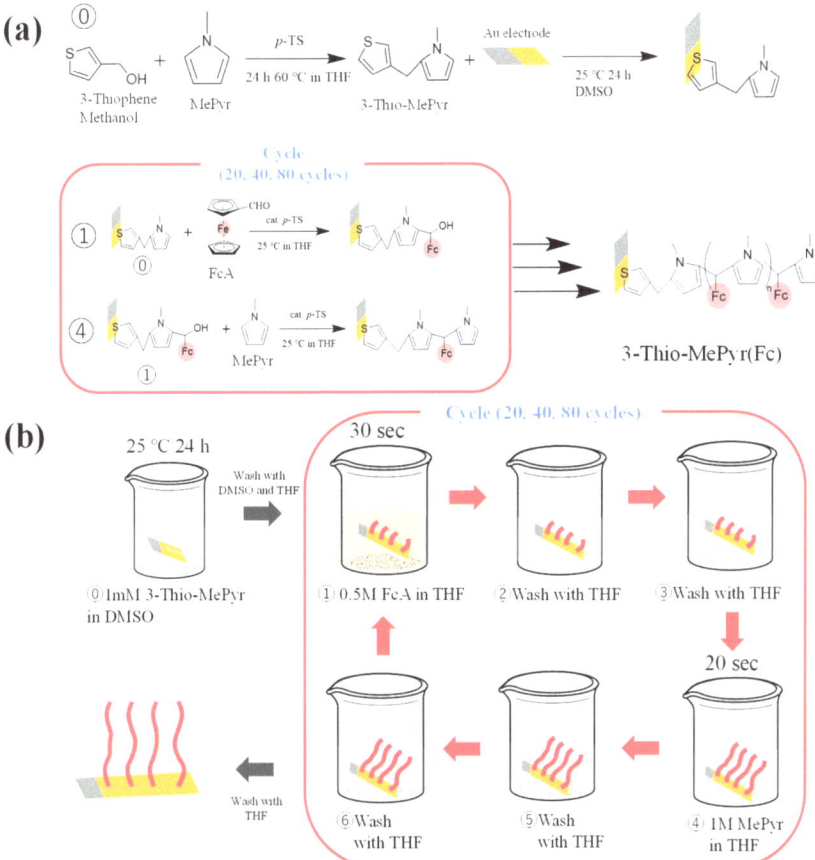

Figure 2. (a) Synthesis of redox-active polymer on a gold electrode using automatic sequential polymerization equipment; (b) preparation of polymer gold electrodes using automatic sequential polymerization equipment.

The electrochemical measurements were conducted using a three-electrode system, with the polymer gold electrode serving as the working electrode, Ag/AgCl as the reference electrode, and a Pt wire as the counter electrode. Cyclic voltammograms of the polymer gold electrodes were recorded in a 0.1 M TBuAPF$_6$ propylene carbonate solution, and the results are presented in Figure 3 and Table 1. The redox waves originating from ferrocene were detected in all polymer gold electrodes. Conversely, sequential reactions carried out in the absence of thiophene did not lead to the synthesis of polymers on the gold surfaces. This finding suggests that the successful execution of sequential reactions on the gold electrode was facilitated by the SAM, as depicted in Figure 3a, highlighting the importance of incorporating thiophene in the process.

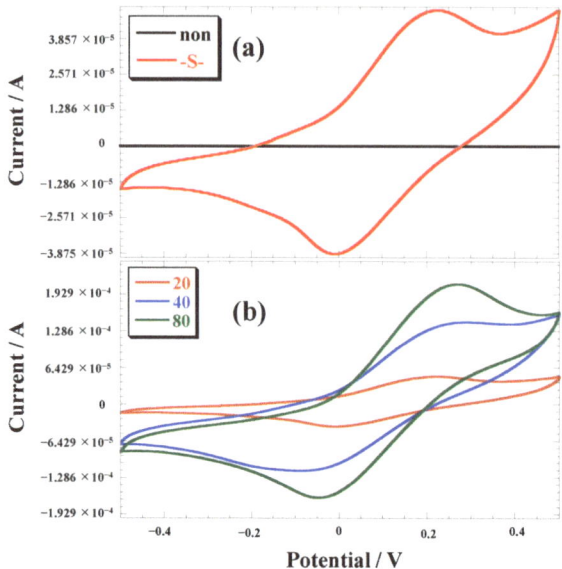

Figure 3. Cyclic voltammograms of polymer gold electrodes in 0.1 M TBuAPF$_6$ propylene carbonate solution, (**a**) polymer synthesis on the gold electrode with (-S-) and without (non) thiophene, (**b**) polymer gold electrodes of syntheses with 20, 40, and 80 cycles, reference electrode: Ag/Ag$^+$, at a sweep rate of 50 mV/s.

Table 1. Electrochemical properties of samples.

Sample	Redox Potential/V	Peak to Peak/V	Charges Recorded at 1 s/$\times 10^{-4}$ C
20	0.11	0.231	2.17
40	0.107	0.361	4.78
80	0.111	0.313	7.14
Fc	0.035	0.430	-

The current value progressively increased with an increasing number of cycles, indicating gradual polymerization of the polymer from the electrode surface (Figure 3b). Notably, the width of the oxidation and reduction peaks in the redox wave of the polymer electrode was narrower than that of ferrocene in solution (Figure 4; CV measurement was performed on the ferrocene solution utilizing a gold electrode with the same dimensions of 1.0 cm × 1.0 cm as the gold electrode used for the polymerization process). This can be attributed to the binding of the ferrocene-based polymer to the electrode through the SAM, resulting in a more confined electrochemical environment. The cyclic voltammograms of the species adsorbed on electrodes typically exhibit sharp peaks. However, in this particular system, the redox wave of the adsorption system did not exhibit sharp peaks. This deviation from expected behavior can be explained by the redox reaction of the Fc groups in the polymer, which involves the incorporation of counterions that diffuse through the polymer to maintain electrical neutrality. It is likely that the current observed in the experiment was influenced by the diffusion of these counterions.

Figure 5 illustrates the electrode charge transfer for each of these polymer electrodes, as measured during the 1 s chronocoulometric measurements. The corresponding charges recorded at 1 s are listed in Table 1. The amount of charge accumulated on the polymer electrodes produced by different cycles was found to vary. In addition, consistent with the CV results, the amount of accumulated charge increased with the number of cycles. This observation strongly suggests that the polymer grows on the gold electrode as the number of cycles increases.

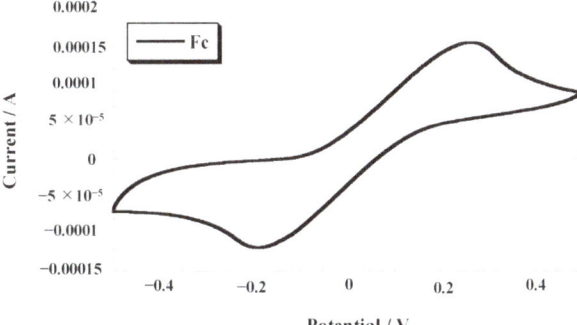

Figure 4. Cyclic voltammogram of Fc in 0.1 M TBuAPF$_6$ propylene carbonate solution, reference electrode: Ag/Ag$^+$ at a sweep rate of 50 mV/s.

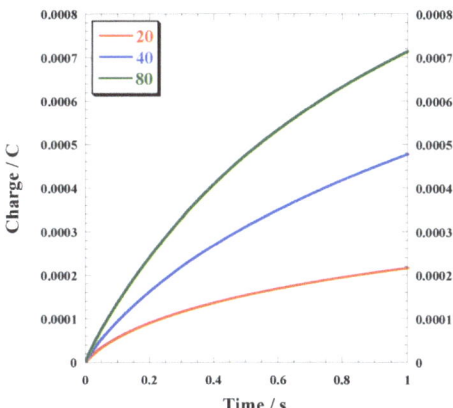

Figure 5. Chronocoulometry of polymer gold electrodes in 0.1 M TBuAPF$_6$ propylene carbonate solution at 0.2 V taken for 1 s.

4. Conclusions

In conclusion, we have successfully demonstrated the synthesis of a redox-active polymer on a gold electrode using a step-by-step approach, employing an automatic sequential polymerization apparatus. By controlling the cycle of the sequential operation, we achieved precise control over the degree of polymerization of the polymer grown on the gold electrode. Furthermore, our observations revealed a direct correlation between the number of cycles and amount of accumulated charge, indicating the potential for further optimization and enhancement of the electrochemical properties of the polymer. These findings highlight the potential of redox-active polymers synthesized on gold electrodes for high-performance energy storage applications. Our ongoing research focuses on the development of redox capacitors with increased energy densities, achieved through improved sequential cycle numbers and electrode design.

Supplementary Materials: The following supporting information can be downloaded at https://www.mdpi.com/article/10.3390/polym15173517/s1. Video S1: One cycle of the step-by-step synthesis process.

Author Contributions: Conceptualization, H.-X.G. and H.A.; Investigation, H.-X.G., D.T. and Y.T.; Equipment design, J.K.; Project administration, H.A.; Writing—original draft, H.-X.G. All authors have read and agreed to the published version of the manuscript.

Funding: This research received no external funding.

Institutional Review Board Statement: Not applicable.

Data Availability Statement: Not applicable.

Acknowledgments: We thank M. Ishikawa (Kansai University) for his significant advice. We also thank T. Yanai for his help during the early part of this study.

Conflicts of Interest: The authors declare no conflict of interest.

References

1. Nishide, H.; Oyaizu, K. Toward Flexible Batteries. *Science* **2008**, *319*, 737–738. [CrossRef] [PubMed]
2. Song, Z.-P.; Zhou, H.-S. Towards sustainable and versatile energy storage devices: An overview of organic electrode materials. *Energy Environ. Sci.* **2013**, *6*, 2280–2301. [CrossRef]
3. Akerlund, L.; Emanuelsson, R.; Renault, S.; Huang, D.; Brandell, D.; Stromme, M.; Sjodin, M. The Proton Trap Technology—Toward High Potential Quinone-Based Organic Energy Storage. *Adv. Energy Mater.* **2017**, *7*, 1700259. [CrossRef]
4. Emanuelsson, R.; Sterby, M.; Stromme, M.; Sjodin, M. An All-Organic Proton Battery. *J. Am. Chem. Soc.* **2017**, *139*, 4828–4834. [CrossRef] [PubMed]
5. Oyaizu, K.; Nishide, H. Radical Polymers for Organic Electronic Devices: A Radical Departure from Conjugated Polymers? *Adv. Mater.* **2009**, *21*, 2339–2344. [CrossRef]
6. Sato, K.; Ichinoi, R.; Mizukami, R.; Serikawa, T.; Sasaki, Y.; Lutkenhaus, J.; Nishide, H.; Oyaizu, K. Diffusion-Cooperative Model for Charge Transport by Redox-Active Nonconjugated Polymers. *J. Am. Chem. Soc.* **2018**, *140*, 1049–1056. [CrossRef] [PubMed]
7. Tamura, K.; Akutagawa, N.; Satoh, M.; Wada, J.; Masuda, T. Charge/Discharge Properties of Organometallic Batteries Fabricated with Ferrocene–Containing Polymers. *Macromol. Rapid Commun.* **2008**, *29*, 1944–1949. [CrossRef]
8. Mondal, S.; Yoshida, T.; Maji, S.; Ariga, K.; Higuchi, M. Transparent Supercapacitor Display with Redox-Active Metallo-Supramolecular Polymer Films. *ACS Appl. Mater. Interfaces* **2020**, *12*, 16342–16349. [CrossRef]
9. Oyaizu, K.; Ando, Y.; Konishi, H.; Nishide, H. Nernstian Adsorbate-like Bulk Layer of Organic Radical Polymers for High-Density Charge Storage Purposes. *J. Am. Chem. Soc.* **2008**, *130*, 14459–14461. [CrossRef]
10. Janoschka, T.; Hager, M.D.; Schubert, U.S. Powering up the Future: Radical Polymers for Battery Applications. *Adv. Mater.* **2012**, *24*, 6397–6409. [CrossRef]
11. Karlsson, C.; Suga, T.; Nishide, H. Quantifying TEMPO Redox Polymer Charge Transport toward the Organic Radical Battery. *ACS Appl. Mater. Interfaces* **2017**, *9*, 10692–10698. [CrossRef] [PubMed]
12. Choi, W.; Harada, D.; Oyaizu, K.; Nishide, H. Aqueous Electrochemistry of Poly(vinylanthraquinone) for Anode-Active Materials in High-Density and Rechargeable Polymer/Air Batteries. *J. Am. Chem. Soc.* **2011**, *133*, 19839–19843. [CrossRef]
13. Oka, K.; Furukawa, S.; Murao, S.; Oka, T.; Nishide, H.; Oyaizu, K. Poly(dihydroxybenzoquinone): Its high-density and robust charge storage capability in rechargeable acidic polymer–air batteries. *Chem. Commun.* **2020**, *56*, 4055–4058. [CrossRef] [PubMed]
14. Yun, T.G.; Hwang, B.I.; Kim, D.; Hyun, S.; Han, S.M. Polypyrrole–MnO_2-Coated Textile-Based Flexible-Stretchable Supercapacitor with High Electrochemical and Mechanical Reliability. *ACS Appl. Mater. Interfaces* **2015**, *7*, 9228–9234. [CrossRef] [PubMed]
15. Kim, J.; Kim, J.H.; Ariga, K. Redox-Active Polymers for Energy Storage Nanoarchitectonics. *Joule* **2017**, *1*, 739–768. [CrossRef]
16. Wu, Z.-Z.; Liu, Q.-R.; Yang, P.; Chen, H.; Zhang, Q.-C.; Li, S.; Tang, Y.-B.; Zhang, S.-Q. Molecular and Morphological Engineering of Organic Electrode Materials for Electrochemical Energy Storage. *Electrochem. Energy Rev.* **2022**, *5*, 26. [CrossRef]
17. Barbosa, J.C.; Marijuan, A.F.; Dias, J.C.; Gonçalves, R.; Salado, M.; Costa, C.M.; Mendez, S.L. Molecular design of functional polymers for organic radical batteries. *Energy Storage Mater.* **2023**, *60*, 102841. [CrossRef]
18. Rehman, A.U.; Afzal, A.M.; Iqbal, M.W.; Ali, M.; Wabaidur, S.M.; Al-Ammar, E.A.; Mumtaz, S.; Choi, E.H. Highly efficient and stable layered AgZnS@WS2 nanocomposite electrode as superior charge transfer and active redox sites for energy harvesting device. *J. Energy Stor.* **2023**, *71*, 108022. [CrossRef]
19. Bain, C.D.; Whitesides, G.M. Molecular-Level Control over Surface Order in Self-Assembled Monolayer Films of Thiols on Gold. *Science* **1988**, *240*, 62–63. [CrossRef]
20. Laibinis, P.E.; Whitesides, G.M. Self-assembled monolayers of n-alkanethiolates on copper are barrier films that protect the metal against oxidation by air. *J. Am. Chem. Soc.* **1992**, *114*, 9022–9028. [CrossRef]
21. Schlenoff, J.B.; Li, M.; Ly, H. Stability and Self-Exchange in Alkanethiol Monolayers. *J. Am. Chem. Soc.* **1995**, *117*, 12528–12536. [CrossRef]
22. Murugan, P.; Krishnamurthy, M.; Jaisankar, S.N.; Samanta, D.; Mandal, A.B. Controlled decoration of the surface with macromolecules: Polymerization on a self-assembled monolayer (SAM). *Chem. Soc. Rev.* **2015**, *44*, 3212–3243. [CrossRef]
23. Yi, R.-W.; Mao, Y.-Y.; Shen, Y.-B.; Chen, L.-W. Self-Assembled Monolayers for Batteries. *J. Am. Chem. Soc.* **2021**, *143*, 12897–12912. [CrossRef]

24. Guo, H.-X.; Yoshida, Y.; Aota, H. Structure-controlled polymers prepared by pseudo-living addition-condensation polymerization and their application to light harvesting. *Chem. Commun.* **2016**, *52*, 11819–11822. [CrossRef]
25. Guo, H.-X.; Aota, H. Light-harvesting and Electron Transfer in a Structure-controlled Polymer for Artificial Photosynthetic Antenna-reaction Centers. *Chem. Lett.* **2017**, *46*, 191–193. [CrossRef]
26. Park, K.-S.; Schougaard, S.B.; Goodenough, J.B. Conducting-Polymer/Iron-Redox- Couple Composite Cathodes for Lithium Secondary Batteries. *Adv. Mater.* **2007**, *19*, 848–851. [CrossRef]

Disclaimer/Publisher's Note: The statements, opinions and data contained in all publications are solely those of the individual author(s) and contributor(s) and not of MDPI and/or the editor(s). MDPI and/or the editor(s) disclaim responsibility for any injury to people or property resulting from any ideas, methods, instructions or products referred to in the content.

Article

Nanocarbon Black and Molybdenum Disulfide Hybrid Filler System for the Enhancement of Fracture Toughness and Electromechanical Sensing Properties in the Silicone Rubber-Based Energy Harvester

Md Najib Alam, Vineet Kumar, Taemin Jeong and Sang-Shin Park *

School of Mechanical Engineering, Yeungnam University, 280, Daehak-ro, Gyeongsan 38541, Republic of Korea; mdnajib.alam3@gmail.com (M.N.A.); vineetfri@gmail.com (V.K.); jjtt00mm@naver.com (T.J.)
* Correspondence: pss@ynu.ac.kr

Citation: Alam, M.N.; Kumar, V.; Jeong, T.; Park, S.-S. Nanocarbon Black and Molybdenum Disulfide Hybrid Filler System for the Enhancement of Fracture Toughness and Electromechanical Sensing Properties in the Silicone Rubber-Based Energy Harvester. *Polymers* 2023, 15, 2189. https://doi.org/10.3390/polym15092189

Academic Editor: Marcin Masłowski

Received: 5 April 2023
Revised: 1 May 2023
Accepted: 3 May 2023
Published: 5 May 2023

Copyright: © 2023 by the authors. Licensee MDPI, Basel, Switzerland. This article is an open access article distributed under the terms and conditions of the Creative Commons Attribution (CC BY) license (https://creativecommons.org/licenses/by/4.0/).

Abstract: Recently, hybrid fillers have been found to be more advantageous in energy-harvesting composites. This study investigated the mechanical and electromechanical performances of silicone rubber-based composites made from hybrid fillers containing conductive nanocarbon black (NCB) and molybdenum disulfide (MoS_2). A hybrid filler system containing only 3 phr (per hundred grams of rubber) MoS_2 and 17 phr NCB provided higher fracture strain, better tensile strength, and excellent toughness values compared to the 20 phr NCB-only-filled and 5 phr MoS_2-only-filled rubber composites. The chemical cross-link densities suggest that NCB promoted the formation of cross-links, whereas MoS_2 slightly reduced the cross-link density. The higher mechanical properties in the hybrid filler systems suggest that the filler particles were more uniformly distributed, which was confirmed by the scanning electron microscope study. Uniformly distributed filler particles with moderate cross-link density in hybrid filler systems greatly improved the fracture strain and fracture toughness. For example, the hybrid filler with a 17:3 ratio of NCB to MoS_2 showed a 184% increment in fracture toughness, and a 93% increment in fracture strain, compared to the 20 phr NCB-only-filled composite. Regarding electromechanical sensing with 2 kPa of applied cyclic pressure, the hybrid filler (17:3 CB to MoS_2) performed significantly better (~100%) than the 20 phr NCB-only compound. This may have been due to the excellent distribution of conducting NCB networks and piezoelectric MoS_2 that caused symmetric charging–discharging in the toughened hybrid composite. Thus, hybrid composites with excellent fatigue resistance can find dynamic applications, such as in blood pressure measurement.

Keywords: silicone rubber; nanocarbon black; molybdenum disulfide; nanocomposites; energy harvester

1. Introduction

Recently, silicone rubber has become increasingly popular in various electronic, medical, and soft robotic applications, either directly or in combination with reinforcing fillers [1–6]. Silicone rubber is a highly dielectric material, and cannot transmit electricity. It exhibits a very high actuation value in the presence of electrical fields, considering its application in the field of soft robotics [7]. Inversely, it exhibits fluctuations in the electrical dipoles upon mechanical deformation, and converts mechanical energy to electrical energy. These properties make it valuable for applications in electromechanical pressure, strain, and other sensors [8,9].

To fabricate a soft and flexible electromechanical sensor, silicone rubber must be compounded with a conducting filler. Among various nanofillers, carbon black (CB), nano graphite (NG), and carbon nanotubes (CNTs) are the most suitable that can sufficiently improve the mechanical and electrical properties of silicone rubber [6]. CB is of particular interest because of its isotropic filler structure, excellent reinforcing properties, and good electrical properties. Owing to their isotropic shape, the composites are very stable, with fatigue in all directions; they are suitable for dynamic applications such as tires. Thus, electrodes made of CB in silicone rubber should maintain a uniform conductivity, which is necessary for electromechanical sensing.

Recently, instead of using a single filler, a hybrid filler was used to make rubber composites to reduce filler percolation and achieve synergistic mechanical and electrical properties [10]. The hybridization of different-dimensional fillers in rubber could be useful for improving filler dispersion, reducing filler percolation, and improving the stretchability of hybrid filler networks. These advantages make hybrid-filler-based rubber composites tougher than single-filler-based rubber composites. Furthermore, CB has recently been hybridized with silica [11], nano clay [12], graphene [13], and carbon nanotubes [14,15] in different polymers.

After the significant achievements of two-dimensional (2D) graphene in science and technology, many other 2D nanomaterials, such as hexagonal boron nitride, graphitic carbon nitride, and transition metal dichalcogenides (TMDs), have been investigated as alternatives to graphene [16,17]. Among the different TMDs, MoS_2 has gained significant interest because of its electronic properties that are complementary to those of graphene. Unlike graphene, MoS_2 has an intrinsic bandgap that enables its use in semiconducting electronic and optoelectronic applications [18,19]. Single-layered MoS_2 has an extraordinary fracture strength of ~23 GPa and an elastic modulus of 300 GPa, which are similar to those of chemically reduced graphene [20]. Moreover, MoS_2 is a piezoelectric material [21]. Therefore, MoS_2 can be used as a reinforcing filler. Different polymer composites [22–26], including rubber composites [27–31], have been fabricated using modified and unmodified MoS_2 particles for mechanical, microwave absorption, and tribology applications. Tang et al. [31] hybridized CB and 2D-MoS_2 particles to reinforce natural rubber, and found about a 50% enhancement in the tensile modulus from a single CB-loaded composite. Although 2D MoS_2 remarkably improves the properties of rubber composites, few studies have investigated silicone rubber composites for improving its toughness and its applications in electromechanical energy harvesting.

Reinforcing filler, with a high aspect ratio such as CNT, can simultaneously improve the electrical and mechanical moduli, but can significantly reduce the fracture strain [32]. Hence, a low aspect ratio with an electrically conductive filler could be useful to make a soft and stretchable energy harvester for low-energy sensing applications. This research aimed to improve the mechanical properties, especially the fracture strain and fracture toughness, of silicone rubber composites that can find suitable applications in energy harvesting devices for low mechanical energy sensitivity. Nano carbon black (NCB) is a very useful reinforcing filler for all types of rubber, owing to its considerable mechanical and electrical properties. NCB provides good electrical conductivity at 15–20 phr of filler amounts in silicone rubber [33]. In rubber-based flexible piezoresistive strain sensors, the sensitivity depends highly on the gauge factor. At low deformation, the gauge factor is significantly low. However, at higher deformation, the gauge factor could be enhanced, but the stability of the piezoresistive strain sensor gradually decreases. Ding et al. prepared flexible electrodes for energy harvesting devices, considering carbon black and polydimethylsiloxane rubber have low energy loss and high durability [34]. Due to the isotropic structures of the filler particles and good filler–polymer interactions, the fatigue life of the rubber composites may be very high [35]. Although the mechanical modulus at this filler level was sufficient, the fracture toughness could be improved by increasing the fracture strain value. Toughness and stretchability are two important factors for most stretchable electronic applications. Hence, the present study investigated the effects of two-dimensional MoS_2 particles on NCB-reinforced room-temperature-vulcanized (RTV) silicone rubber composites, for tough but stretchable energy harvesting applications. The mechanical properties and fatigue under cyclic deformation were studied to determine the importance of hybrid-filler-based rubber composites. The composites were tested as electrodes in a flexible-type capacitor such as an energy harvesting device. Owing to the effects on the mechanical and piezoelectric properties of different amounts of MoS_2, the hybrid composites can be synergized.

The synergistic effects were described, followed by filler distribution and filler–polymer interactions in hybrid filler systems.

2. Materials and Methods

2.1. Materials

The silicone rubber was condensation-cured one-component room-temperature vulcanized (RTV)-silicone rubber (SR, grade KE-441) purchased from Shin-Etsu Company, Tokyo, Japan. The catalyst (CAT-RM) for room-temperature vulcanization was provided by the same company. The RTV-thinner was purchased from Shin-Etsu Company, Tokyo, Japan. Molybdenum disulfide (particle size < 2 µm) was purchased from Sigma-Aldrich. Nanocarbon black (NCB, Conductex SC Ultra grade, electrical conductivity = 2.5 S/cm [36]) was purchased from Saehan Silichem Corporation, Ltd., Seoul, Republic of Korea. All materials were used as received.

2.2. Characterization of Fillers

Primary characterization of the fillers was performed using an X-ray diffractometer (XpertPro-PANanalytical-Diffractometer) with an X-ray wavelength for CuKα (0.154 nm), with scanning Bragg angles (2θ) from 10° to 80°. Field emission scanning electron microscopy (FE-SEM, S-4800, Hitachi, Japan) was used to investigate the morphology of the filler structures.

2.3. Preparation of Rubber Nanocomposites

The different amounts of materials used for fabricating the rubber nanocomposites are provided in Table 1. The required amounts of filler(s) (in phr) were added to 100 g of RTV silicone rubber and mechanically mixed using a stirrer for 10 min. After homogeneous mixing, 2 phr vulcanizing initiator was added and thoroughly mixed for 1 min. The compounded rubber was poured into a mold and maintained for 24 h under compressed conditions. The vulcanisates were stored in a refrigerator to protect them from further curing. Before measuring the chemical and mechanical properties, the samples were removed and maintained for 24 h at room temperature (~25 °C).

Table 1. Formulation table of the silicone rubber-based composites.

Formulation	RTV-SR (phr)	NCB (phr)	MoS$_2$ (phr)	Vulcanizing Agent (phr)
SR-unfilled	100	-	-	2
SR/20-NCB	100	20	-	2
SR/19-NCB/1-MoS$_2$	100	19	1	2
SR/17-NCB/3-MoS$_2$	100	17	3	2
SR/15-NCB/5-MoS$_2$	100	15	5	2
SR/5-MoS$_2$	100	-	5	2

2.4. Measurement of Cross-Link Density

The cross-link densities of the silicone rubber composites were measured using the equilibrium swelling method [37] after 7 days of immersion in toluene.

$$V_c = -\left[\ln(1 - V_r) + V_r + \chi V_r^2\right] / V_s V_r^{1/3}$$

where V_c is the cross-link density, V_r is the volume fraction of rubber in the swollen sample, $\chi = 0.465$ is the interaction parameter between toluene and silicone rubber [32], and $V_s = 106.2$ is the molar volume of toluene. The V_r values were calculated as follows:

$$V_r = (w_r/d_r)[(w_r/d_r) + (w_s/d_s)]$$

where w_r is the weight of the rubber, d_r is the density of the rubber, w_s is the weight of the swelled toluene, and d_s is the density of toluene.

2.5. Mechanical and Hysteresis Properties

The compressive and tensile mechanical properties were investigated using a universal testing machine (UTM, LLOYD, United Kingdom) with a 1 kN load cell. Cylindrical samples (h = 10 mm × d = 20 mm) were used for analysis of the compressive mechanical properties. For the tensile properties, dumbbell-shaped test specimens were used according to ISO 37, Type 2 (gauge length = 25 cm). The average of four tests was calculated for each reported value. For the hysteresis test, a cylindrical sample was used with 30% dynamic compressive strain over 100 cycles.

2.6. Filler Distribution Studies

The filler distribution was characterized using scanning electron microscopy (SEM). The level of filler dispersion was investigated, followed by an energy-dispersive spectroscopy (EDS) mapping technique fitted with the SEM instrument.

2.7. Fabrication of Energy Harvesting Device

Electrodes of 0.1 mm thickness were painted on both sides of the 1 mm thick unfilled elastomer slab, following the method described previously [38]. NCB (20 phr) or hybrid filler (20 phr of NCB: MoS_2 at 17:3 ratio) was used to fabricate electrode composites, in addition to 60 phr of thinner and 100 phr of RTV silicone rubber. The painted electrodes were then coated with protective silicone rubber layers.

3. Results and Discussion

3.1. Crystal Structure and Morphology of Filler

The characteristic X-ray diffraction (XRD) plots of MoS_2 and NCB are shown in Figure 1a,b, respectively. The different major peaks in Figure 1a with 2θ values of 14.402°, 32.631°, 33.459°, 35.833°, 39.510°, 44.233°, 49.781°, 55.992°, 58.236°, 60.33°, 72.831°, and 76.009° correspond to the 002, 100, 101, 102, 103, 104, 105, 106, 110, 008, 203, and 116 crystal planes, respectively, for the hexagonal crystal system of pure molybdenum disulfide according to the reference (JCPDS card no. 00-024-0513). Similarly, the different major peaks in Figure 1b with 2θ values of 24.029°, 43.731°, 44.533°, and 47.723° correspond to the 009, 104, 015, and 018 crystal planes, respectively, of the rhombohedral crystal system of NCB, according to the card reference (JCPDS card no. 01-074-2328). Figure 1c shows an SEM image of plate-like MoS_2 layers stacked together to form larger particles. The SEM image in Figure 1d shows spherical particles aggregated to form branched structures of NCB, which are responsible for electrical conductivity.

3.2. Curing Properties

The cross-link densities of the different rubber composites after 24 h of curing are shown in Figure 2. NCB enhanced the cross-link density, and MoS_2 slightly reduced the cross-link density, compared to unfilled rubber. The hybrid filler-containing vulcanisates showed cross-link densities that were between the NCB and MoS_2-only compounds. The surface chemistry of the filler particles may have been the reason for the different cross-linking densities in the different filler systems. The cross-link density in condensation-cured silicone rubber depends mainly on the curing catalyst and filler surface functionalities. Since MoS_2 particles are platelet-shaped without hydrolyzable functional groups, they may act as a barrier to moisture, which may influence condensation-cured silicone rubber with a lower cross-link density [39,40]. In contrast, NCB promotes the cross-linking of condensation-cured silicone rubber because of its hydrolyzable surface functional groups [41].

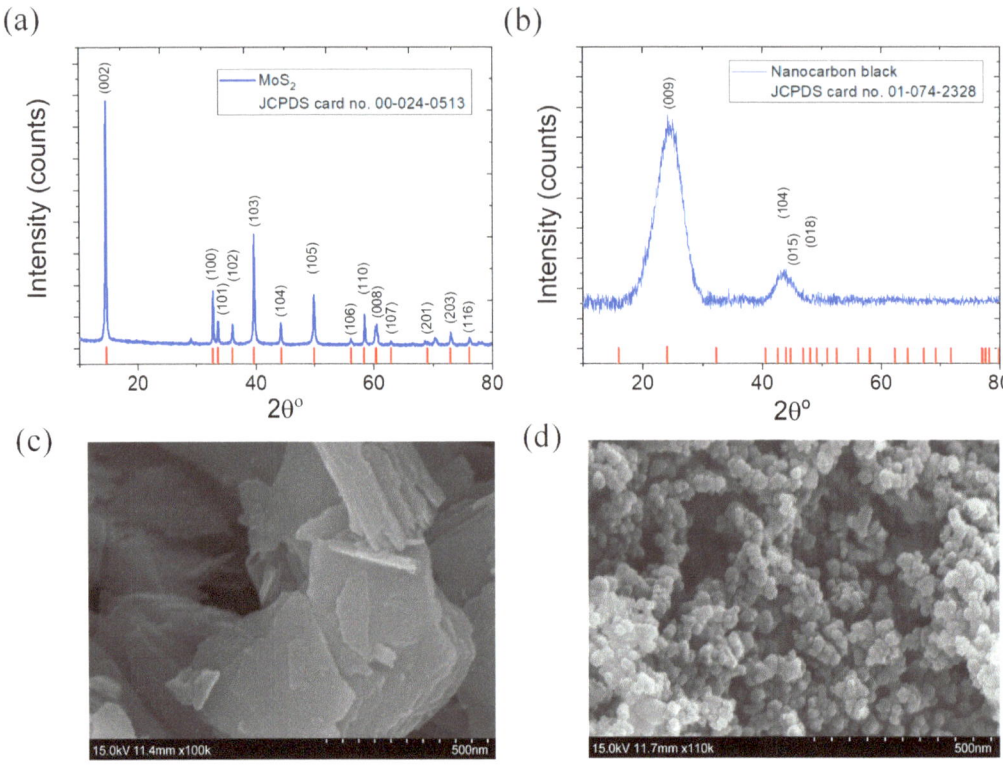

Figure 1. XRD plots of (**a**) MoS$_2$ and (**b**) nanocarbon black; SEM images of (**c**) MoS$_2$ and (**d**) nanocarbon black.

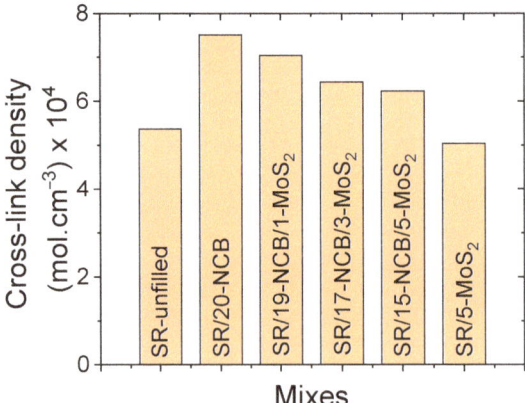

Figure 2. Cross-link densities of different vulcanizates.

3.3. Mechanical Properties of Rubber Nanocomposites

3.3.1. Compressive Mechanical Properties

The compressive stress–strain curves for the different rubber composites are shown in Figure 3a. From the different curves, it is evident that the compressive stress increased with slightly higher values after 15% of compressive strain. This could be due to the generation of filler percolation and polymer chain packing at higher compressive strains [42]. The addition of NCB significantly improved the compressive stress at 35% strain. Little improvement in the mechanical properties was observed with the addition of MoS_2, unlike in the case of the unfilled rubber. This could be due to the cure retardation of sulfur-based MoS_2 particles in silicone rubber [39,40]. The highest compressive stress at 35% of deformation was obtained for the SR/17-NCB/3-MoS_2 hybrid composite, which was much higher than that for the unfilled rubber. The elastic modulus values of the different composites are shown in Figure 3b. The enhanced compressive stress and elastic modulus of the SR/17-NCB/3-MoS_2 hybrid composite over those of the SR/20-NCB composite may indicate an improved filler distribution in the composite matrix [43]. The elastic modulus obtained in the compressive mode was much higher than the elastic modulus determined in the tensile mode in the present study. It is believed that the filler networks have a significant effect on the compressive modulus compared to the tensile modulus. In the compressive mode of detection, the filler particles became closer to agglomerate, which resulted in filler percolation, and improved the mechanical strength of the composite. On the other hand, filler particles became separated by the tensile strain that reduced the filler percolation, and filler networks showed slightly different behavior to the elastic modulus obtained via the tensile mode. The higher elastic modulus of SR/17-NCB/3-MoS_2 than that of SR/20-NCB indicates more homogeneously distributed filler particles that require higher energy for packing at a higher compressive strain.

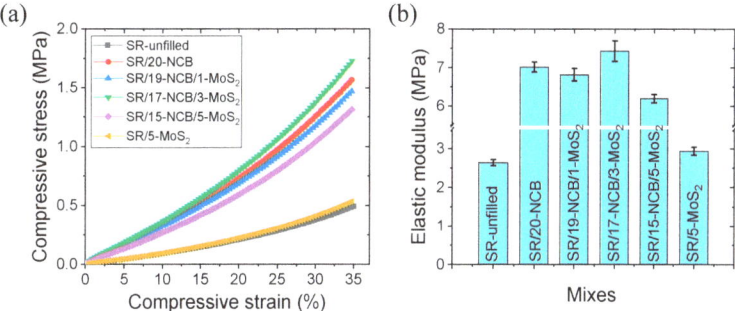

Figure 3. Compressive mechanical properties of different composites; (**a**) compressive stress–strain and (**b**) compressive elastic modulus.

3.3.2. Tensile Mechanical Properties

The different tensile mechanical properties are shown in Figure 4a–e. From the stress–strain curves in Figure 4a, it is evident that hybrid filler-containing composites provide better stretchability, and the area under the curve is also higher than those of the single and unfilled rubber composites. This could be due to the higher flexibility of the rubber matrix, followed by controlled cross-links and strong filler–polymer interactions through NCB. Thus, in the presence of MoS_2, more flexible NCB filler networks may be formed that can withstand larger deformations. From Figure 4b of the elastic modulus chart, it is evident that the addition of MoS_2 to the NCB-containing composites resulted in a varied elastic modulus with increasing MoS_2 content. The SR/20-NCB composite showed the highest elastic modulus in tensile mode, which could be due to the higher reinforcing efficiency of carbon black that restricted the movement of the rubber chains. At 1 phr of MoS_2 in the hybrid (SR/19-NCB/1-MoS_2), the decrease in the elastic modulus was mainly due to the sum of the reduced chemical cross-links and insufficient filler dispersion. At 3 phr of MoS_2 (SR/17-NCB/3-MoS_2), strong and mutual dispersion [43] of fillers may have occurred, which resulted in a further increase in the elastic modulus with a subsequent increase in the tensile strength, fracture strain, and fracture toughness, as shown in Figure 4c–e. Another reason for the improved tensile strength in hybrid filler systems may be the increase in the effective surface area owing to better filler dispersion, which improves the van der Waals interactions between the polymer and filler particles [44]. Further additions of MoS_2 may result in reduced mutual filler dispersion and a reduced number of polymer cross-links, as shown by the reduced elastic modulus in Figure 4b. Thus, MoS_2 in the hybrid filler controls curing and improves mutual filler dispersion, which provides long-range filler–polymer connectivity. At similar modulus values in different composites, higher toughness indicates that the sample can withstand larger deformations. Hence, for stretchable electronic devices, a higher toughness is necessary to achieve higher durability. Table 2 compares the tensile properties of a few silicone rubber composites where CB is the main filler material [45,46]. From this table, it can be concluded that CB/MoS_2 hybrid filler obtained overall good mechanical properties. Moreover, the fracture toughness was highly improved, which could be beneficial for improving the electromechanical energy harvesting, which was not reported in the literature [45,46].

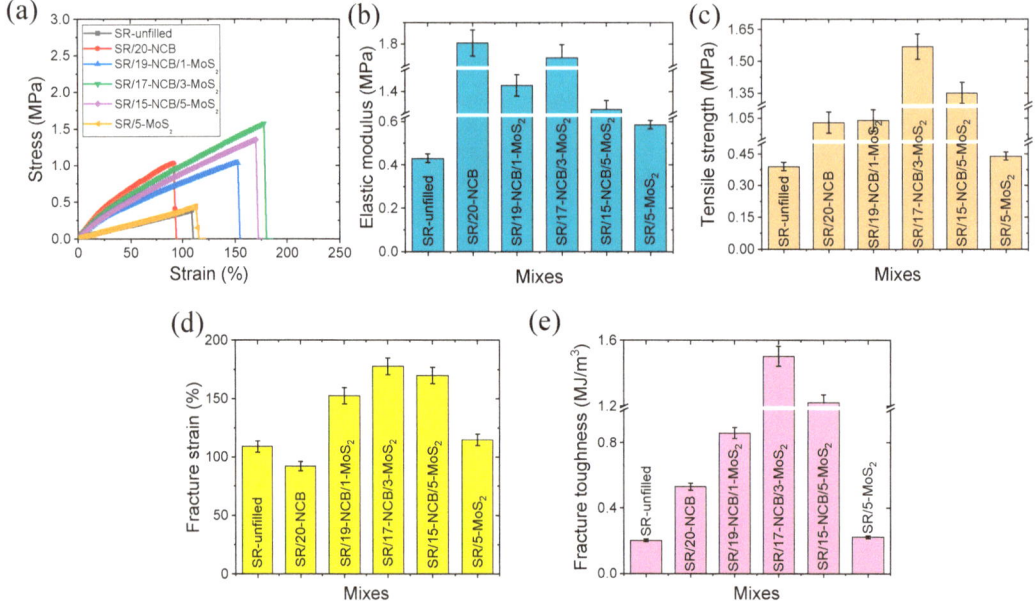

Figure 4. Tensile mechanical properties of different composites; (**a**) stress–strain, (**b**) elastic modulus, (**c**) tensile strength, (**d**) fracture strain, and (**e**) fracture toughness.

Table 2. Comparison of tensile properties of silicone rubber composites with carbon black as the main filler material.

Composites	Filler (Grade)	Amount of Filler	Tensile Strength (MPa)	Elongation at Break (%)	References
SR/CB (N990)	CB (N990)	20 phr	0.61	93.5	[45]
SR/CB (Vulcan XC-72)	CB (Vulcan XC-72)	20 phr	3.6	257.3	[45]
SR/CNT/CB (BP 2000)	CNT/CB (BP 2000)	5.76 vol%	4.5	211	[46]
SR/17-NCB/3-MoS_2	CB (Conductex SC Ultra)/MoS_2	20 phr	1.57	177.67	This study

3.3.3. Hysteresis Losses on Dynamic Loading–Unloading Cycles

The hysteresis losses in the filled compounds were mainly due to the breakdown of the filler structures and rubber networks [47]. Up to a certain deformation, the expended energy was used to break down the filler structures. After the complete breakdown of the filler structure, the expended energy caused the breakdown of the rubber networks [47]. The hysteresis loss increased up to a certain level, and then decreased because the breakdown of the filler network was proportional to the first-order deformation, while the breakdown rubber network was proportional to the second-order deformation [47]. Thus, a lower hysteresis loss signifies bonding stability. Figure 5a,b show the variations in load values of 100 cycles of cyclic compression and relaxation for up to 30% deformation of the rubber composites. From Figure 5c, it can be seen that the area under the first cyclic deformation is higher for the NCB-only compound than for the hybrid filler-loaded SR/17-NCB/3-MoS$_2$ composites. In addition, Figure 5a,b show that the highest load value at maximum deformation reduced more rapidly, with increasing cycles for the SR/20-NCB composite compared to that for the hybrid filler-based SR/17-NCB/3-MoS$_2$ composite. A higher number of inelastic changes, such as the permanent deformation of filler structures, indicates higher fatigue than the elastic change of filler networks [47]. In this respect, it seems that the hybrid filler containing the SR/17-NCB/3-MoS$_2$ composite is more flexible, and returns to quick equilibrium networks with higher fatigue resistance than the only NCB-filled SR/20-NCB composite.

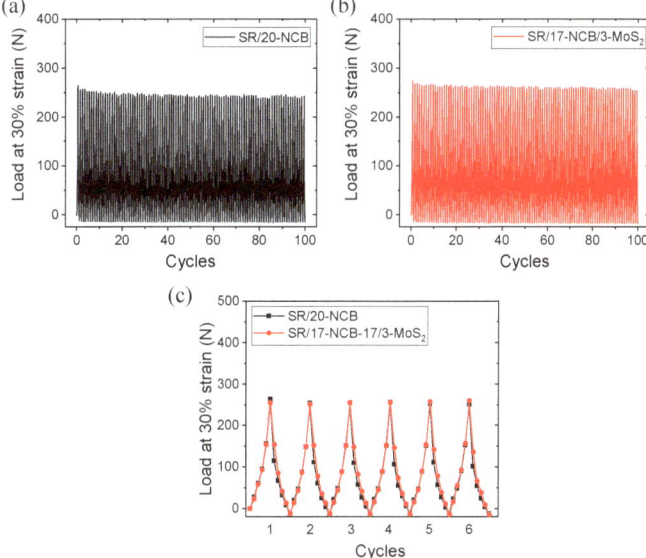

Figure 5. Hysteresis losses in rubber composites; (**a**) SR/20-NCB with 100 deformation cycles, (**b**) SR/17-NCB/3-MoS$_2$ with 100 deformation cycles, and (**c**) SR/20-NCB & SR/17-NCB/3-MoS$_2$ with first 6 deformation cycles.

3.4. Filler Dispersion

The filler dispersion was investigated using SEM analysis. The SEM images in Figure 6a–c show that a more homogeneous filler distribution was possible in the SR/17-NCB/3-MoS$_2$ composite (Figure 6b) compared with the higher MoS$_2$-containing SR/15-NCB/5-MoS$_2$ composite (Figure 6c), as reflected by the mechanical properties having lower values than those of the highly dispersed fillers in the SR/17-NCB/3-MoS$_2$ compound. Moreover, the NCB was well dispersed in the SR/17-NCB/3-MoS$_2$ composite (Figure 6b) compared to the SR/20-NCB composite (Figure 6b).

The reduced filler distribution in SR/15-NCB/5-MoS$_2$ can be confirmed from the EDS mapping shown in Figure 7a–f. Figure 7a shows the area under investigation. While the carbon, silicon, and oxygen elements in Figure 7b–d show homogeneity in the matrix, the MoS$_2$ particles are distributed heterogeneously, as shown in Figure 7e–f. Figure 7e–f also show that nano-range filler dispersion was possible along with microdispersion. The increased nano-level filler dispersion may be due to the increased filler–filler mechanical interactions during mixing [43]. Thus, an optimum filler ratio could be the best for obtaining improved reinforcing properties.

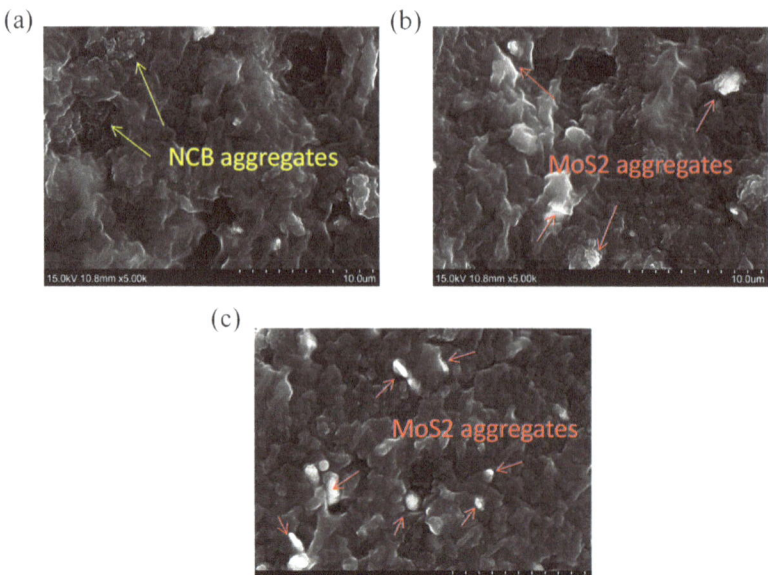

Figure 6. SEM images of rubber composites; (**a**) SR/20-NCB, (**b**) SR/17-NCB/3-MoS$_2$, and (**c**) SR/15-NCB/5-MoS$_2$.

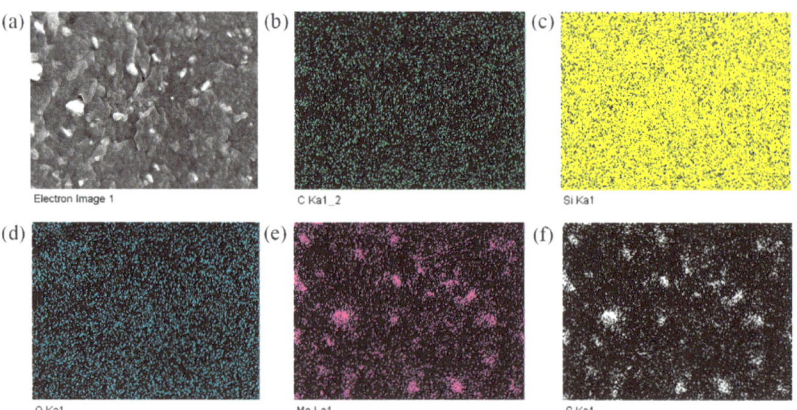

Figure 7. Elemental mapping of SR/15-NCB/5-MoS$_2$ composite; (**a**) SEM image of mapping area, (**b**) carbon, (**c**) silicon, (**d**) oxygen, (**e**) molybdenum, and (**f**) sulfur.

From the mechanical properties, cross-link densities, and SEM analysis, it was evident that MoS$_2$ has a small reinforcing effect on silicone rubber. However, due to its lubricating properties, it can promote the distribution of NCB particles that have greater interactions with silicone rubber. Thus, in hybrid filler systems, more homogeneously distributed fillers showed improved physical interactions with the silicone rubber matrix, and largely enhanced the fracture toughness in the hybrid composites. At 17 phr NCB and 3 phr MoS$_2$, the hybrid composites showed excellent filler distribution of both NCB and MoS$_2$, which is beneficial for improving the toughness value. It is believed that in highly tough rubber composites, the stress distribution is more homogenous and could result in higher electromechanical sensitivity, as found in the later section.

3.5. Electromechanical Sensing Performance of Energy Harvester

To study the effect of MoS_2 on the capacitance-based electromechanical performance, two specimens, one with 20 phr CB (SR/20-NCB) and the other with 20 phr hybrid filler containing a 17:3 CB to MoS_2 ratio (SR/17-NCB/3-MoS_2), were prepared according to the method described above (Section 2.7). The samples were tested for up to 5000 cycles of repeated dynamic loading and unloading in a machine, and the results are shown in Figure 8a–d. With the same applied cyclic pressure up to 2 kPa, the output voltage was much higher for the SR/17-NCB/3-MoS_2 composite than the SR/20-NCB controlled composite. The decrease in voltage efficiency with increasing cycles may be due to the permanent breakdown of conducting NCB filler networks in the controlled composite, while a similar or increasing efficiency in the SR/17-NCB/3-MoS_2 composite may indicate better retention of the conducting filler networks [48,49], which is also evident from Figure 5a,b, with a lower stress softening tendency for the hybrid composite. The energy output was ~100% higher in the SR/17-NCB/3-MoS_2 composite compared to the SR/20-NCB controlled composite as an electrode. This could be due to the improved CB filler dispersion aided by MoS_2, and the piezoresistive effect of the electronic band gap of MoS_2 under strain [50,51] that may enhance the output voltage difference. Moreover, due to the higher toughness in the SR/17-NCB/3-MoS_2 composite, the stress distribution was more homogeneous throughout the matrix than the SR/20-NCB controlled composite. The approximately 100% higher efficiency of SR/17-NCB/3-MoS_2 with only 3 phr MoS_2 compared to the control composite indicates the role of the electronic band gap of MoS_2 in the sensitivity of the electromechanical energy harvesting performance. It was also evident from Figure 8b,d that the SR/17-NCB/3-MoS_2 composite had very uniform sensitivity compared to the SR/20-NCB controlled composite as an electrode. Hence, two-dimensional MoS_2 could be very useful in combination with conducting fillers for capacitance-type energy harvesters for electromechanical sensing applications [52–54]. Although the peak sharpness is much better in Figure 8d than in Figure 8b, this could be further enhanced by increasing the conductivity and the elasticity of the composite. Since viscoelastic materials undergo typical stress relaxation behavior and have a slow stress relaxation rate, charging and discharging are consequently not fast, and may reduce the sharpness of the peaks.

When electrical conducting filler disperses in dielectric rubber, such as in silicone rubber, a capacitor can be produced. Since the capacitance depends on the electrode surface area and the distance between electrodes, mechanical deformation can change the capacitance, followed by changing the electrode surface area and the distance between the electrodes [55]. Due to the deformation, the capacitance value of the capacitor is changed, and the charging–discharging results as negative and positive output voltages. In the hybrid filler system, it is believed that the conducting networks are distributed homogeneously in the rubber matrix, which enhances the capacitance. Since a higher capacitance belongs to a higher charge, hence the amplitudes of output voltages become higher. Moreover, in a hybrid filler system, due to the piezoelectric behavior of MoS_2, some additional potential gradients may be generated, and can further increase the efficiency of the capacitor sensor. Such a type of capacitance-based sensing is a very low-energy process, and can be useful as a pressure sensor in health monitoring applications [55]. For example, under normal body conditions, the systolic (16 kPa) to diastolic (11 kPa) pressure difference is 5 kPa, which is much higher than the applied pressure ($\leq$2 kPa) in this experiment. Hence, this composite can easily detect and measure blood pressure.

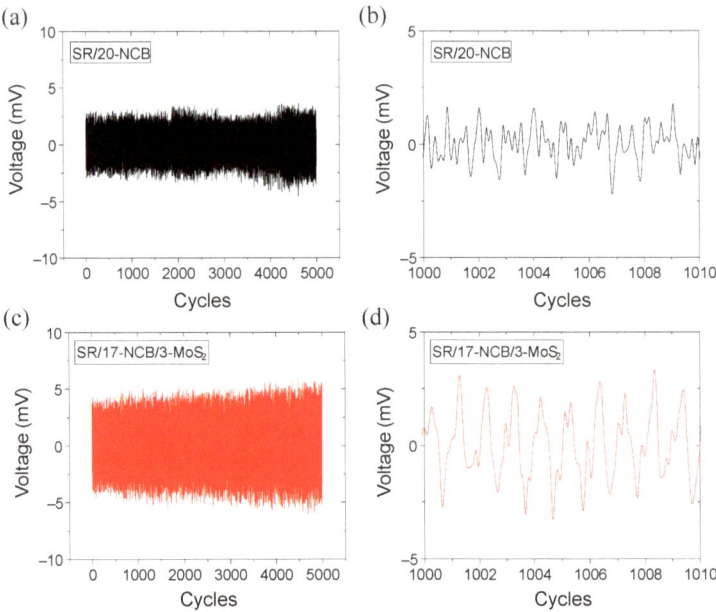

Figure 8. Output voltages in energy harvester with increasing cycles for selective composites as electrodes; (**a**,**b**) SR/20-NCB and (**c**,**d**) SR/17-NCB/3-MoS$_2$.

4. Conclusions

This study examines the mechanical and electromechanical sensing performance of conducting NCB and MoS$_2$ hybrid fillers in RTV silicone rubber. In examining the mechanical properties and fatigue properties, it was evident that a suitable ratio of NCB and MoS$_2$ in silicone rubber can provide excellent fracture toughness with improved tensile strength and fracture strain compared to unfilled and single-filler systems. Such improvements in the mechanical properties could be due to the mutually interacting fillers for excellent filler dispersion in the rubber matrix. Owing to the controlling effect of MoS$_2$ on excessive cross-links in the silicone rubber matrix, the hybrid composites showed higher elongation properties, and maintained better conductivity of the NCB networks in the rubber composites. Thus, the hybrid filler with 17 phr CB and 3 phr MoS$_2$ in silicone rubber provides improved mechanical properties and excellent efficiency for electromechanical sensing. A nearly 100% higher energy harvesting efficiency was obtained with ≤2 kPa of applied pressure for the 17:3 ratio of NCB to MoS$_2$ hybrid compared to NCB only at 20 phr, which confirmed better electromechanical sensitivity of the hybrid composite. The improved sensitivity of the hybrid filler-loaded composite could be attributed to higher charging and discharging during capacitance change followed by the piezoelectricity of the two-dimensional MoS$_2$ particles during mechanical deformation. Hence, 2D MoS$_2$ could be a fascinating hybrid component filler with nanocarbon black for the future development of rubber nanocomposites with advanced mechanical and electromechanical energy harvesting and sensing applications.

Author Contributions: Conceptualization, M.N.A.; methodology, M.N.A., V.K. and T.J.; validation, M.N.A., V.K., T.J. and S.-S.P.; formal analysis, M.N.A., V.K. and T.J.; investigation, M.N.A.; data curation, M.N.A. and V.K.; writing—original draft preparation, M.N.A.; writing—review and editing, M.N.A., V.K., T.J. and S.-S.P.; visualization, M.N.A. and S.-S.P.; supervision, S.-S.P.; project administration, S.-S.P.; funding acquisition, S.-S.P. All authors have read and agreed to the published version of the manuscript.

Funding: This research is funded by the Korea Government (MOTIE, 2022, The Competency Development Program for Industry Specialist) (Grant no. P0002092).

Institutional Review Board Statement: Not applicable.

Data Availability Statement: Data will be available based on request to the corresponding author.

Acknowledgments: This research was supported by the Korea Institute for Advancement of Technology (KIAT).

Conflicts of Interest: The authors declare no conflict of interest.

References

1. Shit, S.C.; Shah, P. A review on silicone rubber. *Natl. Acad. Sci. Lett.* **2013**, *36*, 355–365. [CrossRef]
2. Wolf, M.P.; Salieb-Beugelaar, G.B.; Hunziker, P. PDMS with designer functionalities—Properties, modifications strategies, and applications. *Prog. Polym. Sci.* **2018**, *83*, 97–134. [CrossRef]
3. Liu, J.; Yao, Y.; Li, X.; Zhang, Z. Fabrication of advanced polydimethylsiloxane-based functional materials: Bulk modifications and surface functionalizations. *Chem. Eng. J.* **2021**, *408*, 127262. [CrossRef]
4. Lo, T.Y.; Krishnan, M.R.; Lu, K.Y.; Ho, R.M. Silicon-containing block copolymers for lithographic applications. *Prog. Polym. Sci.* **2018**, *77*, 19–68. [CrossRef]
5. Lv, X.; Tang, Y.; Tian, Q.; Wang, Y.; Ding, T. Ultra-stretchable membrane with high electrical and thermal conductivity via electrospinning and in-situ nanosilver deposition. *Compos. Sci. Technol.* **2020**, *200*, 108414. [CrossRef]
6. Kumar, V.; Alam, M.N.; Manikkavel, A.; Song, M.; Lee, D.J.; Park, S.S. Silicone rubber composites reinforced by carbon nanofillers and their hybrids for various applications: A review. *Polymers* **2021**, *13*, 2322. [CrossRef]
7. Schmitt, F.; Piccin, O.; Barbé, L.; Bayle, B. Soft robots manufacturing: A review. *Front. Robot. AI* **2018**, *5*, 84. [CrossRef]
8. Pang, C.; Lee, G.Y.; Kim, T.I.; Kim, S.M.; Kim, H.N.; Ahn, S.H.; Suh, K.Y. A flexible and highly sensitive strain-gauge sensor using reversible interlocking of nanofibres. *Nat. Mater.* **2012**, *11*, 795–801. [CrossRef]
9. Huang, X.; Yin, Z.; Wu, H. Structural Engineering for High-Performance Flexible and Stretchable Strain Sensors. *Adv. Intell. Syst.* **2021**, *3*, 2000194. [CrossRef]
10. Luo, X.; Yang, G.; Schubert, D.W. Electrically conductive polymer composite containing hybrid graphene nanoplatelets and carbon nanotubes: Synergistic effect and tunable conductivity anisotropy. *Adv. Compos. Hybrid Mater.* **2022**, *5*, 250–262. [CrossRef]
11. Rattanasom, N.; Saowapark, T.; Deeprasertkul, C. Reinforcement of natural rubber with silica/carbon black hybrid filler. *Polym. Test.* **2007**, *26*, 369–377. [CrossRef]
12. Liu, Y.; Li, L.; Wang, Q. Effect of carbon black/nanoclay hybrid filler on the dynamic properties of natural rubber vulcanizates. *J. Appl. Polym. Sci.* **2010**, *118*, 1111–1120. [CrossRef]
13. Zhang, H.; Wang, C.; Zhang, Y. Preparation and properties of styrene-butadiene rubber nanocomposites blended with carbon black-graphene hybrid filler. *J. Appl. Polym. Sci.* **2015**, *132*, 41309. [CrossRef]
14. Socher, R.; Krause, B.; Hermasch, S.; Wursche, R.; Pötschke, P. Electrical and thermal properties of polyamide 12 composites with hybrid fillers systems of multiwalled carbon nanotubes and carbon black. *Compos. Sci. Technol.* **2011**, *71*, 1053–1059. [CrossRef]
15. Gao, J.; He, Y.; Gong, X.; Xu, J. The role of carbon nanotubes in promoting the properties of carbon black-filled natural rubber/butadiene rubber composites. *Results Phys.* **2017**, *7*, 4352–4358. [CrossRef]
16. Xu, M.; Liang, T.; Shi, M.; Chen, H. Graphene-like two-dimensional materials. *Chem. Rev.* **2013**, *113*, 3766–3798. [CrossRef] [PubMed]
17. Wan, C.; Zhou, L.; Xu, S.; Jin, B.; Ge, X.; Qian, X.; Xu, L.; Chen, F.; Zhan, X.; Yang, Y.; et al. Defect engineered mesoporous graphitic carbon nitride modified with AgPd nanoparticles for enhanced photocatalytic hydrogen evolution from formic acid. *Chem. Eng. J.* **2022**, *429*, 132388. [CrossRef]
18. Perkins, F.K.; Friedman, A.L.; Cobas, E.; Campbell, P.M.; Jernigan, G.G.; Jonker, B.T. Chemical vapor sensing with monolayer MoS_2. *Nano Lett.* **2013**, *13*, 668–673. [CrossRef] [PubMed]
19. Zong, X.; Yan, H.; Wu, G.; Ma, G.; Wen, F.; Wang, L.; Li, C. Enhancement of photocatalytic H_2 evolution on CdS by loading MoS_2 as cocatalyst under visible light irradiation. *J. Am. Chem. Soc.* **2008**, *130*, 7176–7177. [CrossRef]
20. Castellanos-Gomez, A.; Poot, M.; Steele, G.A.; Van Der Zant, H.S.; Agraït, N.; Rubio-Bollinger, G. Elastic properties of freely suspended MoS_2 nanosheets. *Adv. Mater.* **2012**, *24*, 772–775. [CrossRef] [PubMed]
21. Zhu, H.; Wang, Y.; Xiao, J.; Liu, M.; Xiong, S.; Wong, Z.J.; Ye, Z.; Ye, Y.; Yin, X.; Zhang, X. Observation of piezoelectricity in free-standing monolayer MoS_2. *Nat. Nanotechnol.* **2015**, *10*, 151–155. [CrossRef]
22. Zhou, K.; Liu, J.; Zeng, W.; Hu, Y.; Gui, Z. In situ synthesis, morphology, and fundamental properties of polymer/MoS_2 nanocomposites. *Compos. Sci. Technol.* **2015**, *107*, 120–128. [CrossRef]
23. Wang, D.; Song, L.; Zhou, K.; Yu, X.; Hu, Y.; Wang, J. Anomalous nano-barrier effects of ultrathin molybdenum disulfide nanosheets for improving the flame retardance of polymer nanocomposites. *J. Mater. Chem. A* **2015**, *3*, 14307–14317. [CrossRef]
24. Eksik, O.; Gao, J.; Shojaee, S.A.; Thomas, A.; Chow, P.; Bartolucci, S.F.; Lucca, D.A.; Koratkar, N. Epoxy nanocomposites with two-dimensional transition metal dichalcogenide additives. *ACS Nano* **2014**, *8*, 5282–5289. [CrossRef]

25. Zhou, K.; Jiang, S.; Shi, Y.; Liu, J.; Wang, B.; Hu, Y.; Gui, Z. Multigram-scale fabrication of organic modified MoS$_2$ nanosheets dispersed in polystyrene with improved thermal stability, fire resistance, and smoke suppression properties. *RSC Adv.* **2014**, *4*, 40170–40180. [CrossRef]
26. Wang, X.; Kalali, E.N.; Wang, D.Y. An in situ polymerization approach for functionalized MoS$_2$/nylon-6 nanocomposites with enhanced mechanical properties and thermal stability. *J. Mater. Chem. A* **2015**, *3*, 24112–24120. [CrossRef]
27. Tang, Z.; Wei, Q.; Guo, B. A generic solvent exchange method to disperse MoS$_2$ in organic solvents to ease the solution process. *Chem. Commun.* **2014**, *50*, 3934–3937. [CrossRef]
28. Tsai, C.Y.; Lin, S.Y.; Tsai, H.C. Butyl rubber nanocomposites with monolayer MoS$_2$ additives: Structural characteristics, enhanced mechanical, and gas barrier properties. *Polymers* **2018**, *10*, 238. [CrossRef] [PubMed]
29. Fuming, K.; Xincong, Z.; Jian, H.; Xiaoran, Z.; Jun, W. Tribological properties of Nitrile Rubber/UHMWPE/Nano-MoS$_2$ water-lubricated bearing material under low speed and heavy duty. *J. Tribol.* **2018**, *140*, 61301. [CrossRef]
30. Geng, H.; Zhao, P.; Mei, J.; Chen, Y.; Yu, R.; Zhao, Y.; Ding, A.; Peng, Z.; Liao, L.; Liao, J. Improved microwave absorbing performance of natural rubber composite with multi-walled carbon nanotubes and molybdenum disulfide hybrids. *Polym. Adv. Technol.* **2020**, *31*, 2752–2762. [CrossRef]
31. Tang, Z.; Zhang, C.; Wei, Q.; Weng, P.; Guo, B. Remarkably improving performance of carbon black-filled rubber composites by incorporating MoS$_2$ nanoplatelets. *Compos. Sci. Technol.* **2016**, *132*, 93–100. [CrossRef]
32. Alam, M.N.; Kumar, V.; Lee, D.J.; Choi, J. Synergistically toughened silicone rubber nanocomposites using carbon nanotubes and molybdenum disulfide for stretchable strain sensors. *Compos. Part B Eng.* **2023**, *259*, 110759. [CrossRef]
33. Kumar, V.; Lee, J.Y.; Lee, D.J. Synergistic effects of hybrid carbon nanomaterials in room-temperature-vulcanized silicone rubber. *Polym. Int.* **2017**, *66*, 450–458. [CrossRef]
34. Ding, H.; Zang, W.; Li, J.; Jiang, Y.; Zou, H.; Ning, N.; Tian, M.; Zhang, L. CB/PDMS electrodes for dielectric elastomer generator with low energy loss, high energy density and long life. *Compos. Commun.* **2022**, *31*, 101132. [CrossRef]
35. Li, W.; Liu, X.; Jiang, Y.; Wu, W.; Yu, B.; Ning, N.; Tian, M.; Zhang, L. Extremely high energy density and long fatigue life of nano-silica/polymethylvinylsiloxane dielectric elastomer generator by interfacial design. *Nano Energy* **2022**, *104*, 107969. [CrossRef]
36. Pantea, D.; Darmstadt, H.; Kaliaguine, S.; Roy, C. Electrical conductivity of conductive carbon blacks: Influence of surface chemistry and topology. *Appl. Surf. Sci.* **2003**, *217*, 181–193. [CrossRef]
37. Yang, X.; Li, Z.; Jiang, Z.; Wang, S.; Liu, H.; Xu, X.; Wang, D.; Miao, Y.; Shang, S.; Song, Z. Mechanical reinforcement of room-temperature-vulcanized silicone rubber using modified cellulose nanocrystals as cross-linker and nanofiller. *Carbohydr. Polym.* **2020**, *229*, 115509. [CrossRef] [PubMed]
38. Kumar, V.; Alam, M.N.; Park, S.S. Soft composites filled with iron oxide and graphite nanoplatelets under static and cyclic strain for different industrial applications. *Polymers* **2022**, *14*, 2393. [CrossRef]
39. Moretto, H.-H.; Schulze, M.; Wagner, G. "Silicones" in Ullmann's Encyclopedia of Industrial Chemistry; Wiley-VCH: Weinheim, Germany, 2005.
40. Kochanke, A.; Nagel, J.; Üffing, C.; Hartwig, A. Influence of addition curing silicone formulation and surface aging of aluminum adherends on bond strength. *Int. J. Adhes. Adhes.* **2019**, *95*, 102424. [CrossRef]
41. Kang, M.J.; Heo, Y.J.; Jin, F.L.; Park, S.J. A review: Role of interfacial adhesion between carbon blacks and elastomeric materials. *Carbon Lett.* **2016**, *18*, 1–10. [CrossRef]
42. Martin-Gallego, M.; Bernal, M.M.; Hernandez, M.; Verdejo, R.; López-Manchado, M.A. Comparison of filler percolation and mechanical properties in graphene and carbon nanotubes filled epoxy nanocomposites. *Eur. Polym. J.* **2013**, *49*, 1347–1353. [CrossRef]
43. Alam, M.N.; Kumar, V.; Potiyaraj, P.; Lee, D.J.; Choi, J. Mutual dispersion of graphite–silica binary fillers and its effects on curing, mechanical, and aging properties of natural rubber composites. *Polym. Bull.* **2021**, *79*, 2707–2724. [CrossRef]
44. Liu, Y.; Huang, Y.; Duan, X. Van der Waals integration before and beyond two-dimensional materials. *Nature* **2019**, *567*, 323–333. [CrossRef]
45. Wang, J.; Li, Q.; Wu, C.; Xu, H. Thermal conductivity and mechanical properties of carbon black filled silicone rubber. *Polym. Polym. Compos.* **2014**, *22*, 393–400. [CrossRef]
46. Song, P.; Song, J.; Zhang, Y. Stretchable conductor based on carbon nanotube/carbon black silicone rubber nanocomposites with highly mechanical, electrical properties and strain sensitivity. *Compos. Part B Eng.* **2020**, *191*, 107979. [CrossRef]
47. Kucherskii, A.M. Hysteresis losses in carbon-black-filled rubbers under small and large elongations. *Polym. Test.* **2005**, *24*, 733–738. [CrossRef]
48. Wang, L.; Choi, J. Highly stretchable strain sensors with improved sensitivity enabled by a hybrid of carbon nanotube and graphene. *Micro Nano Syst. Lett.* **2022**, *10*, 17. [CrossRef]
49. Persons, A.K.; Ball, J.E.; Freeman, C.; Macias, D.M.; Simpson, C.L.; Smith, B.K.; Burch, V.R.F. Fatigue testing of wearable sensing technologies: Issues and opportunities. *Materials* **2021**, *14*, 4070. [CrossRef]
50. Amorim, B.; Cortijo, A.; De Juan, F.; Grushin, A.G.; Guinea, F.; Gutiérrez-Rubio, A.; Ochoa, H.; Parente, V.; Roldán, R.; San-Jose, P.; et al. Novel effects of strains in graphene and other two dimensional materials. *Phys. Rep.* **2016**, *617*, 1–54. [CrossRef]
51. Neri, I.; López-Suárez, M.; Gammaitoni, L. Tunable MoS$_2$ strain sensor. *IEEE Instrum. Meas. Mag.* **2020**, *23*, 30–33. [CrossRef]

52. Trung, T.Q.; Lee, N.E. Flexible and stretchable physical sensor integrated platforms for wearable human-activity monitoringand personal healthcare. *Adv. Mater.* **2016**, *28*, 4338–4372. [CrossRef] [PubMed]
53. Zhao, S.; Li, J.; Cao, D.; Zhang, G.; Li, J.; Li, K.; Yang, Y.; Wang, W.; Jin, Y.; Sun, R.; et al. Recent advancements in flexible and stretchable electrodes for electromechanical sensors: Strategies, materials, and features. *ACS Appl. Mater. Interfaces* **2017**, *9*, 12147–12164. [CrossRef] [PubMed]
54. Wang, X.; Liu, X.; Schubert, D.W. Highly sensitive ultrathin flexible thermoplastic polyurethane/carbon black fibrous film strain sensor with adjustable scaffold networks. *Nano-Micro Lett.* **2021**, *13*, 64. [CrossRef] [PubMed]
55. Zang, Y.; Zhang, F.; Di, C.A.; Zhu, D. Advances of flexible pressure sensors toward artificial intelligence and health care applications. *Mater. Horiz.* **2015**, *2*, 140–156. [CrossRef]

Disclaimer/Publisher's Note: The statements, opinions and data contained in all publications are solely those of the individual author(s) and contributor(s) and not of MDPI and/or the editor(s). MDPI and/or the editor(s) disclaim responsibility for any injury to people or property resulting from any ideas, methods, instructions or products referred to in the content.

MDPI AG
Grosspeteranlage 5
4052 Basel
Switzerland
Tel.: +41 61 683 77 34

Polymers Editorial Office
E-mail: polymers@mdpi.com
www.mdpi.com/journal/polymers

Disclaimer/Publisher's Note: The title and front matter of this reprint are at the discretion of the Guest Editors. The publisher is not responsible for their content or any associated concerns. The statements, opinions and data contained in all individual articles are solely those of the individual Editors and contributors and not of MDPI. MDPI disclaims responsibility for any injury to people or property resulting from any ideas, methods, instructions or products referred to in the content.

www.ingramcontent.com/pod-product-compliance
Lightning Source LLC
LaVergne TN
LVHW072334090526
838202LV00019B/2417